T0276210

CAMBRIDGE LIBRARY COLLECTION

Books of enduring scholarly value

Botany and Horticulture

Until the nineteenth century, the investigation of natural phenomena, plants and animals was considered either the preserve of elite scholars or a pastime for the leisured upper classes. As increasing academic rigour and systematisation was brought to the study of 'natural history', its subdisciplines were adopted into university curricula, and learned societies (such as the Royal Horticultural Society, founded in 1804) were established to support research in these areas. A related development was strong enthusiasm for exotic garden plants, which resulted in plant collecting expeditions to every corner of the globe, sometimes with tragic consequences. This series includes accounts of some of those expeditions, detailed reference works on the flora of different regions, and practical advice for amateur and professional gardeners.

Letters on the Elements of Botany

Among the many interests of Swiss philosopher Jean-Jacques Rousseau (1712–78) was botany. These letters 'addressed to a lady' came to the attention of Thomas Martyn, professor of botany at the University of Cambridge, who thought that 'if [they were] translated into English, they might be of use to such … as wished to amuse themselves with natural history'. However, when the translation was done, he 'perceived that the foundation only being laid by the ingenious author, it could be of little service, without raising the superstructure'. Martyn's 1785 publication, of which we have reissued the 1791 third edition, adds notes and corrections to Rousseau's original thirty-two letters, which explain the structure of plants and their ordering in the Linnaean system. Martyn urges the reader not to study it 'in the easy chair at home': it 'can be no use but to such as have a plant in their hand'.

Cambridge University Press has long been a pioneer in the reissuing of out-of-print titles from its own backlist, producing digital reprints of books that are still sought after by scholars and students but could not be reprinted economically using traditional technology. The Cambridge Library Collection extends this activity to a wider range of books which are still of importance to researchers and professionals, either for the source material they contain, or as landmarks in the history of their academic discipline.

Drawing from the world-renowned collections in the Cambridge University Library and other partner libraries, and guided by the advice of experts in each subject area, Cambridge University Press is using state-of-the-art scanning machines in its own Printing House to capture the content of each book selected for inclusion. The files are processed to give a consistently clear, crisp image, and the books finished to the high quality standard for which the Press is recognised around the world. The latest print-on-demand technology ensures that the books will remain available indefinitely, and that orders for single or multiple copies can quickly be supplied.

The Cambridge Library Collection brings back to life books of enduring scholarly value (including out-of-copyright works originally issued by other publishers) across a wide range of disciplines in the humanities and social sciences and in science and technology.

Letters on the Elements of Botany

Addressed to a Lady

Jean-Jacques Rousseau
Thomas Martyn

Cambridge
University Press

CAMBRIDGE
UNIVERSITY PRESS

University Printing House, Cambridge, CB2 8BS, United Kingdom

Cambridge University Press is part of the University of Cambridge.
It furthers the University's mission by disseminating knowledge in the pursuit of education, learning and research at the highest international levels of excellence.

www.cambridge.org
Information on this title: www.cambridge.org/9781108076722

© in this compilation Cambridge University Press 2017

This edition first published 1791
This digitally printed version 2017

ISBN 978-1-108-07672-2 Paperback

This book reproduces the text of the original edition. The content and language reflect the beliefs, practices and terminology of their time, and have not been updated.

Cambridge University Press wishes to make clear that the book, unless originally published by Cambridge, is not being republished by, in association or collaboration with, or with the endorsement or approval of, the original publisher or its successors in title.

LETTERS

ON THE

ELEMENTS

OF

BOTANY.

LETTERS
ON THE
ELEMENTS
OF
BOTANY.

ADDRESSED TO A LADY.

By the celebrated J. J. ROUSSEAU.

TRANSLATED INTO ENGLISH,
WITH NOTES,
AND TWENTY-FOUR ADDITIONAL LETTERS,
FULLY EXPLAINING THE SYSTEM OF LINNÆUS.

BY THOMAS MARTYN, B. D. F. R. S.
PROFESSOR OF BOTANY
IN THE UNIVERSITY OF CAMBRIDGE.

THE THIRD EDITION,
WITH CORRECTIONS AND IMPROVEMENTS.

LONDON:
PRINTED FOR B. WHITE AND SON,
AT HORACE'S HEAD, FLEET-STREET.
MDCCXCI.

TO

THE LADIES

OF

GREAT BRITAIN:

NO LESS EMINENT

FOR THEIR ELEGANT AND USEFUL

ACCOMPLISHMENTS,

THAN ADMIRED

FOR THE BEAUTY OF THEIR PERSONS:

THIS THIRD EDITION OF THE FOLLOWING

LETTERS

IS, WITH ALL HUMILITY,

INSCRIBED

BY

THE TRANSLATOR AND EDITOR.

THE

TRANSLATOR's PREFACE.

WHEN the Elementary Letters on Botany [a] first presented themselves to me, in turning over the last complete edition of Rousseau's works [b], their elegance and simplicity pleased me enough to make me give them a second more attentive perusal. I then thought that they had considerable merit; and that if they were disembarrassed from the chaos of fifteen quarto volumes, and translated into English, they might be of use to such of my fair countrywomen and unlearned countrymen as wished to amuse themselves with natural history.

When the translation was done, I perceived that the foundation only being laid by the ingenious author, it could be of little

[a] Lettres Elementaires sur la Botanique a Madame de L*. Melanges, tome ii. page 531, &c.

[b] Collection complete des Oeuvres de J. J. Rousseau. Geneve, 1782.

service,

ſervice, without raiſing the ſuperſtructure. This I have attempted; not flattering myſelf that it is executed in Rouſſeau's manner, which is inimitable, but merely with the deſign of being uſeful.

What books can you recommend, that may enable me to acquire a competent knowledge of Botany? is a queſtion that has very frequently been aſked me. To the learned I can readily anſwer, the works of Linnæus alone will furniſh you with all the knowledge you have occaſion for; or, if they are deficient in any point, will refer you to other authors, where you may have every ſatisfaction that books can give you [c]. But I am not very ſolicitous to relieve theſe learned gentlemen from their embarraſſment; they have reſources enough, and know how to help themſelves. As to the unlearned, if I were to ſend them to the tranſlation of Linnæus's works, they would only find themſelves bewildered in an inextricable labyrinth of unintelligible terms, and would only reap diſguſt from a ſtudy, that is, perhaps, more capable of affording

[c] Theſe writings of Linnæus are — *Philoſophia Botanica*, that inexhauſtible mine of elementary knowledge — *Genera Plantarum* — *Species Plantarum* — and *Syſtema Vegetabilium*, which is an epitome of the two laſt.

pleaſure

pleaſure than any other. If I were to bid them ſit down, and ſtudy their grammar [d] regularly; ſo dry and forbidding an outſet might diſcourage the greater number; and few would enter the temple through a veſtibule of ſo unpromiſing an appearance. A language however muſt be acquired; but then it may be done gradually; and the *tædium* of it may, in ſome meaſure, be relieved by carrying on at the ſame time a ſtudy of facts, and the philoſophy of nature. This ſeems to have been Rouſſeau's idea, and I have endeavoured not to loſe ſight of it in my continuation of his eight ingenious letters.

Let an unlearned perſon then, who is deſirous of acquiring ſome knowledge of Botany, begin by taking a few plants with flowers, whoſe parts are ſufficiently viſible, and examine them patiently by the deſcriptions and characters which are given in the following pages. You may perhaps know ſome plants by their names; or if not, you will be unfortunate indeed if you have not a friend who will ſhow you the flower of a lily. If in the courſe of your examination,

[d] In Lee's Introduction, Roſe's Elements, &c.

any term ſhould occur, that is not explained in the page, or mentioned in the index, you may have recourſe to the Dictionary, the Introduction, or the Elements. If you can have patience to go through the firſt ſeven letters, with a plant or two of each natural tribe explained in them; to make yourſelf maſter of the claſſification in the ninth and tenth; and to examine the obvious plants, whoſe characters are given in the twenty following letters, as they occur; I flatter myſelf that you will find little difficulty after that, in determining any plant which you ſhall happen to meet with, by Linnæus's characters, as delivered by his tranſlators[e]: whereas if you had begun with them, I am confident you would have been diſcouraged from proceeding.

Good plates, or figures of plants, will alſo be of conſiderable aſſiſtance: thoſe of Mr. Curtis's *Flora Londinenſis* will ſuffice for moſt of the Britiſh natives: eſpecially as he has accompanied his plates with ample and accu-

[e] A ſyſtem of vegetables, &c. tranſlated from the 13th edition of Linnæus's Syſtema Vegetabilium, by a botanical ſociety at Lichfield.——The Genera Plantarum is ſince alſo tranſlated by the ſame hands.

rate

rate deſcriptions in Engliſh as well as Latin. Mr. Miller's figures to his Gardener's Dictionary, exhibit a great number of the moſt remarkable foreigners. There is indeed no want of ſuch help[f]: but the misfortune is, that theſe books are ſo very expenſive, as to be far beyond the purſe of all but the opulent.

I beg leave to proteſt againſt theſe letters being read in the eaſy chair at home; they can be of no uſe but to ſuch as have a plant in their hand; nor do they pretend to any thing more, than to initiate ſuch as, from their ignorance of the learned languages, are unable to profit by the works of the learned, in the firſt principles of vegetable nature. Botany is not to be learned in the cloſet; you muſt go forth into the garden or the fields,

[f] Cateſby's Carolina. Martyn's Hiſtoria Plantarum Rariorum. Oeder's Flora Danica. Dillenius's Hortus Elthamenſis. Beſler's Hortus Eyſtettenſis. Rheede's Hortus Malabaricus. Rumphius's Herbarium Amboinenſe. Trew's Florum Imagines & Plantæ rariores. Jacquin's Flora Auſtriaca, hortus Vindobonenſis, &c. Ehret's Plantæ rariores. Blackwell's Herbal. Hill's Vegetable Syſtem. Merian's Surinam and European Plants and Inſects. Allionii Flora Pedemontana. Pallas's Flora Roſſica; and Scopoli's Flora Inſubrica—are all very fine works, but coſt an immenſe ſum to purchaſe them.

and there become familiar with Nature herself; with that beauty, order, regularity, and inexhaustible variety which is to be found in the structure of vegetables; and that wonderful fitness to its end, which we perceive in every work of creation, as far as our limited understandings, and partial observations, give us a just view of it.

In the second edition a few mistakes were corrected, and some improvements were made; the principal of these was, a reference at the foot of the page to some authors who have figured the plants. For this purpose I preferred Curtis and Miller: when these failed me, I had recourse to the Flora Danica, &c. and I usually referred to old Gerard, or Morison, or both, for the sake of such as do not possess the more splendid works, and live remote from public libraries.

In this third edition these references are considerably multiplied; and that the plants which are wanted for examination may be the more readily found, the generic names are now first given in the margin, and a running title of the classes and orders is placed at the top of the page.

THE

THE

CONTENTS.

Lately published,

Price 9s. plain, or 18s. coloured,

THIRTY-EIGHT PLATES,

WITH EXPLANATIONS;

Intended to illustrate LINNÆUS's Syſtem of Vegetables, and particularly adapted to the

LETTERS on the ELEMENTS of BOTANY.

By THOMAS MARTYN, B. D. F. R. S.

PROFESSOR OF BOTANY IN THE UNIVERSITY OF CAMBRIDGE.

FLORA DIÆTETICA:

OR,

HISTORY of ESCULENT PLANTS,

Both DOMESTIC and FOREIGN.

IN WHICH

They are accurately deſcribed, and reduced to their LINNÆAN Generic and Specific Names.

WITH

Their ENGLISH NAMES annexed, and ranged under Eleven GENERAL HEADS, viz.

ESCULENT

1 ROOTS,
2 SHOOTS, STALKS, &c.
3 LEAVES,
4 FLOWERS,
5 BERRIES,
6 STONE-FRUIT,
7 APPLES,
8 LEGUMENS,
9 GRAIN,
10 NUTS,
11 FUNGUSES.

AND

A particular Account of the Manner of uſing them; their native Places of Growth; their ſeveral Varieties, and Phyſical Properties: Together with whatever is otherwiſe curious, or very remarkable in each Species.

By CHARLES BRYANT, of NORWICH.

Price Six Shillings in Boards.

INTRODUCTION.

THE principal misfortune of Botany is, that from its very birth it has been looked upon merely as a part of medicine. This was the reaſon why every body was employed in finding or ſuppoſing virtues in plants, whilſt the knowledge of plants themſelves was totally neglected: for how could the ſame man make ſuch long and repeated excurſions as ſo extenſive a ſtudy demands; and at the ſame time apply himſelf to the ſedentary labours of the laboratory, and attendance upon the ſick; which are the only methods of aſcertaining the nature of vegetable ſubſtances, and their effects upon the human body? This falſe idea of Botany, for a long time, almoſt confined the ſtudy of it to medicinal plants, and reduced the vegetable chain to a ſmall number of interrupted links. Even theſe were very ill ſtudied, becauſe the ſubſtance only was attended to, and not the organization. How indeed could perſons be much intereſted in the organical ſtructure of a ſubſtance, of which they had no other idea but as a thing

to

to be pounded in a mortar? Plants were ſearched for, only to find remedies; it was ſimples, not vegetables that they looked after. This was very right, it will be ſaid; may be ſo. Hence nevertheleſs it follows, that, if men were ever ſo well acquainted with remedies, they were very ignorant of plants; and this is all that I have here advanced.

Botany was nothing; there was no ſuch ſtudy; and they who plumed themſelves moſt upon their knowledge of vegetables, had no idea of their ſtructure, or of the vegetable œconomy. Every body knew by ſight five or ſix plants in his neighbourhood, to which he gave names at random; enriched with wonderful virtues, which he took it in his head they poſſeſſed; and each of theſe plants, changed into an univerſal panacea, was alone ſufficient to render all mankind immortal. Theſe plants, transformed into balſams and ointments, quickly diſappeared; and ſoon made room for others, to which new comers, in order to diſtinguiſh themſelves, attributed the ſame effects. Sometimes it was a new plant, decorated with ancient virtues: ſometimes old plants, under new names, ſufficed to enrich new quacks. Theſe plants had a different vulgar name in every province, and they who pointed them out for their drugs, at moſt gave them only thoſe names by which they were known on the ſpot where they lived: thus,

thus, when their recipes travelled into other countries, it was no longer known what plant they ſpoke of; every body ſubſtituted another after his own fancy, without regarding any thing elſe, but giving it the ſame name. Such is the whole art that the Myrepſuſes, the Hildegardiſes, the Suarduſes, the Villanovas, and the reſt of the doctors of that time, employed in the ſtudy of thoſe plants which they treat of; and it would be difficult perhaps for any body to know one of them by the names or deſcriptions which they have given them [a].

At the revival of learning, every thing diſappeared to make room for the works of antiquity; nothing was then either good or true but what was to be found in Ariſtotle or Galen. Inſtead of ſearching for plants where they grew, men ſtudied them only in Pliny and Dioſcorides; and there is nothing ſo frequent in the authors of thoſe

[a] Myrepſus's book is entitled *Antidotarium parvum*. Hildegardis was a lady and an abbeſs; ſhe flouriſhed about 1180, and wrote, among others, a treatiſe entitled *Phyſica Leguminum, Fructuum, Herbarum, &c.* Suardus's book is intitled *Antidotarium*, and was printed at Venice 1551 fol.—Arnoldus de Villanova put together *Regimen Sanitatis Salerni*, printed in 1482, 1484, 1490, 1491, 1493, 1505, 1509, &c. and was author of many other medical and medico-botanical works. He is ſaid to have died in 1313.—But the moſt popular of theſe old works, was *Hortus Sanitatis*, aſcribed to Cuba. See Pulteney's Sketches of the Progreſs of Botany in England, chap. iv.

times, as to find them denying the existence of a plant, for no other reason but because Dioscorides has not mentioned it. These learned plants however must be found in nature, in order to make use of them according to the precepts of their master. They bestirred themselves therefore, they set themselves to search, to observe, to conjecture; and made every effort to find, in the plant which they chose, the characters described in their author; and since translators, commentators, and practitioners, seldom agreed in their choice, twenty names were given to the same plant; and the same name to twenty plants; every man maintaining that his own was the true one, and that all the rest, not being that of Dioscorides, ought to be proscribed. From this conflict indeed it followed at length that more careful researches were made, and some good observations, which deserved not to be forgotten; but at the same time such a chaos of *nomenclature*, that the Physicians and Herbarists no longer understood each other: there was no possibility of communicating their mutual lights; nothing remained but disputes upon words and names; and even every useful enquiry and description was lost, for want of being able to decide what plant each author had spoken of.

Real botanists however began to be formed: such as Clusius, Cordus, Cæsalpinus, Gef-

Gesner[b]; good and instructive books on this subject began to be published, in which already appeared some traces of method[c]. And it has certainly been a loss that these pieces have become useless and unintelligible by the mere discordance of names[d]. But these authors, beginning to unite species and separate genera, according to their own manner of observing the habit and apparent structure, occasioned new inconveniences, and a fresh obscurity; because each author, regulating his nomenclature by his own method, created new genera,

[b] If we follow the order of birth, the arrangement should have been Cordus 1515, Gesner 1516, Cæsalpinus 1519, Clusius 1526: if we range them from the dates of their publications, they should stand thus—Cordus 1535, Gesner 1540, Clusius 1557, Cæsalpinus 1583.

[c] Indeed! some traces only of method in the celebrated work of Cæsalpinus! He who first invented a complete arrangement of plants, and stands unrivalled as the father of method! He to whom every succeeding system-monger owes so many obligations! Though among them all Ray alone confesses it. What Rousseau affirms is true only of the excellent, the illustrious Gesner; the other two thought nothing of arrangement: No, nor the Bauhins, nor any other, till Morison and Ray.

[d] If Rousseau means to speak here concerning the works of the forementioned authors, this is not true. The treatises of Gesner and Clusius are every where referred to, even by Linnæus, and consequently their nomenclature is well known. The principal work of Valerius Cordus is Gesner's History of Plants, which he published in 1561. Cæsalpinus's book is now become rather a matter of respectable curiosity than use.

or ſeparated old ones, as the characters of his own required. So that genera and ſpecies were ſo jumbled together, as to leave ſcarcely any plant without as many names as there were authors who deſcribed it; which made the ſtudy of the nomenclature as tedious as that of the plants themſelves, and frequently more difficult.

At length the two illuſtrious brothers appeared; who alone have done more for the advancement of Botany than all the reſt together who preceded, and even followed them, till Tournefort. Rare geniuſes! whoſe vaſt knowledge and ſolid labours, conſecrated to Botany, render them worthy of that immortality which they have acquired. For, till this part of natural hiſtory falls into oblivion, the names of John and Caſpar Bauhin will live along with it in the memory of mankind [e].

Each of theſe men undertook an univerſal hiſtory of plants: but what more immediately relates to our preſent purpoſe is, that they each of them undertook to join to it a *Synonymy*, or exact liſt of the names that every plant bore in all the writers which preceded them. This labour was become abſolutely neceſſary to enable us to reap any advantage from their obſerva-

[e] John the elder was born at Lyon, in 1541, and died in 1613. Caſpar was not born till 1560, and died in 1624.

tions;

tions; for without that, it was almoſt impoſſible to follow and diſtinguiſh every plant among ſo many names.

The eldeſt almoſt completed this undertaking in three volumes in folio, printed after his death; and he has given ſuch juſt deſcriptions of the plants, that we are rarely deceived in his ſynonyms [f].

The brother's plan was yet more extenſive, as appears by the firſt volume which he publiſhed, and from which we may judge of the immenſity of the whole work, if he had found time to execute it [g]; but, excepting this volume, we have no more than the titles of the reſt in his pinax [h]; and this pinax, the produce of forty years labour, is ſtill the guide to all thoſe who ſtudy

[f] Chabræus was the editor, and Francis Louis de Graffenried, of Bern, was at the expence of the publication. This work derives no excellence from the paper or print. The plates are ſmall and poorly executed; they belonged to Fuchſius, and were purchaſed by the bookſeller for this purpoſe; the editor has not unfrequently put them in wrong places. John Bauhin's Hiſtory however has great intrinſic excellence, for the number of plants well deſcribed, and a judicious compilation of whatever had been done before his time. It is entitled "Hiſtoria Plantarum Univerſalis Auctore Johanne Bauhino Archiatro, &c. Ebrod. 1651."

[g] Theatri Botanici, pars I. Baſil. 1658 and 1663, fol.

[h] Pinax Theatrici Botanici ſive index in Theophraſti, Dioſcoridis, Plinii & botanicorum, qui a ſeculo ſcripſerunt, opera, plantarum circiter 6000 nomina cum ſynonymiis & differentiis. Opus XL annorum. Baſil. 1623 & 1671. 4to.

this ſubject and wiſh to conſult ancient authors[i].

The nomenclature of the Bauhins being formed only from the titles of their chapters, and theſe titles uſually compriſing ſeveral words, hence came the cuſtom of giving, as the names of plants, long ambiguous phraſes; which made this nomenclature not only tedious and embarraſſing, but pedantic and ridiculous. I own there might have been ſome advantage in this, provided their phraſes had been better conſtructed; but being compoſed indifferently of the names of places whence the plants came, of perſons who ſent them, and even of other plants to which they fancied them to bear ſome ſimilitude; theſe phraſes were ſources of new embarraſſment and freſh doubts,

[i] The judicious, the indefatigable Haller, from whoſe judgment there lies no appeal, ſays of Caſpar Bauhin, that he emulated his elder brother in Botany, that he was laborious in collecting, and knew a greater number of plants, being more enriched with them by his ſcholars and friends, but that his judgment was leſs acute; that he admitted too many varieties for ſpecies; that he has repeated the ſame plant under different names; that he was leſs accurate than his brother in his deſcriptions, leſs acquainted with the natural claſſes, and unfortunate, as well as himſelf, in being obliged to divide his time between Anatomy and Botany. Bibl. Botan. I. p. 384.

Haller ſays alſo of this *par nobile fratrum* that for their unwearied diligence they well deſerved to lead the way in a new age of Botany; and accordingly he puts them at the head of the *Collectores* in his ſixth book.

doubts, becauſe the knowledge of one plant required that of ſeveral others to which the phraſe referred, and whoſe names were not better determined than its own.

In the mean time diſtant voyages were inceſſantly enriching Botany with new treaſures; and, whilſt the old names already overloaded the memory, it was neceſſary to invent new ones, for the new plants that were diſcovered. Loſt in this immenſe labyrinth, the botaniſts were obliged to ſeek a thread to extricate themſelves from it; they attached themſelves therefore at laſt ſeriouſly to method; Herman, Rivinus, Ray[k], ſeverally propoſed their own; but the immortal Tournefort carried away the prize from them all[l]; he firſt ranged the whole vegetable kingdom ſyſtematically[m]; and, reforming the nomenclature in part, combined it by his new

[k] The order ſhould have been Ray, Herman, Rivinus. Ray publiſhed his firſt work in 1660, his method in 1682, and even drew up tables for Biſhop Wilkins in 1667, which were printed in the year following. Herman began to write in 1687, and printed his method in 1690. Rivinus publiſhed the firſt part of his method in 1690. Moriſon had before publiſhed his in 1669.

[l] Tournefort firſt publiſhed his ſyſtem in 1697: it was ſpecious, and generally faſhionable, till Linnæus's ſuperſeded it: the plates of generic characters are excellent.

[m] How far this is true may be ſeen in note (k). Tournefort's however may be ſaid to have been the firſt complete regular arrangement; though how it could ever be uſed to good purpoſe, without any characters or deſcriptions of the ſpecies, I do not underſtand.

genera with that of Caſpar Bauhin: but, far from freeing it of its long phraſes, he either added new ones, or loaded the old ones with additions, which his method obliged him to make. The barbarous cuſtom was then introduced of tagging new names to the old ones by a contradictory *qui quæ quod*, making of the ſame plant two diſtinct genera.

For inſtance—'Dens Leonis *qui* Pilo-'ſella folio minus villoſo. Doria *quæ* Ja-'cobœa orientalis limonii folio. Titano-'keratophyton *quod* Lythophyton mari-'num albicans.'

Thus was the nomenclature loaded. The names of the plants became not only phraſes but periods. I ſhall cite one of Plukenet's, to prove that I do not exaggerate. "Gramen myloicophorum caro-"linianum ſeu gramen altiſſimum, pani-"cula maxima ſpecioſa, e ſpicis majoribus "compreſſiuſculis utrinque pinnatis blat-"tam molendariam quodam modo referen-"tibus, compoſita, foliis convolutis mu-"cronatis pungentibus." *Almag* 137[n].

It would have been all over with Botany, if this practice had continued; the nomenclature being now abſolutely inſupportable, could no longer ſubſiſt in this ſtate; and it was become neceſſary either that a reformation ſhould be made, or that

[n] See Linnæus's Critica, and Philoſophia Botanica.

the

the richeſt, the moſt lovely, and the eaſieſt of the three parts of Natural Hiſtory ſhould be abandoned.

At length Linnæus, full of his ſyſtem, and the vaſt ideas which it ſuggeſted to him, formed the project of new-moulding the whole; a taſk which every body felt the neceſſity of, but no one dared to undertake. He did more, he executed it; and, having prepared in his *Critica Botanica* the rules by which it ought to be conducted, he determined the genera of plants in his *Genera Plantarum*, and afterwards the ſpecies in his *Species Plantarum*°; in ſuch a manner, that, by keeping all the old names that agreed with theſe new rules, and new caſting all the reſt, he eſtabliſhed at length a clear nomenclature, founded upon the true principles of the art which he had ſet forth. He preſerved all the ancient genera which were truly natural; he corrected, ſimplified, united, or divided, the reſt as their true characters required. And in forming his names he followed, ſometimes even ſomewhat too ſeverely, the rules which he had laid down.

° The firſt ſketch of Linnæus's ſyſtem was publiſhed in 1735; the laſt edition of Syſtema Vegetabilium in 1784: the Critica Botanica in 1737: the firſt edition of the Genera the ſame year, and the laſt in 1764: the firſt edition of the ſpecies in 1753, the ſecond in 1762 and 1763. See Dr. Pulteney's excellent account of the writings of Linnæus.

With

With respect to the species, descriptions and distinctions were necessary to determine them; phrases therefore remained always indispensable; but, by confining himself to a small number of technical words, well chosen and well adapted, he made good short definitions deduced from the true character of the plant, banishing rigorously all that was foreign to it. For this it was necessary to create a new language for Botany, that would spare the long periphrases of the old descriptions. Complaint has been made that the words of this language are not all to be found in Cicero. This complaint would be reasonable, had Cicero written a complete treatise of Botany. Those words however are all either Greek or Latin, expressive, short, sonorous, and even form elegant constructions by their extreme precision. It is in the constant practice of the art, that we feel all the advantage of this new language, which is as convenient and necessary for Botanists, as that of algebra is for mathematicians.

Hitherto Linnæus had indeed determined the greatest part of known plants, but he had not named them; for defining a thing is not naming it: a phrase can never be a true name, nor can it come into common use. He provided against this defect by the invention of trivial names[p], which

[p] These specific or trivial names appear first in the

which he joined to the generical ones in order to diſtinguiſh the ſpecies. By this contrivance the name of every plant is compoſed only of two words, which alone, when choſen with diſcernment, and applied with propriety, often make the plant better known than the long phraſes of Micheli and Plukenet. To be ſtill better and more regularly acquainted with it, there is the phraſe, which doubtleſs muſt be known, but need not be repeated every time we have occaſion to ſpeak of the object.

Nothing is more pedantic or ridiculous, when a woman, or one of thoſe men who reſemble women, are aſking you the name of an herb or a flower in a garden, than to be under the neceſſity of anſwering by a long file of Latin words that have the appearance of a magical incantation; an inconvenience ſufficient to deter ſuch frivolous perſons from a charming ſtudy offered with ſo pedantic an apparatus.

However neceſſary or advantageous this reform might be, nothing leſs was wanting than Linnæus's profound knowledge to execute it with ſucceſs, and the reputation of this great naturaliſt to make it be univerſally adopted. It met with reſiſtance at firſt, and meets with it ſtill. This could not be otherwiſe; his rivals in the ſame

Pan Suecicus of 1749; but they were brought to perfection in the firſt edition of the *Species Plantarum*, publiſhed four years after.

career

career look upon this adoption as a confeſſion of inferiority which they do not like to make; his nomenclature ſeemed ſo much of a piece with his ſyſtem, that they could not well be ſeparated. And botaniſts of the higher order, who think themſelves obliged through pride not to adopt the ſyſtem of any other, but each man to have his own, will not ſacrifice their pretenſions to the progreſs of an art for which the profeſſors have rarely a diſintereſted fondneſs.

National jealouſies alſo oppoſe the admiſſion of a foreign ſyſtem. People think themſelves obliged to ſupport the famous men of their own country, eſpecially after their death; for even that ſelf-love, which made them ſcarcely bear their ſuperiority whilſt they were alive, is honoured by their glory after they are departed.

The great convenience however of this new nomenclature, and the utility of it, which practice has made known, have cauſed it to be adopted almoſt univerſally throughout Europe, ſooner or later, and even at Paris M. de Juſſieu has eſtabliſhed it in the royal garden; thus preferring public utility to the glory of new-moulding the whole, which the method of natural families, invented by his illuſtrious uncle, ſeemed to require[q].

Not

[q] The royal garden however is certainly arranged by M. de Juſſieu's natural method; which was publiſhed in 1789, under the title of *Genera Plantarum, ſecundum ordines*

Not that the nomenclature of Linnæus is without its faults, or gives no handle to criticiſm; but, till a more perfect one ſhall be found, in which nothing is wanting, it is far better to adopt this than to have none, or to fall again into the phraſes of Tournefort or Caſpar Bauhin. I can even ſcarcely believe that a better nomenclature will in future have ſucceſs enough to proſcribe this, to which the botaniſts of Europe are at preſent ſo wholly accuſtomed; and, having now the double tie of habit and convenience, they will renounce it with ſtill more unwillingneſs than they found in adopting it. In order to bring about ſuch a change, an author muſt be found with credit enough to efface that of Linnæus; one to whoſe authority all Europe would be willing a ſecond time to ſubmit; which appears to me not likely to happen. For if his ſyſtem [r], however excellent it may be, ſhould be adopted by one nation only, it would throw Botany into a new labyrinth, and do it more injury than ſervice.

Even the labour of Linnæus, though immenſe, remains ſtill imperfect, inaſmuch as

dines naturales diſpoſita, juxta methodum in horto regio Pariſienſi exaratam, anno 1774.

[r] He ſhould rather have ſaid *nomenclature* or *language*. It is of no great importance what ſyſtem we adopt, ſo that we all agree to talk the ſame language. That of Linnæus will probably ſtand the teſt of ages, whatever may become of the ſexual ſyſtem.

it does not comprehend all known plants, and is not adopted by all botaniſts without exception; for the writings of ſuch as do not ſubmit to it, require from their readers the ſame labour to ſettle the ſynonyms, as they were forced to take for thoſe which preceded it.

We are obliged to Mr. Crantz, notwithſtanding his rage againſt Linnæus, for having adopted his nomenclature, though he rejected his ſyſtem. But Haller, in his large and excellent work on the Swiſs plants[*], rejects both; and Adanſon does more; for he makes an entire new nomenclature, and furniſhes no information whereby we may refer it to Linnæus's. Haller always quotes the genus, and frequently the ſpecific characters of Linnæus, but Adanſon never quotes either. Haller attaches himſelf to an exact ſynonymy, by which, even when he does not add Linnæus's enunciation of the ſpecies, we may find it at leaſt indirectly by the relation of the ſynonyms. But Linnæus and his books are abſolutely null and void for M. Adanſon and his readers, becauſe the latter gives no information whereby we may connect them. So that we are compelled to chooſe between Linnæus and M. Adanſon,

* Alberti v. Haller Hiſtoria Stirpium Indigenarum Helvetiæ inchoata. Bernæ 1768 folio, in three volumes.

who

who excludes him without mercy; and to throw all the works of one of them into the fire. Or elſe we muſt undertake a new work, which will be neither ſhort nor eaſy, to connect theſe nomenclatures, which offer us no point of union.

Linnæus indeed has not given a complete ſynonymy. For plants known long ſince, he has contented himſelf with quoting the Bauhins and Cluſius, with a figure of each plant. For exotic plants lately diſcovered, he has cited one or two modern authors and the figures of Rheed, Rumphius and ſome others, and has gone no farther. His undertaking did not require of him a more extended compilation, and it is ſufficient that he has given one certain information with regard to every plant which he names[t].

Such is the preſent ſtate of things. Now after this account of it, I would aſk every reader of common ſenſe, how it is poſſible to attach one's ſelf to the ſtudy of plants, and at the ſame time to reject that of the nomenclature? It is juſt as if a man would make himſelf ſkilful in a language, with a determination not to learn the words of it. The names, it is true, are arbitrary, the knowledge of plants has no neceſſary connexion with the nomenclature;

[t] Rouſſeau means to ſpeak here of the *Species Plantarum*, and what he ſays is in general true of that. But in his *Flora Lapponica*, *Suecica*, &c. he has given a much more extenſive ſynonymy.

ture; and it is eaſy to conceive that an intelligent man might be an excellent botaniſt, without knowing a ſingle plant by its name. But that one man alone, without books or any aſſiſtance from communicated information, ſhould become of himſelf even a very moderate botaniſt, is a ridiculous aſſertion to make, and an enterpriſe impoſſible to execute. The queſtion is, whether three hundred years of ſtudy and obſervation ſhould be loſt to Botany, whether three hundred volumes of figures and deſcriptions ſhould be thrown into the fire, whether the knowledge acquired by all the learned, who have conſecrated their purſe, their life, their time, to diſtant, expenſive, painful, and dangerous expeditions, ſhould be uſeleſs to their ſucceſſors, and whether every one ſetting out from nothing, could arrive by himſelf at the ſame knowledge, that a long ſeries of enquiry and ſtudy has ſpread over the maſs of mankind? If not, and if the moſt lovely part of natural hiſtory merit the attention of the curious, let them tell me how we ſhall manage to make uſe of the knowledge heretofore acquired, if we do not begin by learning the language of the writers, and knowing to what objects the names employed by them belong. To admit therefore the ſtudy of botany, and to reject that of the nomenclature, is a moſt abſurd contradiction.

LETTERS

LETTERS ON THE ELEMENTS OF BOTANY; TO A LADY.

LETTER I.

ON THE FRUCTIFICATION AND LILIACEOUS PLANTS.

Dated the 22d of Auguſt, 1771.

I THINK your idea of amuſing the vivacity of your daughter a little, and exerciſing her attention upon ſuch agreeable and varied objects as plants, is excellent; though I ſhould not have ventured to play the pedant ſo far as to propoſe it of myſelf. Since however it comes from you, I approve it with all my heart, and will even aſſiſt you in it; convinced, that at all times of life, the ſtudy of nature abates the taſte for frivolous amuſements, prevents the tumult of the paſſions, and provides the mind with a nouriſhment which is ſalutary, by filling it with an object moſt worthy of its contemplations.

You have begun with teaching your daughter the names of the common plants which you have about you; this was the very thing you ſhould have done. The few plants which ſhe knows by ſight are ſo many points of compariſon for her to extend her knowledge: but they are not ſufficient. You deſire to have a little catalogue of the moſt common plants, with the marks by which they may be known. I find ſome difficulty in doing this for you: that is, in giving you theſe marks or characters in writing, after a manner that is clear, and at the ſame time not diffuſe. This ſeems impoſſible without uſing the language peculiar to the ſubject, and the terms of that language form a vocabulary apart which you cannot underſtand unleſs it be previouſly explained to you.

Beſides, merely to be acquainted with plants by ſight, and to know only their names, cannot but be too inſipid a ſtudy for a genius like yours; and it may be preſumed that your daughter would not be long amuſed with it. I propoſe that you ſhould have ſome preliminary notions of the vegetable ſtructure or organization of plants, in order that you may get ſome real information, though you were to take only a few ſteps, into the moſt beautiful, and the richeſt of the three kingdoms of nature. We have nothing therefore to do yet with the nomenclature, which is but the

the knowledge of a herbariſt. I have always thought it poſſible to be a very great botaniſt without knowing ſo much as one plant by name; and, without wiſhing to make your daughter a very great botaniſt, I think nevertheleſs that it will always be uſeful to her to learn how to ſee, whatever ſhe looks at, well. Do not however be terrified at the undertaking: you will ſoon know that it is not a great one. There is nothing either complicated or difficult in what I have to propoſe to you. Nothing is required but to have patience to begin with the beginning. After that, you may go on no farther than you chooſe.

We are now getting towards the latter ſeaſon, and thoſe plants which are the moſt ſimple in their ſtructure are already paſt. Beſides, I expect you will take ſome time to make your obſervations a little regularly. However in the mean while, till ſpring puts you in a ſituation to begin and follow the order of nature, I am going to give you a few words of the vocabulary to get by heart.

A perfect plant is compoſed of a root, of a ſtem with its branches, of leaves, flower, and fruit, (for in Botany, by fruit, in herbs as well as in trees, we underſtand the whole fabric of the ſeed.) You know the whole of this already, at leaſt enough to underſtand the term; but there is a principal part which requires an examination more at large; I mean the *fructification*, that is, the *flower* and the *fruit*. Let us begin with the flower, which comes firſt. In this part nature has incloſed the ſummary of her work; by this ſhe perpetuates it, and this alſo is commonly the moſt brilliant of all parts of the vegetable, and always leaſt liable to variations.

Lily. Take a lily[a]: I believe you will eaſily find it ſtill in full flower. Before it opens, you ſee at the top of the ſtem an oblong greeniſh bud, which grows whiter the nearer it is to opening; and when it is quite open, you perceive that the white cover takes the form of a baſin or vaſe divided into ſeveral ſegments. This is called the *corolla*, and not the flower, as it is by the vulgar, becauſe the flower is a compoſition of ſeveral parts, of which the corolla is only the principal.

The corolla of the lily is not of one piece, as you eaſily ſee. When it withers and falls, it ſeparates into ſix diſtinct pieces, which are called *petals*. Thus the corolla of the lily is compoſed of ſix petals. A corolla, conſiſting of ſeveral pieces like this, is called a *polypetalous* corolla. If it

[a] *Lilium candidum* of Linnæus, (Pl. 1.) or any of its congeners, (ſee *L. chalcedonicum* & *bulbiferum*, figured, in Curtis's Magazine, 30 and 36.) or almoſt any of the tribe of theſe which are called *liliaceous* flowers, and are, for the greater part, eminently beautiful. As *Amaryllis formoſiſſima*. Curt. Mag. 47.

were all of one piece, like the bell-flower[b] or bind-weeds[c], it would be called *monopetalous*. But to return to our lily.

You will find exactly in the middle of the corolla a ſort of little column riſing from the bottom, and pointing directly upwards. This, taken in its whole, is called the *piſtil* or *pointal:* taken in its parts, it is divided into three; 1, the ſwollen baſe, with three blunted angles, called the *germ* or *ovary*; 2, a thread placed upon this, called the *ſtyle*; 3, the ſtyle crowned by a ſort of capital with three notches: this capital is called the *ſtigma*.

Between the piſtil and the corolla you find ſix other bodies entirely ſeparate from each other, which are called the *ſtamens*. Each ſtamen is compoſed of two parts, one long and thin, by which it is faſtened to the bottom of the corolla, and called the *filament*; the other thicker, placed at the top of the filament, and called *anthera* or *anther*[d]. Each anther is a box which opens when it is ripe, and throws out a yellow duſt, which has a ſtrong ſmell: this is called *pollen* or *farina*.

[b] Campanula rotundifolia *Linnæi*.

[c] Convolvulus ſepium (Pl. 12. f. 3.) & arvenſis, &c. *Linnæi*.

[d] The old Engliſh name of anthera is *ſummit*; Grew called it *ſemet*.—The ſtigma has alſo been named *fibula*.

Such is the general analyſis of the parts which conſtitute a flower. As the corolla fades and falls, the germ increaſes, and becomes an oblong triangular capſule, within which are flat ſeeds in three cells. This capſule, conſidered as the cover of the ſeeds, takes the name of *pericarp*.

The parts here mentioned are found in the flowers of moſt other plants, but in different proportion, ſituation, and number. By the analogy of theſe parts, and their different combinations, the families of the vegetable kingdom are determined: and theſe analogies are connected with others in thoſe parts of the plant which ſeem to have no relation to them. For inſtance, this number of ſix ſtamens, ſometimes only three, of ſix petals or diviſions of the corolla, and that triangular form of the germ, with its three cells, determine the liliaceous tribe; and in all this tribe, which is very numerous, the roots are *bulbs* of ſome ſort or other. That of the lily is *ſquamous*, or compoſed of ſcales; in the aſphodel, it is a number of oblong ſolid bulbs connected together[e]; in the crocus and ſaffron there are two bulbs, one over the other; in the colchicum[f] they are placed ſide by ſide[g].

The

[e] As in the peony, potatoe, &c. Theſe are called by ſome *tuberous* roots.

[f] Or meadow ſaffron.

[g] He might have added that ſome of theſe bulbs are ſolid

The lily, which I have chosen because it is in season; and also on account of the size of the flower and its other parts, is deficient however in one of the constituent parts of a perfect flower, namely the *calyx*, which is that outer green part of the flower usually divided into five parts or composed of five small leaves; sustaining and embracing the corolla at the bottom, and enveloping it entirely before it opens, as you may have remarked in the rose. The calyx which accompanies almost all other flowers, is wanting in the greater part of the liliaceous tribe; as the tulip, the hyacinth, the narcissus, the tuberose, &c. and even in the onion, leek, garlic, &c. which are also liliaceous, though they appear very different at first sight. You will perceive also that in this whole tribe the stems are simple and unbranched, the leaves entire, and never cut or divided: observations which confirm the analogy of the flower and fruit in this family, by that of the other parts of the plants. If you bestow some attention upon these particulars, and make them familiar to you by frequent observations, you are already in a condition to determine, by an attentive

solid like the turnip; others composed of coats, one over another, as in the onion. Linnæus does not allow them to be roots; and indeed it is only their being underground that led former Botanists to call them so. He names them *Hybernacula*, *winter gems* or *buds*, into which the whole plant retires during the cold season.

tentive and continued inſpection of a plant, whether it be of the liliaceous tribe or not; and this without knowing the name of the plant[h]. You ſee that this is not a mere labour of the memory, but a ſtudy of obſervations and facts truly worthy of a naturaliſt[i]. You will not begin by telling your daughter all this at once; and you will be even more cautious, when in the ſequel you ſhall be initiated in the myſteries of vegetation; but you will unveil to her by degrees no more than is ſuitable to her age and ſex, by directing her how to find out things of herſelf, rather than by teaching her[k]. Adieu, my dear couſin; if all this traſh be agreeable to you, I am at your ſervice.

[h] If it ſhould happen to be ſpring when the reader takes up this letter, he may examine the ſnow-drop, crocus, daffodil, narciſſus, crown imperial, tulip, lily of the valley, hyacinth, &c. always taking care, in the garden, to avoid double flowers. See Letter II.

[i] Botany is frequently, but we ſee here how unjuſtly, repreſented as a ſcience which depends wholly upon the memory, as if it were nothing but to get the names of ten thouſand plants by heart.

[k] Rouſſeau takes every occaſion to inculcate this fundamental leſſon of education; and indeed it cannot be inculcated too often. See Letter V.

LETTER II.

ON CRUCIFORM FLOWERS.

The 18th of October, 1771.

SINCE you underſtand ſo well, my dear couſin, the firſt lineaments of plants, though ſo ſlightly marked, as to be able already to diſtinguiſh the liliaceous family by their air; and ſince our little botaniſt amuſes herſelf with corollas and petals, I am going to ſet before you another tribe, upon which ſhe may again exerciſe her little knowledge; with rather more difficulty I own, becauſe the flowers are much ſmaller, and the foliage more varied, but with the ſame pleaſure both on her ſide and on yours; at leaſt if you have as much delight in following this flowery path as I find in tracing it out to you.

When the firſt rays of ſpring ſhall have enlightened your progreſs, by ſhewing you in the gardens hyacinths, tulips, narciſſuſes, jonquils, and lilies of the valley, the analyſis of all which is already known to you, other flowers will ſoon catch your attention, and require of you a new examination; ſuch are ſtocks[l] and rockets[m]. Whenever you find them

[l] Cheiranthus incanus Linnæi. Plate 2.

[m] Heſperis matronalis Linnæi.—Or if theſe are not at hand, wall-flowers, cabbage, turnip, cole-ſeed, muſtard, charlock, radiſh, &c.

them double, do not meddle with them, they are disfigured; or, if you pleaſe, dreſſed after our faſhion: nature will no longer be found among them; ſhe refuſes to reproduce any thing from monſters thus mutilated: for if the moſt brilliant part of the flower, namely the corolla, be multiplied, it is at the expence of the more eſſential parts, which diſappear under this addition of brilliancy.

Stock. Take then a ſingle ſtock gilliflower, or ſtock, as it is vulgarly called, and proceed to the analyſis of the flower: you will perceive immediately an exterior part, which was wanting in the liliaceous flowers, namely the calyx. This conſiſts of four pieces, which we muſt call leaves, leaflets or folioles, having no proper names to expreſs them by, as we have that of petals for the pieces which compoſe the corolla. Theſe four pieces are commonly unequal by pairs; that is, there are two leaflets oppoſite and equal, of a ſmaller ſize, and two others alſo oppoſite and equal, but larger, eſpecially towards the bottom, where they are ſo rounded, as to exhibit a very ſenſible protuberance or bump on the outſide.

In this calyx you will find a corolla compoſed of four petals. I ſay nothing of their colour, becauſe that makes no part of their character. Each of theſe petals is faſtened to the receptacle, or bottom of the calyx, by a narrow pale part, which is called *unguis*, or the *claw* of the petal, and this ſpreads out over the top of the calyx into a large, flat, coloured part, called *lamina*, or the *border* [n].

In the centre of the corolla is one piſtil, long and cylindric, or nearly ſo; chiefly compoſed of a germ ending in a very ſhort ſtyle, and that terminated by an oblong ſtigma, which is *bifid*, that is to ſay, divided into two parts, which are reflex on each ſide.

If you examine carefully the reſpective poſition of the calyx and corolla, you will ſee that each petal, inſtead of correſponding exactly to each leaflet of the calyx, is, on the contrary, placed between two; ſo that it anſwers to the opening which ſeparates them; and this alternate poſition has place in all flowers which have as many petals to the corolla as leaflets to the calyx.

It remains now to ſpeak of the ſtamens. You will find ſix of them in the flower of the ſtock, as in the liliaceous flowers, but not all equal, or elſe alternately unequal, as in thoſe; but you will perceive two oppoſite to each other, ſenſibly ſhorter than the other four which ſeparate them, and which are alſo ſeparate from each other in pairs.

[n] I wonder that Rouſſeau ſays nothing of the regular ſtructure of this corolla, the petals generally ſtanding wide from each other, and forming a figure ſomething like the croſs of the order of St. Louis, whence theſe corollas are called *cruciform*, or *croſs ſhaped*.

I ſhall

I ſhall not enter here into a detail of their ſtructure and poſition: but I give you notice that, if you look carefully, you will find the reaſon why theſe two ſtamens are ſhorter than the other four, and why two leaflets of the calyx are more protuberant, or, as the botaniſts ſpeak, more gibbous, and the other two more flatted.

To finiſh the hiſtory of our ſtock; you muſt not abandon it as ſoon as you have analyſed the flower, but wait till the corolla withers and falls, which it does pretty ſoon; and then remark what becomes of the piſtil, compoſed, as we obſerved before, of the germ, the ſtyle, and the ſtigma. The germ grows conſiderably in length, and thickens a little as the fruit ripens. When it is ripe, it becomes a kind of flat pod, called *ſilique*.

This ſilique is compoſed of two valves, each covering a ſmall cell: and the cells are ſeparated by a thin partition. When the ſeed is ripe, the valves open from the bottom upwards to give it paſſage, and remain faſt to the ſtigma at top. Then you may ſee the flat round ſeeds ranged along each ſide of the partition; and you will find that they are faſtened alternately to right and left by a ſhort pedicle to the ſutures, or each edge of the partition.

I am very much afraid, my dear couſin, that I have fatigued you a little with this long deſcription; but it was neceſſary to give you the eſſential character of the numerous

merous tribe of *cruciform* flowers [o], which forms an entire claſs in almoſt all the ſyſtems of botaniſts: and I hope that this deſcription, which is difficult to underſtand here without a figure, will become more intelligible, when you ſhall have gone through it with ſome attention, having at the ſame time the object before your eyes.

The great number of ſpecies in this claſs [p] has determined botaniſts to divide it into two ſections, in which the flowers are perfectly alike, but the fruits, pericarps, or ſeed-veſſels, are ſenſibly different.

The firſt order comprehends the cruciform flowers with a ſilique, or pod, ſuch as the ſtock, thoſe mentioned in note (m), and the like.

The ſecond contains thoſe whoſe ſeed-veſſel is a *ſilicle*, that is, a ſmall and very ſhort pod, almoſt as wide as it is long, and differently divided within; as whitlow-graſs, mithridate-muſtard, baſtard-creſs, &c. in the fields; and ſcurvy-graſs, horſe-radiſh, candy-tuft, honeſty, &c. in the gardens: though the ſeed-veſſel of the laſt is very large, it is ſtill a ſilicle, becauſe the length exceeds the breadth very little. If none of theſe are known to you, I preſume at leaſt that you are acquainted with the

[o] See note (n).

[p] 287 Species. In the 17th claſs, diadelphia, or two brotherhoods, 695, and in the 19th ſyngeneſia, 1247 ſpecies. Theſe numbers, here and in the ſequel, are given from the 14th edition of *Syſtema Vegetabilium*, by Chevalier Murray.

ſhepherd's-

ſhepherd's-purſe[q], which is ſo common a weed in kitchen gardens. Well then, couſin, this ſhepherd's-purſe is of the cruciform tribe and *ſilicle* branch of it, and the form of the ſilicle is triangular[r]. By this you may form ſome idea of the reſt till they fall into your hands.

But it is time to let you breathe; I will only therefore give you a hint at preſent that in this claſs, and many others, you will often find flowers much ſmaller than thoſe of the ſtock, and ſometimes ſo ſmall that you cannot examine their parts without the aſſiſtance of a glaſs[s]; an inſtrument which a botaniſt cannot do without, any more than he can without a needle, a lancet, or penknife, and a pair of good ſciſſars. Preſuming that your maternal zeal may carry you thus far, I fancy to myſelf a charming picture of my beautiful couſin buſy with her glaſs examining heaps of flowers, a hundred times leſs flouriſhing, leſs freſh, and leſs agreeable than herſelf, Adieu, dear couſin, till the next chapter.

[q] Fl. Dan. t. 729. Curt. Lond. 1. Ger. 276. 1.

[r] The young botaniſt ſhould be advertiſed that theſe ſilicles or little pods differ much in their form: ſome are flat, and round or oval; others are ſpherical or ſpheroidal, (ſee pl. 2. k, l.) and that of ſhepherd's-purſe has a form peculiar to itſelf. Pl. 2. i.

[s] This of the ſmallneſs of the parts in many flowers is an objection that every idle novice makes to the Linnæan ſyſtem, ever trembling leſt any thorn or obſtacle, be it ever ſo minute, ſhould occur in the flowery path: the difficulty however will in great meaſure vaniſh, if he will but have patience to go regularly on his way.

LETTER III.

OF PAPILIONACEOUS FLOWERS.

The 26th of May, 1772.

SINCE you continue, dear coufin, to purfue, with your daughter, that peaceable and delightful ftudy which fills up thofe voids in our time too often dedicated by others to idlenefs, or fomething worfe, with interefting obfervations on nature; I will refume the interrupted thread of our vegetable tribes.

My intention is to defcribe fix of thefe tribes to you firft, in order to render the general ftructure of the characteriftic parts of plants familiar. You have already had two of them; there are four remaining, which you muft ftill have the patience to go through, and after that, quitting for a time the other branches of that numerous race, and going on to examine the different parts of the fructification, we fhall manage fo, that without knowing many plants perhaps, you will at leaft never be in a ftrange country among the productions of the vegetable kingdom.

But I muft inform you, that if you will take books in hand, and purfue the common nomenclature; with abundance of names, you will have few ideas, thofe which you have will be confuſed, and you will not follow properly either my ſteps or thoſe of others; but will have at moſt a mere knowledge of words. I am jealous, dear couſin, of being your only guide in this part of Botany. When it is the proper time I will point out to you the books that you may conſult. In the mean while have patience to read nothing but in that of nature, and to keep wholly to my letters.

Pea. Peas [1] are, at preſent, in full fructification. Seize the moment to obſerve their characters: they are ſome of the moſt curious that Botany affords. One general diviſion of flowers is into regular and irregular. The firſt are they whoſe parts all ſpring uniformly from the centre of the flower, and terminate in the circumference of a circle. This uniformity is the reaſon why when we view flowers of this kind, we do not diſtinguiſh an under from an upper part, nor the right from the left; ſuch are the two tribes which we have already examined. But you will ſee at firſt ſight that the flower of the pea is irregular, that you eaſily diſtinguiſh the longer part of the corolla, which ſhould be at top, from the ſhorter, which ſhould be at bottom; and you know very well, when you hold up the flower to the eye, whether it be in its natural ſituation or not. Thus in examin-

[1] See Plate 3, which is coloured red, to make the flower more conſpicuous.

ing

ing an irregular flower, whenever we ſpeak of the top and the bottom, we ſuppoſe it to be in its natural ſituation.

The flowers of this tribe being of a very particular ſtructure, you muſt not only have ſeveral pea flowers, and diſſect them ſucceſſively, to obſerve all their parts one after another, but you muſt alſo purſue the progreſs of the fructification from the firſt flowering to the maturity of the fruit.

Firſt you will find a *monophyllous* calyx; that is, one of an entire piece, ending in five very diſtinct points, the two wider of which are at top, and three narrower at bottom. This calyx bends towards the lower part, as does alſo the peduncle, or little ſtalk which ſupports it: this peduncle is very ſmall and eaſily moveable; ſo that the flower readily avoids a current of air, and commonly turns its back to the wind and rain.

Having examined the calyx, you may pull it off, ſo as to leave the reſt of the flower entire, and then you will ſee plainly that the corolla is polypetalous.

The firſt piece is a large petal, covering the others, and occupying the upper part of the corolla; it is called the *ſtandard*, or *banner*. We muſt make uſe neither of our eyes nor of common ſenſe, if we do not perceive that this petal is deſigned to protect the other parts of the flower from the principal injuries of the weather. In taking off the ſtandard, you will obſerve, that it is inſerted on each ſide by a little proceſs into the ſide-pieces, ſo that it cannot be driven out of its place by the wind.

The ſtandard being taken off, expoſes to view thoſe two ſide-pieces to which it adhered; they are called the wings. In taking theſe off you will find them ſtill more ſtrongly inſerted into the remaining part, ſo that they cannot be ſeparated without ſome effort. Theſe wings are ſcarcely leſs uſeful in protecting the ſides of the flower, than the ſtandard in covering it.

Taking off the wings, you diſcover the laſt piece of the corolla; this is that which covers and defends the centre of the flower, and wraps it up, eſpecially underneath, as carefully as the three other petals envelope the upper part and the ſides. This laſt piece, which, on account of its form, is called the boat or keel, is, as it were, the ſtrong-box into which nature has put her treaſure, to keep it ſafe from the attacks of air and water.

When you have well examined this petal, draw it gently downwards, pinching it ſlightly by the keel or thin edge, for fear of tearing away what it contains. I am certain you will be pleaſed with the myſtery it reveals when the veil is removed.

The young fruit involved in the boat or keel, is conſtructed in this manner: a cylindric membrane, terminated by ten diſtinct

tinct threads ſurround the germ, or embryo of the legume or pod. Theſe ten threads are ſo many filaments, united below round the germ, and terminated each by a yellow anther, whoſe farina covers the ſtigma which terminates the ſtyle, or grows along the ſide of it: this ſtigma, though yellow with the meal which ſticks to it, is eaſily diſtinguiſhed by its figure and ſize. Thus do theſe ten filaments form alſo about the germ an interior armour, to preſerve it from exterior injuries.

If you examine more curiouſly, you will find that theſe ten filaments are united into one at the baſe, only in appearance. For in the upper part of this cylinder there is a piece or ſtamen which at firſt appears to adhere to the reſt, but as the flower fades and the fruit increaſes, ſeparates and leaves an opening at top, by which the fruit can extend itſelf by opening and ſeparating the cylinder gradually; which otherwiſe, by compreſſing and ſtraitening it all round, would impede its growth. If the flower is not ſufficiently advanced, you will not find this ſtamen detached from the cylinder; but put a fine pin or needle into two little holes which you will ſee near the receptacle, at the baſe of that ſtamen, and you will ſoon perceive the ſtamen with its anther ſeparate from the nine others, which will continue always to form one body, till at length they fade and dry, when the germ becomes a *legume*, and has no longer any occasion for them.

This *legume* is distinguished from the *silique* of the cruciform tribe, by the seeds being fastened to one side only of the case, alternately indeed to each valve of it; but all of them to the same side. You will understand this distinction perfectly if you open the pod of a pea and of a stock at the same time, taking care only to have them before they are quite ripe, that, when the pericarp is opened, the seeds may continue fastened by their proper ligaments to their sutures and their valves [u].

If I have made myself well understood, you will comprehend, dear cousin, what astonishing precautions have been heaped together by nature to bring the embryo of the pea to maturity; and, above all, to protect it, in the midst of the greatest rains, from that wet which is fatal to it, without inclosing it in a hard shell, which would have made it another kind of fruit. The Creator, attentive to the preservation of all beings, has taken great care to protect the fructification of plants from attacks that

[u] In doing this you will also perceive that the legume is unilocular, or has one cell only; whereas you remember that the silique was said to be bilocular. And if you take a ripe legume you will find that it opens by the upper suture, opposite to that to which the seeds are fastened; whereas the silique opens from the bottom upwards by both sutures. Compare Pl. 3. 8. with Pl. 2. h.

may

may injure it; but he ſeems to have doubled his attention to thoſe which ſerve for the nouriſhment of man and animals, as does the greater part of the leguminous or pulſe tribe. The proviſion for the fructification of peas is, in different proportions, the ſame through this claſs. The flowers have the name of *papilionaceous*, from a fancied reſemblance of them to the form of a butterfly *(papilio)*; they have generally a *ſtandard* or *banner*, two *wings*, and a *boat* or *keel*; that is, four irregular petals. But in ſome genera the boat is divided longitudinally into two pieces; and theſe flowers have in reality five petals: others, as clover[v], have all their petals united, and though papilionaceous, are however monopetalous flowers.

The papilionaceous or leguminous plants form one of the moſt numerous and uſeful tribes. Beans, peas, lucerne, ſaintfoin, clover, lupins, lentils, tares or vetches, indigo, liquorice kidney-beans, all belong to it; the character of the laſt is to have the boat ſpirally twiſted, which at firſt ſight might be taken for an accident. There are alſo ſome trees belonging to it; among others that which is commonly called acacia, but which is not the true acacia[w], and many beautiful flowering ſhrubs. But of theſe more hereafter. Adieu, couſin, I wiſh well to every thing that you love.

[v] Trifolium pratenſe *Linnæi.*
[w] Robinia Pſeudacacia *Linnæi.*

LETTER IV.

OF LABIATE AND PERSONATE FLOWERS.

The 19th of June, 1772.

LET us talk of plants, my dear coufin, whilft the feafon for obferving them invites us. Your folution of my queftion concerning the ftamens of cruciform flowers is perfectly right, and fhows that you have underftood me, or rather attended to me; for you have nothing to do but to attend, in order to underftand. You have accounted very well for the fwelling of the two leaflets of the calyx, and the relative fhortnefs of two of the ftamens, in the ftock, by the bending of thefe two ftamens. One ftep more would have led you to the primary caufe of this ftructure; for if you afk once more why thefe ftamens are thus bent, and confequently fhortened, I anfwer that you will find a little gland upon the receptacle, between the ftamen and the germ; and it is this gland which, by throwing the ftamen to a diftance, and forcing it to take a round, neceffarily fhortens it. Upon the fame receptacle are two other glands, one at the foot of each pair of longer ftamens; but being on the outfide of them, between thefe ftamens and the calyx, they do not oblige them to bend, and

and therefore do not ſhorten them : ſo that the two pairs of ſtamens ſtand higher than the two ſingle bent ones ; not becauſe they are longer, but becauſe they are ſtraight. Theſe four glands, or at leaſt veſtiges of them, are more or leſs viſible in almoſt all cruciform flowers, and are much more diſtinct in ſome than in the ſtock[x]. If you aſk me what the glands are for, I anſwer, that they are one of thoſe inſtruments deſtined by nature to unite the vegetable to the animal kingdom, and to make them circulate from one to another. But laying theſe inquiries aſide, in which we anticipate a little too much, let us, for the preſent, return to our tribes of plants.

The flowers which I have hitherto deſcribed to you are polypetalous. I ought perhaps to have begun with the regular monopetalous flowers, which have a much more ſimple ſtructure, but it was this very ſimplicity which diſcouraged me. They conſtitute rather a great nation than a ſingle tribe ; ſo that to comprehend them all under one common mark, we muſt employ characters ſo general and ſo vague, that whilſt we ſeem to ſay ſomething, in effect we ſcarcely ſay any thing. It is better to confine ourſelves within narrower bounds, which we can mark out with more preciſion.

[x] As in arabis turrita, cabbage, muſtard, charlock, radiſh, &c.

Among

Among the irregular monopetalous flowers, there is a tribe whose physiognomy is so marked, that we distinguish the members of it easily by their air. It is that to whose flowers Linnæus has given the name of *ringent*, because they are cut into two lips, the opening of which, whether natural, or produced by a slight compression by the fingers, gives them the air of a gaping mouth. This tribe is divided into two branches: one of *labiate* or ringent flowers, properly so called [y], and the other of *personate* or masked flowers [z]: the Latin word *persona* signifying a mask. The character common to all the tribe is not only a monopetalous corolla, cut into two lips, the upper called the *casque* or *helmet*, the lower, the *beard*; but also four stamens, almost in the same row, distinguished into two pairs, one longer, and the other shorter. The inspection of the object itself will explain these characters better to you than can be done in writing.

Dead Nettle. Let us begin with the labiate flowers. For an example I should willingly give you sage, which is common in almost all gardens: but the singular structure of its stamens, which has occasioned some botanists to separate it from the associates to which it naturally belongs, induces me to look for

[y] Plate 4. f. 1. b.
[z] Plate 4. f. 2. a.

another inſtance [a] in the *white dead-nettle* [b]; which, notwithſtanding its name, has no affinity with nettles, properly ſo called, except in the ſhape of the leaves. This plant is ſo common every where, and continues ſo long in flower, that it cannot be difficult for you to find it [c]. Without ſtopping here to conſider the elegant ſituation of the flowers [d], I will confine myſelf to their ſtructure. The white dead-nettle bears a monopetalous labiate corolla, with the caſque or upper lip arched in order to cover the reſt of the flower, and particularly the ſtamens, which keep, all four of them, very cloſe under cover of its roof. You will eaſily diſcern the longer pair and the ſhorter pair, and in the midſt of them the ſtyle, of the ſame colour, but diſtinguiſhed from them by being forked at the end, inſtead of bearing an anther like the ſtamens. The beard or lower lip bends back, and hangs down, ſo as to let you ſee the inſide of the corolla almoſt to the bottom. In this genus the lower lip is divided

[a] Roſemary, with ſome few others not ſo well known, muſt alſo be avoided, becauſe there are only two ſtamens to the flower.

[b] Lamium album Linnæi. Curtis II. 45. Pl. 4, f. 1.

[c] The largeneſs of the flowers alſo makes it proper for examination; but if the ſmell ſhould be any objection, there is ground-ivy, the other lamiums, betony, hore-hound, baum, ſelf-heal, baum of gilead, &c.

[d] Called verticillate.

length-

lengthwiſe in the middle, but that is not general in this tribe.

If you pull out the corolla, you will take the ſtamens along with it, theſe being faſtened by the filaments to that, and not to the receptacle, whereon the piſtil only will remain. In examining how the ſtamens are faſtened in other flowers, we find them generally attached to the corolla in monopetalous, and to the receptacle, or calyx, in polypetalous flowers: ſo that in the latter caſe one may take away the petals without the ſtamens. From this obſervation we have an elegant, eaſy, and pretty certain rule to know whether a corolla conſiſts of one piece or ſeveral, when it is difficult, as it ſometimes is, to be certain of it immediately.

The corolla, when pulled off, is open at bottom, becauſe it was faſtened to the receptacle, ſo as to leave a circular opening by which the piſtil and what ſurrounds it may grow up within the tube. That which ſurrounds the piſtil in this dead nettle, and all the labiate tribe, is the rudiment of the fruit, conſiſting of four embryos, which become four ſeeds that are naked; that is, without any pericarp or covering: the monophyllous calyx divided into five ſegments ſerving this purpoſe, ſo that the ſeeds, when they are ripe, are detached, and fall to the ground ſeparately. This is the character of the labiate flowers.

The

The other branch or ſection, which is that of the *perſonate* flowers, is diſtinguiſhed from the former; firſt in having the two lips not uſually open, or gaping, but cloſed and joined[e], as you may ſee in the ſnap-dragon[f], a flower not uncommon in gardens; or for want of that, in the toad-flax, a yellow flower with a ſpur, ſo common in the country at this ſeaſon[g]. But a more preciſe and certain character is, that inſtead of having four naked ſeeds at the bottom of the calyx, like the labiate flowers, theſe have a capſule or caſe incloſing the ſeeds, and not opening till they are ripe, in order to diſperſe them. To theſe characters we may add that the greater part of the labiate plants are either ſtrong ſmelling and aromatic, as marjoram, thyme, baſil, mint, hyſſop, lavender, &c. or elſe ſtrong ſmelling and ſtinking, as the dead-nettle, hedge-nettle, cat-mint, black horehound[h], &c. Some few only having little or no ſmell, as bugle, ſelf-heal, and

[e] There are too many exceptions to this, to form a general character, if under the idea of perſonate flowers we include all the plants in the ſecond order of Linnæus's 14th claſs, as Rouſſeau ſeems to do.

[f] Antirrhinum majus Linnæi. Mill. fig. t. 42. pl. 4. f. 2.

[g] Antirrhinum Linaria Linnæi. Curtis I. 47.—It flowers later with us. Moſt of the perſonate tribe flower late.

[h] Here, and in ſome other places, I have taken the liberty of putting plants better known among us, inſtead of thoſe which Rouſſeau has given.

hooded

hooded willow herb: whereas moſt of the plants with perſonate flowers are not odorous, as ſnap-dragon, toad-flax, eye-bright, louſewort, yellow rattle, broom-rape, ivy-leaved toad-flax, round-leaved toad-flax, fox-glove [i], &c. I know of none that have a ſtrong ſmell in this branch but the ſcrophularia, or figwort, which ſmells ſtrong, without being aromatic. Here I am not able to name any but ſuch plants as may perhaps be unknown to you; but you will gradually get acquainted with them, and, whenever you ſee them, you will be able by yourſelf to determine what claſs they belong to. I wiſh you would try to ſettle the branch or ſection by its phyſiognomy; and that you would exerciſe yourſelf in judging at ſight, whether a flower be labiate or perſonate. The exterior form of the corolla may ſuffice to guide you in this choice, which you may verify afterwards by pulling out the corolla, and looking at the bottom of the calyx; for, if you have judged right, the flower which you have named labiate will ſhow you four naked ſeeds, and that which you have named perſonate will ſhow you a pericarp: the contrary would prove that you were miſtaken; and by a ſecond examination of the ſame plant you would prevent a like miſtake another

[i] Some of theſe have the mouth of the corolla gaping. See pl. 4. f. 3.

time.

time[k]. Here, dear coufin, is bufinefs cut out for feveral walks. I fhall not fail to provide fomething for thofe that will fucceed.

[k] This advice will apply in all the other natural claffes. From this paffage it is clear that by labiate flowers Rouffeau underftands all that are included in the firft order; by perfonate flowers all that are in the fecond order of Linnæus's 14th clafs: but many of the flowers in the fecond order have the lips open. Pl. 4. f. 3.

LETTER

LETTER V.

OF UMBELLATE PLANTS.

The 16th of July, 1772.

COMFORT yourſelf, my good couſin, for not having detected the glands in the cruciform flowers. Great botaniſts, and quick-ſighted ones too, have not been more happy. Tournefort himſelf makes no mention of them. They are obvious only in few genera, though we find veſtiges of them in almoſt all; and it is by analyzing ſome of the cruciform flowers, and always obſerving inequalities in the receptacle, and then examining theſe inequalities, that we find out that theſe glands belong to moſt of the genera; and ſuppoſe therefore by analogy that they exiſt in the others, where we do not diſtinguiſh them.

I comprehend that you may not be pleaſed at taking ſo much pains, without knowing the names of the plants which you examine. But I own fairly that it did not enter into my plan to ſpare you that little chagrin. It is pretended that Botany is merely a ſcience of words, which only exerciſes the memory, and teaches the names of plants. For my part, I know not any reaſonable ſtudy which is a mere ſcience of words: and to which of theſe

ſhall

ſhall we give the name of botaniſt, to him who has a name or a phraſe ready when he ſees a plant, but without knowing any thing of its ſtructure; or to him who, being well acquainted with this ſtructure, is ignorant neverthelеſs of the arbitrary name which the plant has in this or that country? If we give our children nothing but an amuſing employment, we loſe the beſt half of our deſign, which is, at the ſame time that we amuſe them, to exerciſe their underſtandings, and to accuſtom them to attention. Before we teach them to name what they ſee, let us begin by teaching them how to ſee. This ſcience, which is forgot in all ſorts of education, ſhould make the moſt important part of it. I can never repeat it often enough; teach them not to pay themſelves in words, nor to think they know any thing of what is merely laid up in their memory.

However, not to play the rogue with you too much, I give you the names of ſome plants, with which you may eaſily verify my deſcriptions, by cauſing them to be ſhown you. For inſtance, if you cannot find a white dead-nettle, when you are reading the analyſis of the labiate or ringent flowers, you have nothing to do but to ſend to an herbariſt for it freſh gathered, to apply my deſcription to the flower; and then having examined the other parts of the plant, in the manner which I ſhall hereafter point out, you will be infinitely better acquainted with the white dead-nettle, than the herbarist who furnished you with it will ever be during his whole life; in a little time, however, we shall learn how to do without the herbarist; but first we must finish the examination of our tribes. And now I come to the fifth, which, at this time, is in full fructification.

Figure to yourself a long stem, pretty straight, with leaves placed alternately upon it, generally cut fine, and embracing at the base, branches which grow from their *alæ*, or *axils*[l]. From the upper part of this stem, as from a centre, grow several pedicles or rays, which spreading circularly and regularly, like the ribs of an umbrella, crown the stem with a kind of basin, more or less open[m]. Sometimes these rays leave a sort of void in the middle, and represent, in that case, more exactly the hollow of a basin: sometimes also this middle is furnished with other rays that are shorter, which, rising less obliquely, form with the others nearly the figure of a half sphere with the convex side uppermost.

Each of these rays is terminated, not by a flower, but by another set of smaller rays, crowning each of the former exactly as the first crown the stem.

[l] The angles formed by a leaf or branch with the stem.

[m] The figure is that of an inverted cone. Pl. 5. f. 1, 2. & pl. 13.

Here

Here then are two ſimilar and ſucceſſive ranks: one of large rays, terminating the ſtem; another of ſmaller rays, like the others; each of them terminating the great ones [n].

The rays of the little umbels are no farther ſubdivided, but each of them is the pedicle to a little flower, of which we ſhall ſpeak preſently.

If you can frame an idea of the figure which I have juſt deſcribed, you will underſtand the diſpoſition of the flowers in the tribe of *umbelliferous* or *umbellate* plants: *umbella* being the Latin word for an umbrella.

Though this regular diſpoſition of the fructification be ſtriking, and ſufficiently conſtant in all the umbellate plants, it is not that however which conſtitutes the character of the tribe. This is taken from the ſtructure of the flower itſelf, which muſt therefore be deſcribed.

But it is expedient, for the ſake of greater clearneſs, to give you in this place a general diſtinction with regard to the relative diſpoſition of the flower and fruit in all plants; a diſtinction which extremely facilitates their methodical arrangement, whatever ſyſtem you adopt for that purpoſe.

The greater number of plants, as the

[n] Linnæus calls the firſt the *univerſal*; and the ſecond ſet the *partial* umbel, or *umbellule*.

pink[o], for inſtance, have the germ incloſed within the flower; theſe are called *inferior flowers*, as incloſing or being below the germ.

Many however have the germ placed below the flower, as in the roſe[p]; for the hep, which is the fruit of it, is that green tumid body which you ſee under the calyx, and this with the corolla crowns the germ, and does not envelope it, as in the former caſe: ſuch are called *ſuperior flowers*, as being above the germ.

The umbellate plants have a ſuperior flower. [q] The corolla has five petals, called regular, though frequently the two outmoſt petals of the flowers at the extremity of the umbel are larger than the three others.

The form of theſe petals varies in the different genera, but it is uſually cordate or heart-ſhaped. They are very narrow next the germ, but gradually widen towards the end, which is emarginate, or ſlightly notched; or elſe they finiſh in a point, which being folded back, gives the petal the air of being emarginate.

Between each petal is a ſtamen, and the anther generally ſtanding out beyond the corolla; the five ſtamens are more viſible

[o] Or jaſmine, roſemary, ſage, borage, primroſe, plum, cherry; all the ringent, cruciform, and papilionaceous tribes; all the compound flowers, &c.

[p] Scabious, honeyſuckle, currant, gooſeberry, elder, ſnow-drop, narciſſu , hawthorn, pear, apple, &c.

[q] See Plate v. f. 5.

than

than the five petals. I make no mention here of the calyx, becauſe it is not very diſtinct in the umbellate plants.

From the centre of the flower ariſe two ſtyles, each furniſhed with its ſtigma, and ſufficiently apparent; theſe are permanent, or continue after the petals and ſtamens fall off, to crown the fruit.

The moſt uſual figure of this fruit is an oblong oval; when ripe it opens in the middle, and is divided into two naked ſeeds faſtened to the pedicle, which, with an art that merits our admiration, divides in two, as well as the fruit, and keeps the ſeeds ſeparately ſuſpended till they fall.

All theſe proportions vary in the different genera, but this is the moſt common order. It requires a very attentive eye to diſtinguiſh accurately objects ſo minute without a glaſs; but they are ſo deſerving of attention, that we cannot regret the trouble of it.

This then is the proper character of the umbellate tribe. A ſuperior corolla, of five petals, five ſtamens, two ſtyles, upon a naked fruit compoſed of two ſeeds growing together.

Whenever you find theſe characters united in one fructification, be ſure that the plant is of this tribe, even though in other reſpects it ſhould have nothing in its arrangement of the order before laid down. And if you ſhould find all this order conformable to my deſcription, and ſee it however contradicted by the examination of the flower, be fure that you are deceived.

For inftance, if it fhould happen that, after having read my letter, you fhould walk out and find an elder in flower, I am almoft certain that at firft fight you would fay, here is an umbellate plant. [r] In looking at it, you would find a large or univerfal umbel, a fmall or partial umbel, little white flowers, a fuperior corolla, and five ftamens; it is certainly an umbellate plant, fay you. But let us fee, let us take a flower.

In the firft place, inftead of five petals, I find a corolla divided into five parts indeed, but all of one piece. Now the flowers of umbellate plants are not monopetalous. There are five ftamens, but I fee no ftyles, and I more often fee three ftigmas than two; more often three feeds than two. Now the umbellate plants have never more or lefs than two ftigmas, and two feeds to each flower. Laftly, the fruit of the elder is a foft berry, and that of the umbellate tribe dry and naked. The elder then is not an umbellate plant.

If now you go back and infpect with more accuracy the difpofition of the flowers, you will fee that the elder has the ftructure of the umbellate tribe only in appearance. Though the principal rays proceed from the fame centre, the fmaller ones are irregular,

[r] See Plate v. f. 4.

and the flowers are borne on a ſecond ſubdiviſion: in ſhort, the whole has not that order and regularity which we find in the umbellate plants. The arrangement of the flowers in the elder is called a *cyme*. Thus by making a blunder ſometimes, we learn to ſee with more accuracy.

Eryngo, on the contrary, has little or nothing the air of an umbelliferous plant, and yet it is one, becauſe it has all the characters of the fructification. If you were by the ſea ſide [5], you would eaſily know it by the bluiſh colour of the leaves, by their prickleneſs, and by the ſmooth membranous conſiſtence of them like parchment. But this plant is uncommon in other ſituations, is rough and untractable, has not beauty enough to make you amends for the wounds it will give you in examining it; and though it were ever ſo beautiful, my little couſin would ſoon be diſguſted at handling ſo ill-humoured a plant. Eryngo.

The umbelliferous tribe is numerous, and ſo natural, that it is very difficult to diſtinguiſh the genera: they are relations, whom we often take for each other, on account of their great reſemblance. To aſſiſt us in diſtinguiſhing them, principal differences are noticed which are ſometimes uſeful, but which we muſt not depend upon too much. The focus of the rays both in the larger or

[5] Eryngo is alſo very common by road-ſides in France, but not with us.

universſal, and in the ſmaller or partial umbel, is not always naked; it is ſometimes ſurrounded with ſmall leaves. This ſet of ſmall leaves or folioles is called the *involucre*. When it is placed at the origin of the univerſal umbel, it is named the univerſal involucre; and when at the origin of the partial umbel, it is named the partial involucre. This gives riſe to three ſections of umbellate plants.

1. Thoſe which have both involucres.

2. Thoſe which have partial involucres only.

3. Thoſe which have neither.

There ſeems a fourth diviſion wanting of thoſe which have an univerſal involucre only; but there is no genus which is conſtantly ſo.

Your aſtoniſhing progreſs, my dear couſin, and unwearied patience, have emboldened me ſo much, that not regarding your ſufferings, I have ventured to deſcribe the umbellate plants, without fixing your eyes upon any model, which muſt needs have rendered your attention much more fatiguing. I am certain, however, that, reading as you do, after you have looked over my letter once or twice, an umbellate plant in flower will not eſcape you; and at this ſeaſon you cannot fail finding many, both in the gardens and the fields.

Moſt of them have their little flowers white. As the carrot, chervil, parſley, hemlock,

hemlock, fool's parſley, angelica, cow-parſnep, water-parſnep, burnet ſaxifrage, pig-nuts, cow-weed, &c [t].

Some, as fennel, dill, parſnep, have yellow flowers; there are ſome few with reddiſh flowers, but none of any other colour.

Here, you will tell me, may be a good general notion of umbellate plants; but how will all this vague knowledge enſure me from confounding fool's parſley with true parſley or chervil, which you have mentioned all together? [u] The meaneſt kitchen-maid will know more of this matter than we with all our learning. You are right. But, however, if we begin with obſervations in detail, we ſhall ſoon be overwhelmed with the number of them; our memory will abandon us, and we ſhall be loſt the firſt ſtep we make in this vaſt region; whereas if we begin with knowing the great roads well, we ſhall ſeldom be loſt in the by-paths, and ſhall always find our way again without much trouble. Let us, however, admit an exception in favour of the utility of the object, and let us not expoſe ourſelves, whilſt we are analyzing the vegetable kingdom, to eat fool's parſley with our meat, or in our ſoup, through mere ignorance.

This plant, which is ſo common a weed

[t] Here, and in other places, I ſet down the names of Hudſon's Flora.

[u] See Pl. v. f. 1, 2, 3.

in

in gardens, is of the umbellate tribe, as well as parſley and chervil. It has a white flower as well as they [v], it is in the ſame ſection with the latter, among thoſe which have the partial, and not the univerſal involucre ; it is ſo like them in its foliage that it is not eaſy to mark the difference in writing. But here follow characters ſufficient to prevent you from being miſtaken.

Fool's Parſley.

You muſt conſider theſe plants when they are all in flower ; for in that ſtate only they have their proper character. The fool's parſley (æthuſa cynapium) has under every partial umbel an involucre of three narrow, long, pointed folioles, all placed on the outer part of the umbel, and hanging down ; whereas the folioles of the partial umbels in the chervil ſurround it entirely, and grow equally on every ſide : and as to parſley, it has only a few ſhort folioles, fine almoſt as hairs, and diſtributed indifferently at the baſe of both umbels.

When you are very certain of the fool's parſley in flower, you will confirm yourſelf in your judgment by ſlightly bruiſing and ſmelling its foliage ; for the diſagree-

[v] The flower of parſley is yellowiſh. But the flowers appear yellow in many of the umbellate plants, from the germ and anthers being ſo, though the corolla is white. *Rouſſeau.*—The germ and anthers alſo are frequently large in proportion to the ſize of theſe minute flowers, and the corolla eaſily falls off, eſpecially with wet.

able

able venomous ſmell will no longer ſuffer you to confound it with parſley or chervil, which have both rather a pleaſant ſmell. Very certain at length, not to make a miſtake, you will examine theſe three plants together and ſeparately in every ſtate, and in all their parts, eſpecially in their foliage, which accompanies them more conſtantly than the flower; and by this examination compared and repeated, till you have acquired certainty at ſight, you will be able to know and diſtinguiſh them without the leaſt trouble. Thus does ſtudy bring us to the very door of practice; after which the latter confers the facility of knowing things.

Take breath, dear couſin, for this is an unconſcionable letter; and yet I dare not promiſe you more diſcretion in the next; after that, however, we ſhall have nothing before us but a path bordered with flowers. You deſerve a garland for the cheerfulneſs and perſeverance with which you have condeſcended to follow me through theſe briars, without being diſcouraged at their thorns.

LETTER VI.

OF COMPOUND FLOWERS.

May the 22d, 1773.

THOUGH there be ſtill, dear couſin, a great deal wanting to complete our idea of the five former tribes of plants, and I have not always known how to adapt my deſcriptions to the underſtanding of our young botaniſt; I flatter myſelf however that I have given you ſuch an idea of them, as to enable you, after ſome months herbarization, to render the *air*, *port*, or *habit* of each tribe familiar to you: ſo, that when you ſee a plant, you may conjecture nearly whether it belong to one of theſe five tribes, and to which; provided always that by an analyſis of the fructification, you afterwards ſee whether you may not have been deceived in your conjecture. The umbellate plants, for inſtance, have thrown you into ſome embarraſſment, from which however you may eaſily eſcape when you pleaſe, by means of the hints which I ſubjoined to my deſcriptions. In ſhort, carrots and parſneps are ſo common, that nothing is eaſier in the middle of ſummer than for the gardener to ſend you one or other of them in flower out of the kitchen garden. Now from the mere view of an umbel,

bel, and the plant which bears it, you muſt acquire ſo clear an idea of the umbellate tribe, that you will rarely be deceived at firſt ſight, whenever you meet with one. This is all that I have hitherto pretended; for we have nothing to do yet with genera and ſpecies; and I repeat it once more, that it is not the nomenclature of a parrot which I wiſh you to acquire, but a real ſcience, and one of the moſt delightful ſciences that it is poſſible to cultivate. I go on therefore to our ſixth tribe before I take a more methodical road. It may perhaps at firſt embarraſs you as much, if not more than the umbellate plants. But my deſign at preſent is nothing more than to give you a general notion of it, eſpecially as we have ſtill plenty of time, before the generality of theſe plants are in full flower; and the interval, well employed, will ſmooth thoſe difficulties againſt which we have not ſtrength to contend.

Take one of thoſe little flowers which, at this ſeaſon, cover all the paſtures, and which every body knows by the name of *daiſy*. [w] Look at it well; for by its appearance, I am ſure you will be ſurpriſed when I tell you, that this flower, which is ſo ſmall and delicate, is really compoſed of between two and three hundred other flowers, all of them perfect; that is, hav- Daiſy.

[w] Plate 6. f. 1.

ing

ing each its corolla, germ, piſtil, ſtamens, and ſeed; in a word, as perfect in its ſpecies as a flower of the hyacinth or lily. Every one of thoſe leaves which are white above and red underneath, and form a kind of crown round the flower, appearing to be nothing more than little petals, are in reality ſo many true flowers; and every one of thoſe tiny yellow things alſo which you ſee in the centre, and which at firſt you have perhaps taken for nothing but ſtamens, are real flowers. If your fingers were already exerciſed in botanical diſſections, and you were armed with a good glaſs, and plenty of patience, I might convince you of the truth of this; but at preſent you muſt begin, if you pleaſe, by believing me on my word, for fear of fatiguing your attention upon atoms. However, to put you at leaſt in the way, pull out one of the white leaves from the flower; you will think at firſt that it is flat from one end to the other; but look carefully at the end by which it was faſtened to the flower, and you will ſee that it is not flat, but round and hollow in form of a tube; and that a little thread ending in two horns iſſues from the tube; this thread is the forked ſtyle of the flower, which, as you now ſee, is flat only at top.

Now look at thoſe little yellow things in the middle of the flower, and which, as I have told you, are all ſo many flowers; if the flower be ſufficiently advanced, you will

will ſee ſeveral of them open in the middle, and even cut into ſeveral parts.

Theſe are monopetalous corollas, which expand, and a glaſs will eaſily diſcover in them the piſtil, and even the anthers with which it is ſurrounded. Commonly the yellow florets towards the centre are ſtill rounded and cloſed. Theſe however are flowers like the others, but not yet open; for they expand ſucceſſively from the edge inwards. This is enough to ſhow you, by the eye, the poſſibility that all theſe ſmall affairs, both white and yellow, may be ſo many diſtinct flowers; and this is a conſtant fact. You perceive, neverthleſs, that all theſe little flowers are preſſed, and incloſed in a calyx, which is common to them all, and which is that of the daiſy. In conſidering then the whole daiſy as one flower, we give it a very ſignificant name, when we call it a *compound flower*. Now there are many genera and ſpecies of flowers formed, like the daiſy, of an aſſemblage of other ſmaller flowers, contained in a common calyx. This is what conſtitutes the ſixth tribe, of which I propoſed to treat, namely, that of the *compound flowers*.

Let us begin by avoiding all ambiguity with regard to the word flower, which we may do in the preſent caſe by reſtraining it to the compound flower[x], and giving the

[x] Pl. 6. f. 1. a.

name

name of *floſcules* or *florets* [y] to the little component flowers; but in the midſt of this verbal preciſion let us not forget that each of theſe florets is a genuine flower.

You have obſerved two ſorts of florets in the daiſy: the yellow ones, which occupy the middle or diſk of the flower, and the little white tongues or ſtraps which ſurround them.

The former are ſomething like the flowers of the lily of the valley, or hyacinth in miniature: and the latter bear ſome reſemblance to thoſe of the honeyſuckle. We ſhall leave to the firſt the name of *florets* [z]; and to diſtinguiſh the ſecond we ſhall call them *ſemi-florets* [a]: for in reality they have a little the air of monopetalous flowers gnawed off on one ſide, and having ſcarcely half the corolla remaining.

Theſe two ſorts of florets are combined in the compound flowers in ſuch a manner, as to divide the whole tribe into three ſections, very diſtinct from each other.

The firſt ſection conſiſts of thoſe which are entirely compoſed of ſemiflorets, both in the middle and circumference; theſe are called *ſemi-floſculous flowers*, and the whole is always of one colour, which is generally yellow. Such is the common dandelion [b],

[y] Pl. 6. f. 1. c. e. f. 2. b. f. 3. b.

[z] Pl. 6. f. 1. e & f. 3. b.

[a] Linnæus alſo calls theſe *ligulate* florets, from *ligula* a ſtrap. Pl. 6. f. 1. c. & f. 2. b.

[b] Pl. 6. f. 2.

the lettuce and ſowthiſtle; the ſuccory and endive, which have blue flowers; the ſcorzonera, ſalſafy, &c.

The ſecond ſection comprehends the *floſculous flowers*, or ſuch as are compoſed of florets only: [c] theſe are alſo commonly of one colour; as immortal flowers, burdock, wormwood, mugwort, thiſtles, and artichoke, which is nearly allied to them: it is the calyx of this that we ſuck, and the receptacle that we eat, whilſt it is yet young, before the flower opens, or is even formed. The choke, which we take out of the middle, is an aſſemblage of florets which are beginning to be formed, and are ſeparated from each other by long hairs fixed in the receptacle.

The third ſection is of flowers compoſed of both theſe. They are always ſo arranged that the florets occupy the centre of the flower, and the ſemi-florets the circumference, as you have ſeen in the daiſy. [d] The flowers of this ſection are called *radiate*. Botaniſts have given the name of *ray* to the ſet of ſemi-florets which compoſe the circumference; and of *diſk* to the area or centre of the flower occupied by the florets. This name of diſk is ſometimes given to the ſurface of the receptacle in which all the florets and ſemi-florets are fixed. In the radiate flowers the diſk is often of one co-

[c] Pl. 6. f. 3.
[d] Pl. 6. f. 1. & Pl. 26.

lour, and the ray of another; there are, however, genera and ſpecies in which both are alike.

Let us endeavour now to fix in your mind an idea of a *compound flower*. The common clover is in blow at this ſeaſon; [e] the flower is purple: if you ſhould take one in hand, ſeeing ſo many little flowers aſſembled, you might be tempted to take the whole for a compound flower. You would however be miſtaken; in what? ſay you. Why, in ſuppoſing that an aſſemblage of many little flowers is ſufficient to conſtitute a compound flower: whereas, beſides this, one or two parts of the fructification muſt be common to them all; ſo that every one muſt have a part in it, and no one have its own ſeparately: theſe two parts in common are the calyx and receptacle. The flower of the clover indeed, or rather the group of flowers, which has the appearance of being but one flower, ſeems at firſt to be placed upon a ſort of calyx; but remove this pretended calyx a little, and you will perceive that it does not belong to the flower, but that it is faſtened below it to the pedicle that bears it. This then is a calyx only in appearance; but in reality it belongs to the foliage, not to the flower; and this ſuppoſed compound flower is only an aſſemblage of very ſmall leguminous or

[e] Pl. 6. f. 4.

papi-

papilionaceous flowers, each of which has its diſtinct calyx, and they have nothing common to them but their being faſtened to the ſame pedicle. Vulgarly all this is taken for one flower; it is a falſe idea however, or if we muſt look upon it as ſuch, we muſt not at leaſt call it a compound, but an *aggregate* or *capitate* flower, or a *head* of flowers; and theſe terms are ſometimes ſo applied by botanical writers.

This, dear couſin, is the moſt ſimple and natural notion I can give you of this numerous claſs of compound flowers, and the three ſections into which it is ſubdivided. I now come to the ſtructure of the fructifications peculiar to this claſs, and this perhaps will bring us to determine the character of it with more preciſion.

The moſt eſſential part of a compound flower is the receptacle [f]; upon which are placed firſt the florets and ſemi-florets, and then the ſeeds which ſucceed them. This receptacle, which forms a diſk of ſome extent, makes the centre of the calyx, as you may ſee in the dandelion, which we will here take as an inſtance. The calyx in this tribe is commonly divided into ſeveral parts, down to the baſe, that it may cloſe, open again, and turn back, as it does during the progreſs of the fructification, without being torn. The calyx of the dandelion is formed of two rows of folioles, inſerted into each

[f] Pl. 6. f. 1. b. & 26. e.

other; and the folioles of the outer row turn back and curl downwards towards the pedicle, whilſt the folioles of the inner row continue ſtraight, to ſurround and hold in the ſemi-florets compoſing the flower.

One of the moſt common forms alſo of the calyx in this claſs is the *imbricate*, or that which is made up of ſeveral rows of folioles, lying over each other like tiles on a roof. The artichoke, blue-bottle, knapweeds, and ſcorzoneras, may ſerve as inſtances of imbricate calyxes.

The florets and ſemi-florets incloſed within the calyx are placed very thick upon the diſk or receptacle in form of a quincunx, or the checks upon a cheſs-board. Sometimes they touch each other without any thing interpoſed between them; ſometimes they are ſeparated by partitions of hairs, or ſmall ſcales, which continue faſt to the receptacle after the ſeeds are fallen. You are now in the way to obſerve the differences of calyxes and receptacles: we will go on then to the ſtructure of florets, and ſemi-florets, beginning with the former.

A floret [g] is a monopetalous flower, commonly regular, with the corolla divided at top into four or five parts. The five filaments of the ſtamens are faſtened to the tube of this corolla: they are united at top into a little round tube, which ſurrounds the piſtil, and this tube is the five anthers

[g] Pl. 6. f. 1. e, f. 3. b.—Pl. 25. f. 2. c. Pl. 26. d.

united

united circularly into one body. This union of the anthers, according to modern botanifts, forms the effential character of compound flowers, and belongs to their florets only, exclufive of all others. If therefore you find feveral flowers upon the fame difk, as in the fcabioufes and teafels, unlefs the anthers are united in a tube round the piftil, and the corolla ftands upon one naked feed, fuch flowers are not florets, nor do they form a compound flower [h]. On the contrary, whenever you find in a fingle flower the anthers thus united, and a fuperior corolla on a fingle feed, this flower, though fole, is a genuine floret, and belongs to the compound tribe; for it is better thus to take the character from a precife ftructure than from a deceitful appearance.

The piftil has the ftyle generally longer than the floret, above which it rifes through the tube formed by the anthers. It is moft frequently terminated at top by a forked ftigma, the two curling horns of which are very vifible. The piftil does not reft upon the receptacle any more than the floret, but both upon the germ, which ferves them as a bafe, and grows and lengthens as the floret withers, becoming in time a longifh feed, remaining faftened to the receptacle till it is ripe: then it falls, if it be naked; or the wind wafts it to a diftance if it be crowned with an egret of feathers or hairs; and the

[h] See Pl. xi. f. 1.

receptacle remains quite naked in ſome genera, but is furniſhed with ſcales or hairs in others.

The ſtructure of the ſemi-florets [i] is like that of the florets; the ſtamens, the piſtil, and the ſeed, are arranged almoſt in the ſame manner; only in the radiate flowers there are many genera, wherein the ſemi-florets of the ray are apt to be abortive, either becauſe they have no piſtils, or becauſe thoſe which they have are barren: in ſuch caſes the flower ſeeds only by the florets in the middle [k].

In the whole compound claſs the ſeed is always ſeſſile, that is, it bears immediately upon the receptacle without any intermediate pedicle. But there are ſeeds in which the down or egret which crowns them is ſeſſile [l]; and others in which it is faſtened to the ſeed by a pedicle [m]. You underſtand that the uſe of this down is to ſpread the ſeeds about to a diſtance, by giving the air more hold upon them.

To theſe irregular imperfect deſcriptions I ſhould add that the calyx has generally the property of opening when the flower expands; of cloſing when the florets fall off, in order to confine the young ſeed, and to hinder it from falling before it is ripe; and,

[i] Pl. 6. f. 2. b. Pl. 25. f. 1. b. Pl. 26. c. and Pl. 27. f. 2. e.

[k] Sunflower.

[l] Thiſtles, artichoke. See Pl. 25. f. 2. c.

[m] Lettuce, dandelion. See Pl. 25. f. 1. d.

laſtly,

laſtly, of opening again and turning quite back to give a larger area to the ſeeds which increaſe in ſize as they grow ripe. You muſt often have ſeen the dandelion in this ſtate, when children gather it, to blow off the down that forms a ball round the reverted calyx.

To underſtand this claſs well, you muſt follow the flowers from before their expanſion to the full maturity of the fruit; and in this ſucceſſion you will ſee transformations and a chain of wonders, which will keep every ſenſible mind that obſerves them in a continual admiration. One flower proper for theſe obſervations is the ſunflower, which is radiate; as are alſo ox-eye, Chineſe aſter, and many others, which are the ornament of the borders in autumn. I have already ſaid that there are thiſtles for the floſculous, and ſcorzonera and dandelion for the ſemifloſculous flowers. All theſe are large enough to be diſſected, and ſtudied with the naked eye, without fatiguing yourſelf too much.

I will not trouble you at preſent any more upon the tribe or claſs of compound flowers. I tremble already at having abuſed your patience too much by details which would have been clearer if I had known how to make them ſhorter; but it is impoſſible for me to avoid the difficulty ariſing from the ſmallneſs of objects. Adieu, dear couſin.

LETTER VII.

OF FRUIT TREES.

HERE, dear coufin, you have the names of thofe plants which you fent me laft. I have put a mark of interrogation to thofe which I had any doubt of, becaufe you had not taken care to put the leaves with the flower, and they are often neceffary to determine the fpecies, efpecially to fo flender a botanift as I am. When you arrive at Fourriere you will find moft of the fruit-trees in flower; and I remember you requefted fome directions from me upon this article. At prefent I can only give you fome hints upon the fubject, becaufe I am very bufy; and yet I would not have you lofe the feafon for this examination.

You muft not, my dear friend, give more importance to Botany than it really has; it is a ftudy of pure curiofity, and has no other real ufe than that which a thinking fenfible being may deduce from the obfervation of nature and the wonders of the univerfe.

Man has changed the nature of many things to convert them better to his own ufe; in that he is not to be blamed; but then it is neverthelefs true that he has often disfigured them, and that when he thinks he is ftudying nature in the works of his own hands, he is frequently miſtaken. This error is found above all in civil ſociety; but it has a place alſo in gardens. The double flowers, which we admire ſo much in our borders and beds, are but monſters, deprived of the power of producing their like; a power with which nature has endowed every organized being. Fruit-trees are ſomewhat in the ſame caſe, by being ingrafted; you may plant the pips or ſeeds of pears and apples of the beſt ſorts, but they will produce nothing but wildings. To know then the pear and the apple of nature, you muſt not look for them in orchards, but in woods. The fleſh or pulp is not ſo large and ſucculent, but the ſeeds ripen better, multiply more, and the trees are vaſtly bigger, and more vigorous. But I am entering on a ſubject that would carry me too far: let us return to the orchard.

Our fruit-trees, though ingrafted, preſerve all the botanical characters which diſtinguiſh them; and it is by an attentive conſideration of theſe characters, as well as by the transformation of the graft, that we aſcertain there being but one ſpecies of pear, for inſtance, under a thouſand different names, by which the ſhape and taſte of their fruits has cauſed them to be diſtinguiſhed into ſo many pretended ſpecies, which are at bottom, but varieties: nay more, the pear and apple are only two ſorts or ſpecies of the ſame kind or genus, and their only characteriſtic

racteriſtic difference is, that the ſtalk of the apple enters into a hollow in the fruit, and that of the pear is faſtened to the narrow part of a fruit a little lengthened out [n]. In the ſame manner the different ſorts of cherries are nothing but varieties of the ſame ſpecies; all the plums are but one ſpecies of plum; nay the genus of *prunus* or plum contains three principal ſpecies; the plum properly ſo called, the cherry and the apricot, which alſo is only a ſpecies of plum. Thus when the learned Linnæus, in dividing the genus into its ſpecies, has enumerated the domeſtic plum, the plum cherry, and the plum apricot [o]; ignorant people have laughed at him, but obſervers have admired the juſtneſs of his arrangement.

The fruit-trees belong moſtly to a numerous tribe, which has a character not difficult to ſeize; the ſtamens, which are many in number, inſtead of ariſing from the receptacle, are faſtened to the calyx, [p] either immediately, or with the corolla, which is

[n] Nor is this always conſtant, ſome pears having the common ſhape of the apple. It is extremely difficult to find any permanent differences between fruits, which are diſtinguiſhed by every body at firſt ſight. We may add, however, that the corollas of the pear are white, thoſe of the apple red on the outſide: the apple alſo has a firmer pulp, and none of thoſe tubercles which ſome ſorts of pear have: and, laſtly, the leaves of the pear are very ſmooth; thoſe of the apple more rounded, leſs ſerrated, and villous underneath.

[o] 1. Prunus domeſtica. 2. Prunus Ceraſus. 3. Prunus Armeniaca. The fruit-trees are figured by Duhamel.

[p] Pl. 18. f. 1. c. and f. 2.

polypetalous,

polypetalous, and conſiſts commonly of five petals. The following are characters of ſome of the principal genera.

The pear, comprehending alſo the apple and the quince, has the calyx monophyllous, divided into five ſegments; the corolla of five petals faſtened to the calyx, about twenty ſtamens, all faſtened likewiſe to the calyx. The germ is inferior, and there are five ſtyles. The fruit, as every body knows, is fleſhy, and has five cells containing the ſeeds.

The genus plum, comprehending the apricot and cherry, as was before obſerved, and alſo the laurel, has the calyx, corolla, and ſtamens, nearly as in the pear. But the germ is ſuperior, or within the corolla; and there is but one ſtyle. The fruit is rather watery than fleſhy, and contains a ſtone.

The genus almond, including the peach and nectarine, is almoſt like the plum, but the germ has a down upon it, and the fruit, which every body knows is ſucculent in the peach, and dry in the almond, incloſes a hard ſtone, which is rough and full of cavities [q]

All this is very roughly ſketched out, but I hope contains enough to amuſe you for the preſent. Adieu, dear couſin.

[q] Beſides thoſe mentioned above, this claſs, called *icoſandria* by Linnæus, contains other fruits, as the pomegranate, ſervice, medlar, raſpberry, ſtrawberry, &c.

LETTER

LETTER VIII.

OF MAKING A HORTUS SICCUS, OR HERBARIUM.

April the 11th, 1773.

THE earth, dear coufin, begins to put on its green robe, the trees to bud, the flowers to open; fome are even already paft; an inftant of delay would be the lofs of a whole year for Botany: I proceed then without farther preamble.

I fear we have hitherto treated our fubject in too abftract a way, by not having applied our ideas to determinate objects: it is a fault which I have been guilty of, efpecially in the umbellate tribe. If I had begun by fetting one of them before your eyes, I fhould have fpared you a very fatiguing application to an imaginary object, as well as a very difficult defcription to myfelf, and fuch as a fingle look would have fupplied. Unfortunately, at a diftance to which the law of neceffity reftrains me, I am not able to deliver the objects into your hand; but provided each of us can fee with the fame eyes, we fhall underftand one another very well, when we relate what we fee. The whole difficulty is, that the indication muft come from you; for to fend you dried plants from

from hence, would be doing nothing. To know a plant well you muſt begin with ſeeing it growing. A *hortus ſiccus*, or *herbarium*, by which Latin terms we call a collection of dried plants, may ſerve to put us in mind of the plants we have once known; but it gives us only a poor knowledge of thoſe we have never ſeen before. You therefore muſt ſend me ſuch plants as you wiſh to know, and have gathered yourſelf; and it is my buſineſs to name, claſs, and deſcribe them; till by comparative ideas, become familiar to your eye and your underſtanding, you arrive at claſſing, arranging, and naming, by yourſelf, thoſe which you ſee for the firſt time: and this is the ſcience which diſtinguiſhes the true botaniſt from the mere herbariſt or nomenclator. My deſign then here is to teach you how to prepare, dry, and preſerve plants, or ſpecimens of plants, in ſuch a manner as that they may be eaſily known and determined. In a word, I propoſe to you to begin a *hortus ſiccus*. Here is a deal of buſineſs preparing at a diſtance for our little botaniſt: for at preſent, and for ſome time to come, the addreſs of your fingers muſt ſupply the weakneſs of hers.

Firſt, here is ſome proviſion to be made; namely, five or ſix quires of gray paper, and almoſt as many of white, of the ſame bigneſs, pretty ſtrong and well ſized, without which the ſpecimens would rot in the

gray

gray paper, the plants, or at leaſt the flowers, would loſe their colour, and this, of all the parts, is that by which they are moſt eaſily known, and which it is moſt pleaſant to ſee in a collection of dried plants[r]. It were alſo to be wiſhed that you had a preſs of the ſame ſize with your paper, or at leaſt two pieces of board well planed, between which you may keep your papers and ſpecimens, preſſed by ſtones or any other weight, with which you may load the upper plank. When you have made theſe preparations, you muſt obſerve the following rules, in order to prepare your plants ſo as to preſerve them and know them again.

The preciſe time to gather your plant is when it is in full flower, or rather when ſome of the flowers begin to fall, to give place to the fruit, which begins to make its appearance. It is at this time, when all parts of the fructification are viſible, that you muſt endeavour to gather the plant in order to dry it.

Small plants may be taken whole with their roots, which muſt be bruſhed, that no earth may remain. If the earth be wet, it muſt either be dried, that it may be bruſhed, or elſe the root muſt be waſhed; but in this caſe you ſhould wipe it well, and dry it before you put it into the papers, without which it would infallibly rot and injure the

[r] See Dr. Withering's Arrangements of Britiſh Plants, edit. 2. introd. p. 45.

plants near it. You need not, however, preſerve the roots, unleſs they have ſome remarkable ſingularities; for in moſt plants the branching fibrous roots are ſo alike, that it is not worth the trouble. Nature, which has done ſo much for elegance and ornament, in the form and colour of plants, in whatever ſtrikes our ſight, has deſtined the roots entirely to uſeful functions; becauſe being concealed within the earth, to give them an agreeable ſtructure, would have been to hide a light under a buſhel.

Trees and all great plants can only be had by ſpecimens: but then that ſpecimen ſhould be ſo well choſen, as to contain all the conſtituent parts of the genus and ſpecies, that it may ſuffice to know and determine the plant from whence it is taken. It is not ſufficient that all the parts of the fructification are diſtinguiſhable, which would be enough to determine the genus; but the character of the foliation and ramification alſo muſt be ſufficiently viſible; that is, the origin and form of the leaves and branches, and even, as much as may be, ſome portion of the main ſtem itſelf; for, as you will ſee in the ſequel, all this ſerves to diſtinguiſh the ſpecies of the ſame genus, which are perfectly alike in the flower and fruit. If the branches are too thick, they may be made thinner, by cutting them with a ſharp knife nicely underneath, as much as may be, without cutting and

and mutilating the leaves. There are botanists who have the patience to slit the bark, and draw the wood out so nicely, that when the bark is united again, the branch seems to be entire though the wood is gone: by which means there are none of those inequalities and bumps, which spoil and disfigure a collection, and give a bad form to the plants. Where the flowers and leaves do not come out at the same time, or grow too far distant from each other, you will take a little branch in flower, and another in leaf, and placing them together on the same leaf of your book, you thus have before you different parts of the same plant, sufficient to give you a complete knowledge of it. As to plants where you find only the leaves, the flower being either past or not yet come, you must wait with patience till they show their faces, to be fully acquainted with them. A plant being no more certainly to be known by its foliage than a man by his clothes.

Such is the choice that you should make in what you gather: you must have a choice also as to the time in which you do it. Plants gathered in the morning before the dew is off, or in the evening when it is damp, or in the day-time when it is wet, will not keep. You must absolutely choose a dry season, and even then, the driest and hottest time of the day, which in summer is between eleven in the morning and five in the afternoon. Even then, if you find the leaſt moiſture on them, you muſt not take them, for they will certainly not keep.

When you have gathered your ſpecimens, you muſt bring them home as ſoon as you can, quite dry, to put and arrange them in your papers. For this purpoſe you lay down at leaſt one ſheet of gray paper, upon this half a ſheet of white paper, and then your plant, taking great care that all the parts of it, eſpecially the leaves and flowers, are well opened, and laid out in their natural ſituation. If the plant be a little withered, without being too much ſo, it will generally ſpread out better upon the paper, with the fingers and thumb. But there are rebellious plants which ſtart up on one ſide, whilſt you are ranging them on the other. To prevent this inconvenience, I have leads, halfpence, and farthings, which I place upon thoſe parts that I have juſt put in order, whilſt I am arranging the reſt, ſo that when I have done, my plant is almoſt covered with theſe pieces, which keep it in its proper ſituation. Then you place another half ſheet of white paper upon the firſt, preſſing it with your hand, to keep the plant in the poſition you have given it, bringing your left hand that preſſes gradually forward, and at the ſame time taking away the leads, &c. with your right; then put another ſheet of gray paper upon the ſecond white paper, all the while preſſing the plant, leſt it loſe the poſition you have given it: upon the gray paper place another half ſheet of white, as before; upon this another plant arranged and covered like the former, till you have placed your whole harveſt, which ought not to be too numerous at once; both that your taſk may not be too laborious, and that your paper may not contract too much humidity during the drying; which would infallibly ſpoil your plants, unleſs you haſtened to change the papers with the ſame attention as before; this, however, is what you muſt do from time to time, till your ſpecimens have taken their bent, and are all very dry.

Your pile of plants and papers thus arranged, muſt be put into the preſs, without which your plants will not be flat and even; ſome are for preſſing them more, others leſs; experience will teach you this, as well as how often the papers ſhould be changed, without taking unneceſſary pains. Laſtly, when your plants are quite dry, put each of them ſeparately into a ſheet of paper, one upon another, without other papers between, for which there is no occaſion, and you will thus begin a *hortus ſiccus*, which will continually increaſe with your knowledge, and at length contain the hiſtory of all the vegetation of the country. Take care always to keep your collection very cloſe, and a little preſſed; without which the plants, however dry they might be, will attract the humidity of the air, and again get out of form.

Now the uſe of all theſe pains is to arrive at a knowledge of each particular plant, and to underſtand one another well when we talk of them.

For this purpoſe you muſt gather two ſpecimens of each plant ; one larger to be kept, the other ſmaller to ſend me. You muſt number them carefully, ſo that both great and little ſpecimen ſhall always have the ſame number. When you have a dozen or two of ſpecies thus dried, you will ſend them to me in a little parcel by the firſt opportunity. I will ſend you back their names and deſcriptions; by means of the numbers you will know them in your collection, and after that in their natural ſtate, wherein, I preſume, you firſt examined them. This is the certain way to make as ſecure and rapid a progreſs as you can, at a diſtance from your guide.

P. S. I forgot to tell you that the ſame papers may ſerve over and over again, provided you take care to air and dry them well. I ſhould alſo add here, that your *hortus ſiccus* muſt be kept in the drieſt part of the houſe, and rather on the firſt than the ground-floor.

LETTER IX.

EXPLANATION OF THE CLASSES IN THE LINNÆAN SYSTEM.

March the 25th, 1774.

I Have received all your packets very ſafe, and cannot but admire the neatneſs with which you have arranged your plants; the care you have taken in having all the parts neceſſary to determine both the genus and ſpecies in your ſpecimens; and the brilliancy of colour in moſt of the flowers. All this ſerves to ſhow how much better the female fingers are adapted to ſuch operations than ours. I am pleaſed alſo to hear that our little botaniſt had ſo large a ſhare in laying out and drying theſe plants, which I ſhall carefully preſerve as a memorial of the induſtry and adroitneſs of both. But what gives me the moſt pleaſure is, to ſee that you have remarked, with ſo much ſucceſs in general, to which of the natural claſſes your plants belong: ſo that I am well convinced you have profited by my leſſons, and have paid a due attention to my letters.

What reward, dear couſin, can I give you for your unwearied patience and perſeverance in following me through ſo much abſtract

abſtract matter, when your curioſity muſt needs have been piqued, and your deſire of being acquainted with the rank and names of the beautiful objects which you gathered arranged, and dried, with ſo much affection, muſt have been awakened? I have now, in ſome degree, endeavoured to content you, by the paper which accompanies this, containing the names of all the plants in your packets, placed after the numbers which you have put to them in your collection: ſo that to the common objects which you knew by rote, you are now enabled to add a conſiderable number, whoſe acquaintance you will value more, becauſe you know them, upon thorough examination. You have therefore ſo many more points to reſt upon; but this is not ſufficient; you cannot be a botaniſt till you are able to help yourſelf, to caſt me off entirely, and to find out a plant with which you are unacquainted. All this, however, will ſtill require ſome time and patience; and as you remember that you are not to take any more ſteps in this kingdom than are agreeable, you will inform me when you are tired.

Such information I propoſe now to convey to you by degrees: and having initiated you by ſhowing how you may determine the claſs of ſome plants, I will now open the whole myſtery, and inſtruct you how to determine the claſs of them all. To do this you muſt learn a ſyſtem; in which, however, you are not to expect that all vegetables are arranged in natural claſſes, ſuch as I have hitherto explained to you, but after an artificial method, the order of nature not being in all points yet unveiled to our mortal eyes. Your pains, however, will not have been thrown away; becauſe I promiſe you that our artificial ſyſtem ſhall preſerve the natural tribes which you have ſtudied ſo well.

Do not ſuffer yourſelf to be terrified at the word *ſyſtem*. I promiſe you there ſhall be little difficulty in it to you who have patience and attention; and as little parade of hard words as poſſible, only allowing me to name my claſſes and orders [r]. The ſyſtem I propoſe to you is not the French one by Tournefort, which is very beautiful, and has great merit; but the Swediſh one by Linnæus. I prefer this, becauſe it is moſt complete, and moſt in faſhion.

You are ſo well acquainted with all the conſtituent parts of the fructification, that you need not be told what the ſtamens and piſtils are. Linnæus has founded his claſſes upon the former, and many of his orders upon the latter of theſe. But at preſent

[r] The Engliſh ſtudent will find great advantage in poſſeſſing many elementary books, explaining all the terms, in his own language. Now alſo he has Linnæus's ſyſtem of vegetables and genera tranſlated. Hudſon's Flora Anglica, and Withering's arrangement, connect the Engliſh names with thoſe of Linnæus.

the

the claſſes will furniſh you with ſufficient employment.

I ſuppoſe you take a plant in hand that is in full flower; the firſt thing you have to ſee is, whether the flowers are complete or perfect, that is, have both ſtamens and piſtils: if ſo, view the ſtamens well, in order to diſcover whether they are entirely ſeparate from the piſtil and each other from top to bottom, or united in ſome part or other; if they are ſeparate, of the ſame, or an indeterminate length, and leſs in number than twenty, then the number alone will ſuffice to determine the claſs; and thoſe which have one ſtamen will belong to the firſt claſs entitled *monandria*; thoſe with two ſtamens to the ſecond, *diandria*; thoſe with three to the third, *triandria*, and ſo on to the tenth, entitled *decandria*[s]. Theſe are Greek names, and ſome of them not ſhort ones: ſince however they are only four-and-twenty in all, you will indulge me ſo far in time as to have them by heart. The flowers for examination ſhould be gathered as nearly as poſſible in their natural ſtate; for many of thoſe which are cultivated in gardens undergo ſtrange transformations, and either loſe the ſtamens and piſtils entirely, or acquire an additional number. The firſt claſſes, which have but few ſtamens, are not ſo liable to change as

[s] Plate 7. to 16. with pl. 5. & 1.

thoſe which have many. Thus the number in the three claſſes already mentioned is not variable; nor in the fourth claſs, *tetrandria*. In the fifth, *pentandria*, ſome plants have more than their proper quota of ſtamens to the flower, at leaſt when cultivated in gardens; but this is a very numerous claſs, and it is no wonder if we find ſome few irregular among ſo many. To ſecure you in ſome meaſure againſt miſtakes on this and other occaſions, I muſt obſerve, that nature in general carries a certain proportion through all the parts of the ſame work; and therefore if you have a flower which has a calyx divided into five ſegments, and a corolla conſiſting of five petals, or divided into five parts; if you count ſix or ſeven ſtamens, be ſure all is not right, and take the pains to inſpect ſome other flowers of the ſame ſpecies, before you determine. I dare affirm ſuch examination will convince you that your flower belongs to the fifth claſs, *pentandria*, in which the natural number of ſtamens is five. In the ſixth claſs, *hexandria*, whoſe beautiful flowers have ſix ſtamens, I do not obſerve ſo conſiderable a variation as one might expect in plants that are ſo much the objects of culture; you will however frequently count more than ſix ſtamens in the flowers of the tulip. The flowers of the claſs *heptandria* ſhould have ſeven ſtamens; but you will often find thoſe of the horſe-cheſnut

chesnut faulty in this respect. As you will also some flowers in the three following classes, *octandria* which has eight, *enneandria* which has nine, and *decandria* which has ten stamens, as the names all imply. With a little attention however to the proportion of the parts, and by a repetition of your examination where any doubt arises, you will find these ten classes easy to determine.

No flowers being known at present that have constantly and regularly eleven stamens, the eleventh class in the system of Linnæus contains those which have twelve; and is therefore entitled *dodecandria* [u]. But the genera which have this precise number being few; and, as I observed before, the number being uncertain when the stamens are many, all plants are comprehended in this class that have any number of stamens, from eleven to nineteen inclusive, provided they are disunited.

All plants that have more separate stamens than these belong to one of the two following classes. Here then you must take in another consideration, besides the number of the stamens, to determine in which of these two classes you are to search for your plant. This consideration is, the *situation* of the stamens; which in the class *icosandria*, is either on the calyx or corolla [v], and

[u] Plate 17.
[v] Plate 18.

in the thirteenth, *polyandria*, on the baſe or receptacle of the flower [w]. This difference of ſituation is only to be attended to in thoſe flowers which have many ſtamens; for you will frequently obſerve in the fifth claſs that the monopetalous flowers have the ſtamens growing out of the corolla; but this circumſtance has nothing to do in determining their claſs. The twelfth claſs has its name *icoſandria*, from the flowers in it having uſually twenty ſtamens or thereabouts, at leaſt in the greater part of the genera: this circumſtance, however, is not to determine the claſs; but all plants which have many ſtamens, that is, more than nineteen, faſtened either immediately, or mediately by means of the claws of the petals, to the calyx, are to be referred to the claſs *icoſandria*. To aſſiſt you farther in diſtinguiſhing the flowers of this from thoſe of the following claſs, it may be remarked that the calyx in this is monophyllous or all of one piece, and concave; and the corolla is fixed by its claw or ſmall end into the calyx, inſtead of the baſe or bottom of the flower, as it generally is in the other claſſes.

When on the contrary you find more than nineteen detached ſtamens in the ſame flower, with a piſtil or piſtils, and ſituated on the baſe or receptacle of the flower, that plant muſt belong to the claſs *polyandria*,

[w] Plate 19.

ſignifying

ſignifying many ſtamens, and the ſtamens may vary in number from twenty to a thouſand in the different genera. Theſe alſo either have a polyphyllous calyx, that is, conſiſting of ſeveral folioles, generally five, or none at all; though ſometimes it falls off, as in the poppy when the flower opens.

We have hitherto ſuppoſed you to find all the ſtamens of the ſame length, or nearly ſo; or if not, ſtill we preſume that you have not found a certain regular and determinate proportion in their lengths. Now, on the contrary, we ſuppoſe you to take up a flower which has an appearance of regularity in its whole ſtructure; and that, on an attentive examination, you diſcover four ſtamens, not all equal in length, but ranged in one row, and the inner pair ſhorter than the outer one. This plant will probably belong to the fourteenth claſs, the name of which is *didynamia* [x], ſignifying that two of the ſtamens are ſtronger than the others. Here you will immediately perceive that you are got among your old acquaintance, for it will ſtrike you that all the flowers which have the character juſt deſcribed are either labiate or perſonate, and therefore that you were miſtreſs of the claſs *didynamia*, before you knew that it had this Greek name [y]. All then that I need ſay to you is, that Linnæus makes the eſſential character

[x] Plates 20. & 4.

[y] See Letter IV.

to consist, in the proportional arrangement of four stamens above expressed, accompanied with one pistil, and invested with an irregular monopetalous corolla.

There is yet another class of these plants with proportional stamens, which, though you do not know it by the dreadful long name *tetradynamia*, is however one of your first acquaintance under the gentler appellation of cruciform flowers [z]. These, you remember, have four stamens longer than the other two: this is the classical character, and hence its name. For the other distinctive marks by which this class is readily known at first sight, you have them at your fingers ends.

You are now in possession of all those classes which have the stamens free, separate, disunited. If a flower that has both stamens and pistils should present itself, in which you find the stamens united at bottom, it certainly belongs to one of the three next classes: and if, on the contrary, they are united at top, that is, the anthers form one body, it will belong to the nineteenth class.

In the sixteenth class, called *monadelphia* [a], the filaments are united so as to form one regular membrane at bottom, whilst they are distinct at top. Of this character you have a clear and convincing instance in

[z] See Letter II. Plates 21. & 2.
[a] Plate 22.

that

that very common plant the mallow. In ſome others, however, of this claſs, the character is not ſo evident, and without a careful inſpection of the flowers to the very bottom, you might eaſily be tempted to give them to another claſs. Obſerve then farther, that the flower has always a calyx, and frequently a double one: that the corolla conſiſts of five heart-ſhaped petals: that the receptacle of the fruit, as it is called, or the column to which the ſeeds are faſtened, projects above them in the centre of the flower: that the germs ſurround this in a ring: that all the ſtyles are united at bottom and form one body with the receptacle, but are divided at top into as many threads as there are germs: and that theſe germs grow into a kind of capſule divided into as many cells as there are piſtils, or conſiſting of the ſame number of arils, which are looſe coats covering each ſeed ſeparately, and not eaſily falling from it.

In the ſeventeenth claſs, *diadelphia*, the filaments are united at bottom: not however into one, but two bodies. Theſe flowers alſo have but one piſtil; the fruit is a *legume* or pod; and if I add that the flowers are papilionaceous, you will immediately diſcover that this is another claſs with which you are perfectly acquainted, and with the form of whoſe flowers you were ſo much delighted [b].

[b] See Letter III. Plates 23. & 3.

In the eighteenth claſs the filaments are united in three or more bundles, and the name of it is *polyadelphia*[c]. The union being generally at the bottom only, without extending up the filaments, and the flowers having no diſtinguiſhing character, you muſt pull out the ſtamens, in order to be certain that the plant belongs to this claſs. The names of the three laſt-mentioned claſſes ſignify literally one, two, and three brotherhoods.

If inſtead of the filaments being joined at bottom, they are free and diſtinct, but the anthers are connected together, ſo as to form one body, then your plant will be found in the claſs *ſyngeneſia*. But the flowers in this claſs being ſmall, and the abovementioned circumſtance not being the firſt that will ſtrike an examiner of flowers, it muſt be added that they are *compound*; and this one word is ſufficient to overcome the whole difficulty with you who know theſe flowers at firſt ſight, and have ſo frequently diſſected the florets and ſemi-florets which compoſe them[d].

Though in the four laſt claſſes the ſtamens have been in ſome ſort united, yet both in theſe, and in all the former, they have been found detached from the piſtil, ſo at leaſt as that the one may be taken off from the plant without the other. But what if a

[c] Plate 24.

[d] See Letter VII. and Plates 25. to 29. & Pl. 6. Syngeneſia ſignifies *congeneration*, or union of the anthers.

flower

flower ſhould occur to you in which you are unable to do this, but you find on the contrary that the ſtamens grow upon the piſtil itſelf? Then, I anſwer, it belongs to a claſs entitled *gynandria* [e], which is the twentieth in the ſyſtem of Linnæus, and derives its name from this peculiar circumſtance, by which it ſtands inſulated as it were, and detached from all the others. From the poſition of the piſtils in this claſs, ariſes a ſingularity in the appearance and ſhape of the flowers in moſt of the genera; and ſometimes the receptacle is lengthened out in form of a ſtyle, and bears both ſtamens and piſtils upon it[f].

Hitherto you have been concerned with ſuch plants only as have flowers which I call complete or perfect, becauſe they have both ſtamens and piſtils. But a plant perhaps may have occurred to your obſervation in which you have found theſe parts always in ſeparate, diſtinct flowers. In this caſe I beg leave to coin two words, and to call thoſe which have only the ſtamens *ſtaminiferous*, and thoſe which have only the piſtils *piſtilliferous* flowers. Now when you find theſe, and theſe only on the ſame tree or plant, that tree or plant belongs to the twenty-firſt claſs in the arrangement of Linnæus, called by him *monœcia* [g], a term ſignifying one houſe: the

[e] Plate 30.

[f] As in the common Arum, Curtis, Lond. 2. Mill. fig. 52. 1. J. Mill. illuſtr. Ger. 834. 1.

[g] Plate 31.

flowers

flowers of different kinds being produced in the ſame habitation, or on the ſame individual plant. Whereas in the following claſs, theſe ſtaminiferous and piſtilliferous flowers are not merely ſeparate from each other, but are always found on diſtinct plants of the ſame ſpecies, and in other reſpects ſo alike, as not to be diſtinguiſhed when they are out of flower. The name of this claſs therefore is *diœcia* [h], ſignifying two houſes, and implying that incomplete flowers only are found in different habitations, or on ſeparate trees or plants, never on the ſame.

There remains now only one poſſible caſe to provide for, in the arrangement of conſpicuous flowers, which is this. Suppoſe you find ſome flowers that are complete, and at the ſame time others which bear only ſtamens or piſtils, on the ſame plant with the complete flowers, or on different plants of the ſame ſpecies; there is a claſs, namely, the twenty-third, provided for the reception of ſuch plants, and it is entitled *polygamia* [i], from this variety in the flowers [k].

For plants with inconſpicuous flowers, as being of leſs conſideration, there is only one claſs provided, and that is called *cryptoga-*

[h] Plate 32.

[i] Plate 33.

[k] Thunberg, and ſome others, have ſunk the four claſſes from Gynandria to Polygamia, melting the ſpecies into other claſſes. I ſhall not diſpute the propriety or convenience of this reformation: but it is my deſign to explain the ſyſtem of Linnæus, as the great author himſelf delivered it.

mia,

mia[1], from the circumſtance of the fructification being concealed, or not obvious to our eyes. For the flowers in the moſt perfect of theſe are hardly to be diſtinguiſhed without a glaſs, and in many not even with it; nay, the moſt acute obſervers have not detected flowers in them all, though in all probability there is no vegetable without them. They will be eaſily known from plants with conſpicuous flowers, by their ſingular ſtructure; as you will readily acknowledge when I inform you that the objects of this, the loweſt claſs of vegetables, are ferns, moſſes, ſea-weeds, and funguſes: and therefore when we talk of inconſpicuous flowers, we do not mean to include ſuch as are deſtitute of a magnificent corolla, but ſuch only as have not the ſtamens and piſtils viſible to the naked eye. But you are too good an obſerver to require ſuch admonitions. By this time you are doubtleſs ſufficiently fatigued, as well as myſelf, with all this dry matter; and what is worſe, you have not learnt to find out one plant: but patience, we are in the way, and have made great progreſs, though we are not arrived at the end of our journey. We will ſoon make another long ſtage, unleſs you tell me you have enough, and in that caſe I promiſe to trouble you no more with this traſh: if it does not amuſe and even intereſt you, throw it at once into the fire.

[1] Plate 35 to 38.

LETTER X.

EXPLANATION OF THE ORDERS IN THE LINNÆAN SYSTEM.

May the 1ſt, 1774.

PRESUMING, dear couſin, that you have already examined abundance of ſpring flowers, and determined their claſſes, upon the inſtructions contained in my laſt letter, I ſhall proceed in this to give you the characters of the orders, or diviſions of the claſſes. If you were to proceed at once to the examination of the ſpecies, all would be confuſion ; juſt as if you attempted to eſtimate a vaſt mixt multitude, fluctuating in tumultuary diſorder : but if you have patience to make a regular progreſs ; to throw this multitude into large bodies, to ſubdivide theſe into ſmaller ones, and theſe again into others ſo ſmall as to command them well with the eye, you have at length a regular army, which you can number, arrange, and diſcipline at your pleaſure. We will now divide our twenty-four regiments into their reſpective companies. Here I think you will not find ſo much difficulty as in the claſſes : for the orders in the firſt thirteen claſſes are founded wholly upon the *number* of the piſtils, ſo that the chief of your taſk here will be to learn ſo many new terms, which are formed by putting *gynia* inſtead of *andria* to the Greek words ſignifying the numbers: as *monogynia*, one piſtil; *digynia*, two piſtils; and ſo on.

After the firſt thirteen claſſes we no longer uſe the piſtils for the purpoſe of ſubdividing the claſſes into orders. In the claſs *didynamia* it would be nugatory, becauſe you have obſerved that all the flowers of the ringent tribe have one piſtil, and no more. Here then we have recourſe to another circumſtance which anſwers extremely well. For we find that moſt of the plants which have a labiate flower have four naked ſeeds at the bottom of the calyx; and that the perſonate flowers are ſucceeded by a capſule containing many ſmall ſeeds: hence ariſes an elegant, commodious, obvious, and natural diviſion of the fourteenth claſs into two orders, *gymnoſpermia* [m] and *angioſpermia* [n]; the firſt containing all the ringent flowers with four naked ſeeds ripening in the calyx: the ſecond ſuch as have the ſeeds contained in a bilocular pericarp, or ſeed-veſſel of two cells, and faſtened to a receptacle in the middle of it.

In the next claſs, *tetradynamia*, the flowers have alſo one piſtil and no more. Here again it is found convenient to take the fruit

[m] Plate 20. f. 1. & Pl. 4. f. 1.

[n] Plate 20. f. 2. & Pl. 4. f. 2, 3.

for the ſubdiviſion of it into orders. Theſe are called *ſiliculoſa* [o] and *ſiliquoſa* [p], from the form of the fruit, which we call *ſilicle* and *ſilique*; having only the word *pod* current in our language, which will not ſuffice to diſtinguiſh theſe from each other, nor from the pod in the leguminous tribe. The plants of the firſt order then have a *ſilicle* or ſhort roundiſh pericarp; thoſe of the ſecond a *ſilique* or oblong narrow pericarp: both are bilocular; but the ſtructure has been already ſufficiently explained [q].

In the 16th, 17th, and 18th claſſes it is found beſt to take the orders from the number of ſtamens. Here then is no ſort of difficulty; and, what is very pleaſant, you have no new terms to burden the memory.

The chief difficulty, with reſpect to the orders, lies in the claſs *ſyngeneſia*. Tournefort's diviſion of the compound flowers into *floſculous*, *ſemi-floſculous*, and *radiate*, was pretty and obvious; but Linnæus's is abſtruſe and difficult. I will explain it to you however as clearly as I can. You are perfect miſtreſs of a compound flower, and the different ſorts of florets of which it is compoſed [r]. I muſt next inform you, therefore, that what you know by the name of compound, is called by Linnæus a *floſculous* flower; and that he calls the florets, *tubu-*

[o] Plate 2. i, k, l.
[p] Plates 21 & 2.
[q] See Letter II.
[r] See Letter VI.

lous floſcules, and the ſemi-florets, *ligulate* floſcules; this being premiſed, we may uſe the language of Linnæus or Tournefort as we pleaſe. Now if you examine theſe floſcules nicely, you will diſcover that they have ſometimes both ſtamens and piſtil; but you will ſee that others have ſtamens only; others again a piſtil only: and laſtly, ſome have neither ſtamens nor piſtil. The firſt of theſe I call *perfect* [s] floſcules; the ſecond *ſtaminiferous*, the third *piſtilliferous*, and the fourth *neuter* floſcules. All theſe variations are to be found both in the tubulous and ligulate floſcules; and muſt be well attended to, becauſe on theſe variations, aſſiſted by the form of the florets, Linnæus has founded the four firſt orders of this claſs.

Polygamia æqualis [t] is the name of the fir ord er. *Polygamia* is the family name, which this has in common with all the orders except the laſt; it is uſed only in oppoſition to *monogamia*, and implies that there are many florets incloſed within one common calyx; which is your idea of a compound flower. The peculiar name *æqualis* ſignifies equal, regular, or alike, and implies that the whole flower is regular, and that all the component floſcules therefore, whether tubulous or ligulate, are alike; and indeed they are not only ſo,

[s] Perfect at leaſt in appearance, if not always really ſo.
[t] Pl. 6. f. 2. & 25. f. 2.

but likewiſe perfect, or all furniſhed with ſtamens and piſtil; and therefore each followed by a ſeed. If theſe flowers have any ligulate floſcules, all the reſt are ſo; if any tubulous floſcules, all the reſt are ſo likewiſe, except in two genera, Atractylis and Barnadeſia, which have radiate flowers.

In the ſecond order, *polygamia ſuperflua* [u], all the florets of the diſk, centre or middle of the flower are perfect; thoſe of the ray or exterior part piſtilliferous: both of them produce ſeed. Moſt of the flowers in this order are *radiate*, and then they are eaſily known by the circumſtance of having fertile ſeeds both in the diſk and ray: but there are ſome which have tubulous florets only, and appear like the diſk of a radiate flower, as a daiſy would look when ſpoiled of its white ſemi-florets; whence Ray called them *diſcoid* flowers: in theſe however, on an attentive inſpection, you will diſcover that ſome of the outer ones are deficient in ſtamens at leaſt, if not in corolla too. Theſe are by much the largeſt orders, each of them containing almoſt double the number of genera, that are in the three remaining orders of compound flowers taken together.

The third order of theſe compound flowers, or of the claſs *ſyngeneſia*, is entitled *fruſtranea* [v]. The character of the order is, that the florets in the diſk or centre are

[u] Pl. 6. f. 1. & pl. 26. [v] Pl. 27. f. 1.

perfect,

perfect, and produce seed; whilst those of the ray are imperfect, and therefore abortive or frustrate, whence the name. This is a very small order, containing only eight genera; of which seven have radiate flowers, and the eighth, which however is a numerous one, has capitate flowers like the thistles, but differing from them in having either neuter or abortive florets next the calyx, as in the common blue-bottle; in which the neuter floscules distinguish themselves by being much larger than the others; but on examination they are mere corolla, and nothing else.

In the fourth order, *necessaria* [w], the florets in the disk or middle are apparently perfect, but are not really so, and therefore produce no perfect seed; whilst the pistilliferous floscules in the ray or outside of the flower are fertile. All these have radiate flowers, except in two genera, wherein the exterior fertile florets have scarcely any corolla.

In the fifth order, *polygamia segregata* [x], there is a common calyx, as in the foregoing orders; but besides that there is in this order a partial one, including one or more florets, which are thus separated from each other in a manner different from the rest of the orders: and hence the name. By this order the compound approach the aggregate flowers; such as the teasel, scabious, &c.

[w] Pl. 27. f. 2. [x] Pl. 28.

but then theſe have not the character of the claſs ſyngeneſia in the union of the anthers.

The ſixth or laſt order is entitled ſimply *monogamia* [y], becauſe it conſiſts of plants with *ſimple*, not compound flowers, which circumſtance is abundantly ſufficient to diſcriminate this order, provided you attend at the ſame time to the claſſical character.

We have now, dear couſin, happily, I hope, paſſed *the fool's bridge*, and are arrived ſafely on the other ſide, where the way is plain, and we ſhall ſoon get pleaſantly to the end of our ſtage. In ſhort, the orders of the three following claſſes, *gynandria*, *monœcia*, and *diœcia*, being founded upon the ſtamens, and taking their names from the foregoing claſſes, according to the number, and union or diſunion of the ſtamens in the reſpective flowers; there is nothing new to be learnt in any of theſe.

The twenty-third claſs indeed, *polygamia*, has three orders, ariſing from the triple mode in which the three ſorts of flowers may be arranged; either on the ſame plant, on two diſtinct plants, or on three. When the perfect and imperfect flowers are on the ſame plant, the order is entitled *monœcia* [z]. When the perfect flowers are on one plant, and the imperfect ones on a ſecond of the

[y] Pl. 29.—The violets are a good inſtance of this order.

[z] Plate 33. Acer or maple.

ſame

ſame ſpecies, the order is then entitled *diœcia*[a]. And when the perfect flowers are on one plant, ſtaminiferous ones on a ſecond, and piſtilliferous ones on a third, all of the ſame ſpecies; then ſuch plant belongs to an order called *triœcia*[b], ſignifying three houſes; the three ſorts of flowers having three diſtinct habitations.

The laſt claſs having no flowers whoſe parts are diſcernible by the naked eye; and therefore called *cryptogamia:* having alſo many genera in which we are uncertain what the fructification is; many in which we can diſcern no fructification at all: the characters of the orders can no longer be taken from the ſtamens and piſtils. Fortunately the plants of this claſs have a very particular ſtructure, ſerving very well both to aſcertain the claſſical character, and the diviſion of it into four Orders; which are called, I. *Filices*, or Ferns. II. *Muſci*, or Moſſes. III. *Algæ*, or Sea-weeds; and, IV. *Fungi*, or Funguſes.

The *ferns*[c] moſtly have their fructification upon the backs of their leaves. This, when examined by the microſcope, appears to conſiſt of a ſcale ariſing from the leaf, and opening on one ſide; and under that, ſome little balls on pedicles, ſurrounded by an elaſtic ring: in due time the balls burſt,

[a] The aſh is an inſtance of this order.

[b] as in the Fig.

[c] Plate 35.

and

and throw out a fine duſt, which is ſuppoſed to be the ſeed. Linnæus makes the ſcale to be a calyx: and the globules are probably ſo many capſules or pericarps.

The *moſſes* [d] have ſmall threads growing out of the boſoms of the leaves, terminated by a ſmall body, the whole reſembling ſtamens: accompanied by little ſhorter threads ſuppoſed to be piſtils, ſometimes on the ſame plant with the former, and ſometimes on another. The firſt of theſe Linnæus took for anthers, and actually called them ſo; but he ſuſpected them afterwards to be capſules, and ſuch they turn out to be, on a narrower inſpection with greater magnifiers.

Of the *algæ* [e] we know too little about the fructification to give a regular character of the order, which includes not only the ſea-weeds, but the liverworts, &c. theſe have been ranged by others among the moſſes. In the liverworts there are little bodies viſible enough, which are taken for ſtaminiferous and piſtilliferous flowers, diſtinct from each other; but experiments are yet wanting to aſcertain them with preciſion. On the ſea-weeds are little bladders, ſome hollow with hairs within, others filled with a gelly-like ſubſtance; and theſe are ſuppoſed to be the flowers and fruits.

If the *funguſes* [f] have any fructification, it

[d] Plate 36. [e] Plate 37. [f] Plate 38.

is

is imagined to be underneath, in the gills, pores, &c. But I will not detain you with these dregs of vegetable nature, in which you will take no pleasure till you have imbibed an enthusiastic passion for botany.

After the class *cryptogamia* Linnæus has given the *palms*, in a twenty-fifth class, or appendix, without any character. I presume he has thus thrown them into the rear of his system, partly because he could not have ranged this proud set of trees according to his laws, without tearing them from each other; and partly because they have not been examined with sufficient accuracy; you will scarcely have an opportunity of examining this natural class, the most remarkable characters of which are, that the staminiferous flowers are distinct from the pistilliferous, on the same or different individuals; except in one genus, which has complete or perfect flowers accompanied by staminiferous ones on the same individual; all proceeding from a *spathe* or sheath, and growing upon a *spadix*[g]. So that these trees belong to the three last classes of conspicuous flowers in the artificial system.

[g] The *spadix* is the receptacle in this tribe, and has no English name. In another place, Linnæus, in distributing vegetables into nine nations, assigns the first to the palms, calling them Princes of India, bearing their fructification on a *spadix*, within a *spathe*; flowing; remarkable for their prodigious height; distinguished by an unvaried, undivided, perennial trunk; crowned at top by an evergreen bush of leaves; rich in abundance of large, fine fruit.

Thus,

Thus, dear coufin, we have accomplifhed our fecond ftage. And this letter not being of fo unconfcionable a length as the former, I have accompanied it with two tables; one of the claffical characters, and another explaining thofe of the orders: that after reading my diffufe explanation, you may have the whole under your eye at once; and thus perhaps at one view form a better idea of the arrangement of vegetables into claffes and orders, than you could do from many detached pages [h]. We are not yet arrived at fpecific or individual information, but we are on the borders, as I fhall convince you in my next letter. In the mean time you have fufficient employment for your eyes and attention, without doors as well as within: for if you had taken up this trafh of mine only in your dreffing room, you would long fince have thrown it into the fire; if it meets with a better fate, I owe it merely to the beautiful objects which your fair hands have cropt in the garden and fields. Always give the preference to the latter where you can, both for the fake of exercife, and having your plants in their natural ftate. Adieu, dear coufin; continue your kind indulgence to my prate.

[h] See Curtis's beautiful explanation of Linnæus's Syftem of Botany, with coloured plates. And an Illuftration of the Syftem of Linnæus, by John Miller; who has given a plate of one genus in every clafs and order. Lond. 1779, octavo.

SKETCH AND EXPLANATION OF THE ORDERS IN THE SYSTEM OF LINNÆUS.

I. Monandria. *One ſtamen.*
   1. Monogynia. *One piſtil.*
   2. Digynia. *Two piſtils.*

II. Diandria. *Two ſtamens.*
   1. Monogynia. *One piſtil.*
   2. Digynia. *Two piſtils.*
   3. Trigynia. *Three piſtils.*

III. Triandria. *Three ſtamens.*
   1. Monogynia. *One piſtil.*
   2. Digynia. *Two piſtils.*
   3. Trigynia. *Three piſtils.*

IV. Tetrandria. *Four equal ſtamens.*
   1. Monogynia. *One piſtil.*
   2. Digynia. *Two piſtils.*
   3. Tetragynia. *Four piſtils.*

V. Pentandria. *Five ſtamens.*
   1. Monogynia. *One piſtil.*
   2. Digynia. *Two piſtils.*
   3. Trigynia. *Three piſtils.*
   4. Tetragynia. *Four piſtils.*
   5. Pentagynia. *Five piſtils.*
   6. Polygynia. *Many piſtils.*

VI. Hexandria. *Six equal ſtamens.*
   1. Monogynia. *One piſtil.*
   2. Digynia. *Two piſtils.*
   3. Trigynia. *Three piſtils.*
   4. Tetragynia. *Four piſtils.*
   5. Polygynia. *Many piſtils.*

VII. Heptandria. *Seven stamens.*
1. Monogynia. *One pistil.*
2. Digynia. *Two pistils.*
3. Tetragynia. *Four pistils.*
4. Heptagynia. *Seven pistils.*

VIII. Octandria. *Eight stamens.*
1. Monogynia. *One pistil.*
2. Digynia. *Two pistils.*
3. Trigynia. *Three pistils.*
4. Tetragynia. *Four pistils.*

IX. Enneandria. *Nine stamens.*
1. Monogynia. *One pistil.*
2. Trigynia. *Three pistils.*
3. Hexagynia. *Six pistils.*

X. Decandria. *Ten stamens.*
1. Monogynia. *One pistil.*
2. Digynia. *Two pistils.*
3. Trigynia. *Three pistils.*
4. Tetragynia. *Four pistils.*
5. Pentagynia. *Five pistils.*
6. Decagynia. *Ten pistils.*

XI. Dodecandria. *Twelve stamens,* (from 11 to 19.)
1. Monogynia. *One pistil.*
2. Digynia. *Two pistils.*
3. Trigynia. *Three pistils.*
4. Pentagynia. *Five pistils.*
5. Dodecagynia. *Twelve pistils.*

XII. Icosandria. *Twenty stamens,* (on the calyx or corolla.)
1. Monogynia. *One pistil.*
2. Digynia. *Two pistils.*
3. Trigynia. *Three pistils.*

4. Pentagynia.

4. Pentagynia. *Five pistils.*
5. Polygynia. *Many pistils.*

XIII. Polyandria. *Many stamens*, (from 20 to 1000, on the receptacle.)

1. Monogynia. *One pistil.*
2. Digynia. *Two pistils.*
3. Trigynia. *Three pistils.*
4. Tetragynia. *Four pistils.*
5. Pentagynia. *Five pistils.*
6. Hexagynia, *Six pistils.*
7. Polygynia. *Many pistils.*

XIV. Didynamia. *Four stamens, 2 longer and 2 shorter.*

1. Gymnospermia. *Four naked seeds.*
2. Angiospermia. *Seeds inclosed in a pericarp.*

XV. Tetradynamia. *Six stamens, 4 longer and 2 shorter.*

1. Siliculosa. *Pericarp generally roundish, with the style permanent or continuing*, called a silicle.
2. Siliquosa. *Pericarp very long and narrow, called a* silique *or* pod.

XVI. Monadelphia. *One brotherhood; or filaments all connected.*

1. Triandria. *Three stamens.*
2. Pentandria. *Five stamens.*
3. Octandria. *Eight stamens.*
4. Decandria. *Ten stamens.*
5. Endecandria. *Eleven stamens.*
6. Dodecandria. *Twelve stamens.*
7. Polyandria. *Many stamens.*

XVII. Diadelphia. *Two brotherhoods: or filaments in two bodies.*

1 Pentandria.

1. Pentandria. *Five stamens.*
2. Hexandria. *Six stamens.*
3. Octandria. *Eight stamens.*
4. Decandria. *Ten stamens.*

XVIII. Polyadelphia. *Many brotherhoods. filaments in three or more parcels.*

1. Pentandria. *Five stamens.*
2. Dodecandria. *Twelve stamens,*
3. Icesandria. *Twenty stamens.*
4. Polyandria. *Many stamens.*

XIX. Syngenesia. *Congeneration. Anthers united.*

1. Polygamia Æqualis. *All the floscules perfect, and the whole flower regular.*
2. Polygamia superflua. *Perfect floscules in the disk: pistilliferous floscules in the ray: both producing seed.*
3. Polygamia Frustranea. *Floscules in the disk perfect, and producing seed: in the ray imperfect, and without seed.*
4. Polygamia Necessaria. *Floscules in appearance perfect in the disk producing no seed: pistilliferous floscules in the ray producing seed.*
5. Polygamia Segregata. *Many floriferous calyxes contained in one common calyx, and forming one flower.*
6. Monogamia. *Flowers not compound, as in the other orders, but simple, as in all the other classes.*

XX. Gynandria. *Stamens growing on the pistil.*

1. Diandria.

1. Diandria. *Two stamens.*
2. Triandria. *Three stamens.*
3. Tetrandria. *Four stamens.*
4. Pentandria. *Five stamens.*
5. Hexandria. *Six stamens.*
6. Octandria. *Eight stamens.*
7. Decandria. *Ten stamens.*
8. Dodécandria. *Twelve stamens.*
9. Polyandria. *Many stamens.*

XXI. Monoecia. *One house. Imperfect flowers separate on the same plant.*

1. Monandria. *One stamen.*
2. Diandria. *Two stamens.*
3. Triandria. *Three stamens.*
4. Tetrandria. *Four stamens.*
5. Pentandria. *Five stamens.*
6. Hexandria. *Six stamens.*
7. Heptandria. *Seven stamens.*
8. Polyandria. *Many stamens.*
9. Monadelphia. *Filaments united in one.*
10. Syngenesia. *Anthers united.*
11. Gynandria. *Stamens on the pistil.*

XXII. Dioecia. *Two houses. Imperfect flowers on distinct individuals.*

1. Monandria. *One stamen.*
2. Diandria. *Two stamens.*
3. Triandria. *Three stamens.*
4. Tetrandria. *Four stamens.*
5. Pentandria. *Five stamens.*
6. Hexandria. *Six stamens.*
7. Octandria. *Eight stamens.*
8. Enneandria. *Nine stamens.*
9. Decandria. *Ten stamens.*

10. Dodecandria. *Twelve stamens.*
11. Polyandria. *Many stamens.*
12. Monadelphia. *Filaments united in one.*
13. Syngenesia. *Anthers united.*
14. Gynandria. *Stamens on the pistil.*

XXIII. Polygamia. *Perfect flowers, accompanied with one or both sorts of imperfect flowers.*

1. Monœcia. *Perfect and imperfect flowers on the same plant.*
2. Diœcia. *Perfect flowers on one plant, and imperfect on another.*
3. Triœcia. *Perfect flowers on one plant, staminiferous flowers on a second, and pistilliferous flowers on a third.*

XXIV. Cryptogamia. Fructification secret.

1. Filices. *Ferns: bearing seed on the back of the leaves.*
2. Musci. *Mosses: having imperfect flowers distinct, and the seeds in a capsule, often covered with a veil.*
3. Algæ. *Having imperfect flowers distinct, and the seeds either like a meal on the leaves or inclosed in bladders.*
4. Fungi. *Having no discernible flowers, but seeds in the gills, pores, cups, &c.*

XXV. Palmæ: *Palms. Flowers on a spadix, in a spathe or sheath: generally staminiferous and pistilliferous distinct.*

LETTER

LETTER XI.

OF THE CLASS MONANDRIA.

June the 10th, 1774.

AT length, dear coufin, I am going to put you in the way of examining plants by yourfelf, and determining the genus and fpecies, as you have before done the clafs and order. You have been already initiated in my firft letters; but now I fhall proceed in more form, and prefent you with one plant or more of each clafs; explaining to you as we go along fome others of the natural claffes; which form, or are contained in the artificial ones.

The firft clafs, *Monandria*, in the Syftem of Linnæus is a very fmall one; comprifing, as you have feen already, in the fecond table which I fent you, but two orders. There are alfo but eighteen *genera* in it, and forty-four fpecies. Very few of thefe plants are natives of Europe; and the Indian forts are not eafy to be met with, at leaft in flower, in the beft hot-houfes.

There is a plant, however, not very uncommon in ponds, ditches, and flow muddy Hippuris. ftreams, called *Hippuris*, which is of this clafs, and of the firft order. It has a fingle jointed ftalk, and at each joint is a dozen leaves or more, placed all round in a whorl, which is a form that Linnæus calls *Verticillate*. To each of theſe leaves, cloſe to the ſtalk, belongs a little flower, conſiſting of a ſingle ſtamen and piſtil, one ſeed, and nothing more; for it has neither calyx nor corolla. You will find the ſtamen ſitting on the germ terminated by a bifid anther; and behind this is the ſtyle, which is terminated by a ſtigma tapering to a point. This will be amply ſufficient for you to determine the *Hippuris*[i], which perhaps may not grow near you; and if it does, you muſt not hazard wetting and dirtying yourſelf in a muddy ditch. Since therefore it is abundant in the moat of the neighbouring abbey, I have incloſed ſome ſpecimens of it in my tin pocket caſe, which may ſerve afterwards to bring home your plants freſh and cool, if you are not already provided with ſo neceſſary a thing. If you are not ſtruck with the beauty of the *Hippuris*, you will at leaſt eſteem it for its modeſty and ſimplicity. I have one favour to aſk in return for my tin box and its contents, which is, that whenever you call this plant by its name, you will pronounce the middle ſyllable long, and not ſhort, as

[i] I do not know that this plant has been noticed enough to have a common name in Engliſh. In the books it is called *Female Horſe-Tail* or *Mare's Tail.* Figured in Curtis, Flora Londinenſis. Faſcic. IV. Plate I. and Pl. 7. f. 2. of this work.

many

many do: for I am ſolicitous to pronounce, as well as think, like you. I have ſaid nothing here of the diſtinction between genus and ſpecies, becauſe there is only one ſort of *Hippuris*. I muſt however inform you, once for all, that we invariably take the characters of the genera from the parts of fructification; and thoſe of the ſpecies from the other parts of the plant, particularly the leaves.

There is another plant of this claſs and order, which your gardener may poſſibly have in the hot-houſe. I dare ſay you know it by the upright growth, reedy appearance, and fine ſcarlet flowers. Perhaps you have already found ſome difficulty in determining the claſs and order; for there is no filament, but the anther grows to the edge of a kind of petal, which Linnæus calls the *Nectary:* the ſtyle alſo, which is lance-ſhaped, grows to the ſame petal. The calyx conſiſts of three leaves: the corolla is cut into ſix parts, five erect, and the ſixth reflexed; the ſeeds are contained in a capſule or veſſel of three cells, are round and very hard; whence this plant has the name of *Indian ſhot*. Linnæus calls it *Canna*. Thus much for the *Genus*, of which there are three ſpecies at leaſt; ſome make five. Linnæus has diſtinguiſhed his three ſpecies thus. 1. Canna indica[k]; by its Canna.

[k] This is figured by John Miller, in his Illuſtrations of the Sexual Syſtem:—and in Pl. 7. f. 1. of this work.

ovate leaves, ſharp-pointed towards both ends, and marked with nerves. 2. C. anguſtifolia, *Narrow-leaved Indian ſhot*, by its lance-ſhaped, petiolate leaves, marked alſo with nerves. 3. C. glauca, *Sea-green Indian ſhot*, by its lance-ſhaped petiolate leaves, ſmooth or without nerves[1]. Yours will be one of the two firſt ſpecies, for the laſt has yellow flowers. This order contains ſeveral intereſting plants, ſuch as ginger, cardamom, grain of paradiſe, Arabian coſtus, turmerick, galangale, &c. all which, with *Canna*, belong to a natural tribe entitled *Scitamineæ*, from the Latin word *ſcitum*, which when tacked to *edulium* implies eatables of a pleaſant taſte. They have not only the ſame place in the artificial ſyſtem, but they agree farther in having their ſeeds encloſed in a veſſel below the receptacle, as you perceive plainly it is in the *canna:* the diviſions alſo of the calyx, corolla, and ſeed veſſel, are uſually three.

Short flights are beſt, till you have tried your wings. My next may poſſibly be a little longer, if you give me leave. Adieu for a few days.

[1] The Hortus Kewenſis has only two ſpecies; making Linnæus's ſecond, a variety only of the firſt.

LETTER

LETTER XII.

OF THE CLASS DIANDRIA.

June 17th, 1774.

YOU have ſtarved a week, dear couſin, upon the meagre fare of my laſt: I can now promiſe you more variety, having a larger range and better choice. The ſecond claſs of plants, *diandria*, has 35 genera, and 265 ſpecies.

Linnæus has done every thing in his power to facilitate the inveſtigation of plants; and nothing contributes more to this than the clearneſs and order of his arrangement, and his leading on the ſtudent by regular ſteps from generals to particulars. Thus, after you have ſettled the claſs and order of your plant, you perceive that each order, when numerous, is thrown into ſeveral great diviſions, before you are preſented with the generic characters. This ſhortens your inquiry conſiderably; for, in the firſt order of this claſs, inſtead of having the characters of thirty-five genera to chooſe out of, you have by this means only eight or nine, or perhaps no more than three, or even one. That you may underſtand this the better, I will give you Linnæus's ſubdiviſion of the firſt order of this claſs.

DIANDRIA MONOGYNIA.

1. Flowers inferior, monopetalous, regular, 8 genera.
2. ——— inferior, monopetalous, irregular, with ſeeds incloſed in a veſſel: 9 genera.
3. ——— inferior, monopetalous, irregular, with naked ſeeds: 9 genera.
4. ——— inferior, pentapetalous: 1 genus.
5. ——— ſuperior: 3 genera.

So that if your plant happens to belong to the fourth diviſion, it is determined at once: and in all the reſt your ſearch is much facilitated[m].

In this claſs, though by no means one of the moſt numerous, you will not be at a loſs, either in your garden or in the fields, for examples.

Jaſminum. You are well acquainted with moſt ſorts of jaſmine. Take any of them, and you will perceive immediately that it belongs to the firſt diviſion of the firſt order. Compare as many of the ſpecies as you can meet with in flower, and you will find that they all agree in the characters of it.

[m] It is not neceſſary to be more particular with the Engliſh reader, ſince the botanical ſociety at Lichfield have publiſhed a tranſlation of Linnæus's Syſtem of Vegetables.

But

But other circumſtances are to be found in them all, called *generic characters*: theſe in the preſent caſe are: that the corolla is monopetalous, ſalver-ſhaped [n], and the border divided into five ſegments: the anthers ſmall, and lying within the tube of the corolla: the ſeed-veſſel a berry of two cells: and the ſeeds covered with an *aril* or looſe coat.

Having ſeen in what all the jaſmines agree, to determine the claſs, order, with its diviſions, and genus; now attend to the circumſtances in which they differ, to ſettle the ſix ſpecies. For this the leaves will nearly ſuffice, thus:

1. Leaves pinnate, oppoſite: lobes diſtinct. *Jaſmine officinal.* Curt. Magaz. 31. Pl. 8. f. 2.
2. Leaves pinnate, oppoſite: lobes confluent. *J. Catalonian.*
3. Leaves ternate, oppoſite. *J. Azorian.*
4. Leaves ternate and ſimple, alternate; branches angulate: *J. ſhrubby.*
5. Leaves ternate and pinnate, alternate, acute: branches angulate. *J. dwarf.*
6. Leaves ternate and pinnate, alternate, obtuſe: branches round. *J. ſweet-ſcented.*

[n] If the reader be at a loſs for the meaning of terms, there is no want of books to conſult; ſuch as Lee's and Roſe's Introductions, Berkenhout's Dictionary, Milne's Inſtitutes, &c.

The

The three first have the corolla white; in the three last it is yellow. If you inquire after your favourite *Arabian jasmine*, it belongs to another genus, *Nyctanthes*, because it has the calyx and corolla divided into eight segments. The *Cape jasmine* is of another class, the fifth; and of course has another name, *Gardenia*.

Several other trees and shrubs belong to this same first division. *Privet*, *Phillyrea*, *Olive*, and the *Lilacs*. These have all a quadrifid corolla; and are distinguished by their fruit, which in privet is a *berry* with four seeds; in phillyrea a *berry* with one seed; in olive a *drupe*; in the lilacs a bilocular *capsule*. The common lilac has heart-shaped leaves; a circumstance sufficient to distinguish it from the Persian, which has lance shaped leaves. As to the different colours of the flowers in the first—white, blue, and red, they form but varieties: colour being rarely permanent enough to constitute specific differences.

Veronica. In the second division is a genus, named from a female saint, *Veronica*: it is a very numerous one, containing no less than forty species. Here therefore Linnæus has done with the genus, as he did before with the order—he has thrown it into three principal divisions from the manner of flowering. 1. Such as bear the flowers in spikes. 2. Such as bear them in racemes or bunches. 3. Such as produce them singly.

This

This genus is easily known by the monopetalous, rotate, or wheel-shaped corolla, divided into four segments, the lowest of which is narrower than the rest; and the bilocular, heart-shaped, flatted capsule.

One species is very common among bushes, and in the edges of pastures. Its beautiful blue flowers have doubtless attracted your notice, and in falling off too easily, have given occasion perhaps to a lesson on the short duration of our enjoyments, or the fleeting nature of female charms, to your lovely daughter. If it be not already past flowering, for May is its season, you will find that it belongs to the second division; or even if it be, the oval, wrinkled leaves, indented about the edge, and sitting close to the stalk, together with the weak trailing stems, unless upheld by the bushes, will so clearly point out this humble plant to you, that you cannot well be mistaken [o].

If this species however is out of blow, you will certainly find another [p] in dry pastures or heaths, especially upon old ant-hills: it may perhaps have escaped you; the flowers being small, and of a pale colour; not however without their beauty, on a nearer survey. This belongs to the

[o] Veronica Chamædrys. Wild Speedwell or Germander. Curtis, Lond. I. 2.—Pl. 8. f. 1.

[p] Veronica officinalis. Officinal Speedwell. Curtis, Lond. III. 1.

first

firſt diviſion; having the flowers growing in ſpikes, coming out chiefly from the ſide of the plant, at ſome diſtance from the main ſtem; the leaves are oppoſite, and the ſtalks trail along the ground. It has the trivial name of officinal, becauſe an infuſion of it is ſometimes uſed medicinally.

Other ſpecies are common by the ſides of ditches and brooks, whence they have the name of *Water Speedwell*, or *Brooklime*[q]: theſe are of the ſecond diviſion: and three ſpecies of the third diviſion are abundant among corn, in the ſpring[r].

I know not how it is, but there is a connexion between this claſs and the fourteenth. *Pinguicula* or *Butterwort* has a perſonate flower. Some ſpecies of *Vervain* have two ſtamens, others four of unequal lengths; among the latter is our *common* or *officinal Vervain*[s]; whence ſome authors have removed it to the claſs *didynamia*. *Sage*, *Roſemary*, and others, have labiate flowers, and in every reſpect ſo reſemble the plants of the fourteenth claſs, that they ſhould naturally be placed there; but having only two ſtamens, the artificial ſyſtem

Salvia. ranges them in this claſs. *Sage* ſeems to form the connecting link between the two claſſes; for in this genus are rudiments of

[q] Veronica Becabunga. Curtis, Lond. II. 3. is one of theſe.

[r] Veronica arvenſis Curtis, Lond. II. 2. agreſtis Curtis, Lond. I. 1. hederifolia Curtis, Lond. II. 1.

[s] Curtis, Lond, I. 41.

another

another pair of ſtamens, but without anthers. The ſtructure of the ſtamens in the ſage is ſingular, and merits your obſervation. The two filaments are very ſhort, but two others are faſtened to theſe tranſverſely by the middle; and at one end of theſe laſt is a gland, at the other an anther. This circumſtance diſtinguiſhes the genus from all others, and is called its eſſential character. If you compare the flowers of ſage and roſemary together, you will find them agree in moſt other particulars; but roſemary has not this character: it has very long filaments, bending towards the caſque or upper lip of the corolla.

The genus *Salvia* or *Sage* has no leſs than fifty-two ſpecies. Our common garden ſage [t], of which there are ſeveral varieties, has the flowers growing in ſpikes, the ſegments of the calyx acute, and the leaves of an oblong ovate form, entire, and very ſlightly notched about the edges. There are two ſorts commonly wild in Europe [u], not very unlike each other; but rather *clarys* than *ſages*: You will be at no loſs to know them when you ſee them. To diſtinguiſh them from each other obſerve that *Meadow Clary* [v] has the leaves oblong-heart-ſhaped, and notched about the edges; the

[t] Salvia officinalis *Linnæi*. Pl. 8. f. 3. Ger. 764.

[u] Salvia pratenſis & verbenaca; but the latter only is common in England.

[v] Salvia pratenſis. Ger. 769. 3.

upper

upper ones embracing the ſtalks; the flowers grow in almoſt naked whorls, and the upper lip of the corolla is glutinous. The *Wild Clary*[w] has the leaves ſerrate, ſinuate, and ſmoothiſh: the tube of the corolla very ſmall in compariſon with the calyx, which opens wide.

But enough for our ſecond excurſion, eſpecially as I propoſe that we ſhould take a third very ſoon.

[w] Salvia verbenaca. Ger. 771. 1. The edition of Gerard's Herbal which is quoted here and elſewhere, is that which received the additions of Johnſon, and was printed in 1636.

LETTER XIII.

OF CORN AND GRASSES.

June the 24th, 1774.

I HAVE haſtened this letter, dear couſin, leſt the induſtrious mower ſhould have ſpoiled our harveſt. The brilliancy of the preſent ſeaſon will perhaps have quickened his ſteps: but at the worſt, he will have left you ſome gleanings about the hedges.

The tribe which I now recommend to your examination, is the moſt known and general of any; it is the moſt pleaſant to the eye, and of the moſt extended uſe, ſince it furniſhes man with the beſt portion of his nouriſhment, and at the ſame time is the whole ſupport of many among the beaſts, and of a large proportion of birds. The moſt rigid critic cannot accuſe us of miſpending our time, when we are engaged in the contemplation of ſo uſeful a tribe of plants as that which contains all the different ſpecies of corn and graſſes.

The former being larger, requiring more care and culture, becauſe they are annual, and being immediately neceſſary to the ſupport of man, and the animals about him, in this and many other countries; the ſpecies are

are univerſally known and diſtinguiſhed. But this is not the caſe in the latter; graſs vulgarly forms one ſingle idea; and a huſbandman when he is looking over his incloſure, does not dream that there are upwards of three hundred ſpecies of graſs, of which thirty or forty may be at preſent under his eye. They have ſcarcely had a name, beſides the general one, till within theſe twenty years; and the few particular names that have been lately given, are far from having obtained general uſe: ſo that we may fairly aſſert that the knowledge of this moſt common and valuable tribe of plants is yet in its infancy [x].

Let us not however give more importance to Botany than it really has; but proceed quietly with our own buſineſs. The greater part of the world ſcarcely know that graſs has a flower; or, if they are ſhown

[x] The late excellent Mr. Stillingfleet firſt directed the public attention to *graſſes*; and that moſt reſpectable and uſeful inſtitution, the Society of Arts, &c. has done all in its power to promote an improvement in the culture of them; but without great effect. Nor can much be expected till economical gardens or public farms are inſtituted, for the purpoſe of experiment in this and other parts of huſbandry. It is not enough to tell men of a good thing, and inſtruct them how it may be done; but they muſt actually ſee it put in execution, and be eye-witneſſes of its good effects.—This has lately been done by ſome public-ſpirited gentlemen; particularly by Mr. Coke, of Norfolk. *See Young's Annals.*—Mr. Curtis's Practical Obſervations on the Britiſh Graſſes are highly deſerving of the attention of the public.—See alſo Mr. Swayne's Gramina Paſcua.

it, will coldly aſk, Is this all? And yet graſs not only has a flower, but every conſtituent part of it; which is more than we can ſay of a tulip, and ſome others, that have engroſſed almoſt all the attention of mankind: nay, there is ſuch a variety in the parts, diſpoſition, and manner of flowering, that we have ſufficient marks in the fructification to diſtinguiſh above forty genera.

If you take up a ſpike [y] or panicle [z] of graſs, you may perhaps be diſappointed in your expectation of diſcerning the ſtamens and other parts; be aſſured then that the flower is not yet open, and continue your ſearch till you find one with the parts expanded, the ſlender filaments hanging out, and large, oblong, double anthers playing freely about with the ſlighteſt motion. You will immediately perceive that your graſs, having three of theſe ſtamens, muſt range under the third claſs, *triandria*, provided the flower has a piſtil as well as ſtamens. Searching a little farther, you will eaſily detect two reflex ſtyles, each terminated with a feathered ſtigma: you are at no loſs therefore to determine that your graſs belongs to the ſecond order *(digynia)* of this third claſs [a].

Having thus ſettled the claſs and the order, you will proceed to the other parts of

[y] Pl. 9. f. 1. [z] Pl. 9. f. 2. [a] See Plate 9, b, c.

the flower. The neglected chaff you will find to be double: the outer generally consisting of two leaflets; one large and gibbous, the other smaller and flat; the inner consisting also of two parts or valves, which you may call petals, for this is the corolla, and the former is the calyx. Nay this despised flower has even its *nectary*; which is a little oblong body composed of two leaflets, but so small as to require a glass to discern it well. Grasses have no *pericarp*, but one naked seed, with the shape of which we are well acquainted—it is oblong, and draws to a point towards each end. These characters you will find common to every grass you examine, and also to every species of corn; or however with very few exceptions: this then is called the *classical character*. As these small flowers grow frequently two or more close together, you have only to separate a single flower to avoid confusion in your examination.

But this tribe of plants does not agree in the parts of fructification only, as above described. The whole appearance, the general air, the manner of growth, is the same in all. A simplicity of structure runs through the entire class. Every one has a simple, unbranched, straight, hollow stem, strengthened with knots at certain intervals[b]. There is none but has a single leaf to

[b] Linnæus names it *culmus*.

each

each knot, investing or sheathing the stem to some distance, and then spreading out into a long narrow surface, of equal breadth all the way, till it approaches the end, when it draws off gradually to a point [c]. It is also invariably entire in every species; and without veins or branching vessels, being only marked longitudinally with lines parallel to the sides, and to a nerve or ridge that runs the whole length of it. There is another curious circumstance, almost peculiar to this tribe of plants, and common to them all; namely, that the body of the seed does not split into two lobes, but continues entire [d], till it has accomplished its purpose of giving the young plant its first nourishment, and then rots away: this you may easily observe as corn is springing up; or you may sow a little Canary grass seed, which you have for your birds, in a garden pot in your window, and thus make the observation at home. But though I may indulge you for once, you know I do not encourage this idle domestic manner of observing the operations of nature. You must go abroad and view her seated on her native throne: and in her court you have this advantage, which you will find in no other, that you are gathering health whilst you pay her homage.

[c] Linnæus calls this sort of leaf *linear*.

[d] Such plants are called *monocotyledonous*, the others, *dicotyledonous*.

If you are now miſtreſs of all the circumſtances in which this tribe of plants agree, you may proceed to thoſe in which they differ, and thus ſeparate them firſt into their genera, and then into their ſpecies. But the genera being numerous it may not be inconvenient, as we did once before, to throw the whole tribe into ſome general ſubdiviſions; and that we can eaſily do from the manner in which the flowers are produced—either in a panicle or ſpike; and ſingly, or ſeveral together. Hence we ſhall get four ſubdiviſions:

1. Flowers ſingle — — 14 genera.
2. Flowers two together — 2 genera.
3. Flowers many together — 7 genera.
   Theſe are moſtly panicled: in all, the flowers are irregularly diſpoſed, or *wandering*, as Linnæus calls them.
4. Flowers in a ſpike, with a ſubulate receptacle — — — 6 genera.
   Including wheat, rye, and barley. Oat is in the third diviſion.

Phalaris. Your pot of Canary ſeed, if you do not pull up all the plants to verify what I told you before, will ſerve for an inſtance of the firſt diviſion. When it arrives at a ſtate of perfection, you will obſerve that the two leaves of the calyx are flatted, boat-ſhaped, have a keel running along them, and are equal in length; the corolla is leſs than the

7 calyx,

calyx, and ſhut up within it. This is the character of the genus. It is ſpecifically diſtinguiſhed by the form of the panicle reſembling a ſpike, and being ovate, the chaffs being turgid and hairy, but the keel ſmooth. It is an annual graſs: is found wild in the Canary Iſlands, whence its name of *Phalaris Canarienſis*, and is cultivated in Europe for the food of Canary and other ſmall birds.

Whilſt your Canary-graſs is growing, you muſt go out in ſearch of other inſtances of this firſt diviſion; for I muſt abſolutely inſiſt that you ranſack the neighbouring meadows and paſtures before the furious ſcythe has levelled all their honours.

Meadows of a good quality abound in *Fox-tail graſs*[e], which is indeed one of the earlieſt, as well as the moſt excellent, for hay and feeding cattle. This genus is an exception to one of the general charac ters; for though the calyx has two valves or leaves, the corolla has but one. You will readily diſcover the ſpecies by the cylindric ſhape and hoary appearance of the panicle, which, from its form, you will take for a ſpike, the erectneſs of the ſtalk, and the corollas not being bearded. Alopecurus.

Cat's-tail graſs[f] is another of theſe; the ſpike has not the ſmooth hoary appearance of the laſt, but ſeems rough, and is known Phleum.

[e] Alopecurus pratenſis Linnæi. Stillingfl. t. 9. Curtis, Lond. 5. 5. & obſ. t. 2.

[f] Phleum pratenſe. Lin. Schreber t. 14.

at firſt ſight by the truncated and forked termination of the calyxes, which are alſo linear, and ſit cloſe to the ſtem. The corolla is ſhut up within the calyx. The ſhape of the ſpike is cylindric; the keel of the chaffs is ciliate[g], and the ſtalk is erect. The ſpike of Cat's-tail graſs is ſometimes four inches long in moiſt meadows; in dryer, poorer ſoils, it decreaſes in length, until it dwindles to half an inch; and even leſs in hard barren ground, ſuch as way ſides and heaths. In theſe laſt it cannot raiſe itſelf upright; and the roots not being able to ſpread themſelves freely, grow knotty and bulbous. I mention theſe circumſtances that you may be aware of the changes wrought in plants by ſoil and ſituation; and not ſuppoſe that a new ſpecies preſents itſelf every time you meet with theſe and other ſlight variations. If you tranſplant from the heath into your garden, a dwarf, crooked, knobby-rooted plant, I dare engage that the ſtem will become erect, that the ſpike will lengthen, and the bulbous root change to a fibrous one. It is not however always eaſy to ſay what is a ſpecies, and what a variety only. A great deal of obſervation and experience is neceſſary in many caſes to determine this with preciſion. Moſt varieties indeed are produced by culture, or a change from their

[g] Set with little hairs like eye-laſhes.

native ſoil and ſituation: and, when they regain their natural ſtate, will return to their priſtine form: if this were univerſally ſo, there would be no difficulty to aſcertain the ſpecies from the variety. But it ſometimes happens that when accident has produced a variety, it continues permanent, and having once taſted a poliſhed ſituation, refuſes to return to a ſtate of nature: our teſt therefore is not a certain one.

The ſecond diviſion of the graſſes having only two genera, the diſtinction is eaſy: they are known from the reſt by having two flowers growing together; and from each other by the rudiment of a third flower between the two others, in the *Melica*, of which there is no ſign in the *Aira*.

Of the third diviſion you will find abundance of graſſes ſufficiently common: *Briza* or ladies' hair, *Poa* or meadow graſs, *Feſtuca* or feſcue, *Brome* graſs, oats with all the oat-graſſes, and the reeds. The genera are thus diſtinguiſhed:

Corolla cordate: valves turgid, -	*Briza.*
Corolla ovate: valves rather ſharp,	*Poa.*
Corolla oblong: valves pointed,	*Feſtuca.*
———————: valves bearded below the point, -	*Bromus.*
———————: beard writhed or bent, - - -	*Avena.*
Corolla woolly at the baſe: awnleſs, - - -	*Arundo.*

Briza. The *Brizas*, of which there are five ſorts, are very pretty graſſes; inſomuch that one of them is cultivated in gardens for its beauty and ſingular appearance. They flower early in the month of May, grow in a looſe panicle, the foot-ſtalks of which are ſo ſlender as to be moved by every wind; whence they have obtained the name of *Quaking graſſes*. By theſe circumſtances, and their general air different from their other neighbours, you cannot fail of knowing them. The three ſorts which you are likely to meet with are thus diſtinguiſhed:

1. Spicules [h] triangular: calyx longer than the flower. *Little Briza.* Mor. 8. 6. 47.
2. Spicules ovate: calyx ſhorter than the flower. *Middle Briza.* Mor. 45. Ger. 86. 2.
3. Spicules cordate: 17 flowers. *Great Briza.* Jacq. Obſ. 3. 60.

The ſecond is the ſort which is common in meadows, and the third is that which is cultivated in gardens: in this the flowers grow in a *raceme* rather than a panicle.

Poa. The *Meadow-graſſes* are numerous, there being no leſs than 33 ſorts regiſtered by Linnæus, and ſeveral of them are thrown abundantly from the lap of nature; for

[h] Theſe are the little aſſemblages of flowers, or ultimate ſubdiviſions of the panicle or whole.

perhaps

perhaps they are the beſt of all the graſſes for paſtures, the quantity of their produce being very great, their quality excellent both for green and dry food, and their verdure moſt freſh and pleaſant. But we are not huſbandmen, dear couſin, Botany is our purſuit.

There are four ſorts of *Poa* very common in moſt meadows: which I ſhall diſtinguiſh by the names of 1. Great, 2. Trivial, 3. Narrow leaved, and 4. Annual. They all flower in a looſe branching panicle. The ſtalks of the firſt ſort are generally erect, and throw out runners: the leaves are rather blunt at the end, and the membrane at the bottom is ſhort and blunt: the ſpicules are ovate, and on ſhort footſtalks; the flowers growing cloſe together, moſt commonly five in number. Every part of this graſs is ſmooth. The ſecond ſort is diſtinguiſhed by the leaves being ſharper at the end, and having the membrane at bottom long and pointed: the ſpicules conſiſt of two or three flowers, very ſeldom four. The whole of this ſpecies is rough. The third has the ſtems more erect: the leaves ſharp-pointed and roughiſh, but ſmooth where they ſheathe the ſtalk: the panicle is more erect than the others;

1. Curtis, Lond. II. 5. obſerv. t. 3.
2. Curtis, Lond. II. 6. obſerv. t. 4.
3. Moriſon's hiſt. ſ. 8. t. 5. f. 19.
4. Curtis, Lond. I. 6. Stillingfl. t. 7.

the

the ſpicules on longer foot-ſtalks, with from one to ſix flowers, which are hairy at the baſe. Theſe three are perennial. The fourth is annual, and ſmaller than the others; extremely univerſal, and in flower the greateſt part of the year; it has a very looſe ſpreading panicle growing all on one ſide [i], the lower branches of it often coming out in pairs: the ſpicules producing 3 or 4 flowers: the ſtalk is oblique and compreſſed.

I muſt give you one caution in examining theſe and the reſt of the panicled graſſes, which is this—that you ſhould take them at the time when they are arrived at full maturity; that is, when the panicle is completely expanded, and the flowers ſhow their ſtamens: for, at different periods of their exiſtence, theſe graſſes put on ſuch various appearances, that they have deceived many eminent botaniſts into forming ſeveral ſpecies out of one. To have the hiſtory of a plant complete, we ought to examine it every day during the whole time of its growth. What a work would ſuch a hiſtory of ten thouſand plants form! but the book of nature is inexhauſtible.

Feſtuca. The genus *Feſtuca* or *Feſcue graſs*, though leſs numerous than the laſt, yet contains 19 ſpecies. *Sheep's feſcue* [k] is a well known graſs, always to be found in dry paſtures,

[i] This is what Linnæus calls *Panicula ſecunda*.
[k] Feſtuca ovina. Stillingfl. t. 8.

5 and

and ſheep commons. It has a cloſe contracted panicle, growing on one ſide; the ſpicules having from 3 to 6 flowers; the valves of the flowers are very ſharp pointed, but ſeldom properly awned; the culm is rather ſquare than round, almoſt naked, and the leaves are ſetaceous [l].

Another Feſcue [m], extremely different from the former, grows in watery places, ponds, and ditches. It has a looſe panicle of a conſiderable length, but little branching, growing on one ſide; the branches of the panicle are ſometimes ſingle and ſometimes double; the ſpicules are round, linear, and awnleſs, almoſt an inch long, and preſſed cloſe to the ſtalk; varying in the number of flowers from 9 to 12. The leaves are not round like thoſe of the laſt, but flat; and the culm is very long, procumbent, branching, and flatted. The ſeeds of this being large and ſweetiſh are gathered for the table in Poland and ſome other countries, and appear there under the name of *Manna*.

In this graſs we have another inſtance of the changes wrought by ſoil and ſituation. Three ſpecies having been made out of one, until experiment detected the truth, and informed us that the ſeeds of the *flote* Feſcue ſown in a dry ſoil, become the firſt

[l] Very narrow, like thoſe of ruſhes.

[m] *Feſtuca fluitans*; flote Feſcue. Curtis, Lond. I. 7.

year

year *ſpiked*, and the ſecond *meadow* [n] Feſcue-graſs. Nay *tall* Feſcue, a fourth ſpecies, has ſo many marks in common with the laſt, that it is matter of doubt whether this alſo may not be a variety only [o].

Bromus. The *Bromes* are very nearly allied to the *Feſcues*. They are diſtinguiſhed however by being all bearded, and the beard or awn ſpringing from the back, or below the tip of the chaff: whereas the *Feſcues* are often beardleſs; and when the flowers have a beard, it is an elongation of the chaff itſelf.

No graſs is more common in many paſtures than *Field-Brome graſs*. It has a looſe unbranched panicle: the ſpicules are ovate, the flowers are obtuſe, and the beards are ſtraight. It is an annual plant: and varies ſo much as to have obtained the name of *polymorphus* or *many-formed*. The two principal varieties [p] are, 1. that which has a ſoft down all over the panicles, leaves and ſtalks; with larger, heavier ſpicules; 2. that which is ſmooth all over; with the ſpicules thinner, and not hanging down ſo much, but often rather erect. Between theſe are two other varieties, 1. with the leaves downy, and the panicle almoſt ſmooth;

[n] Feſtuca pratenſis. Curt. obſ. t. 5.

[o] See Hudſon Flora Anglica, edit. 2. p. 47.

[p] Bromus mollis & ſecalinus Linnæi. Mr. Hudſon, after Scopoli, has very judiciouſly made them one, under the title *polymorphus*. Curtis, Lond. I. 8. figures the *mollis*—Moriſon figures this in t. 7. f. 18; and *ſecalinus* in f. 16.

2. with

2. with the lower leaves only a little downy, and the panicle quite ſmooth. Other connecting links may eaſily be remarked by thoſe who are induſtrious in hunting after varieties.

There are thiee very large ſpecies of this genus, to be met with in woods and hedges, but ſeldom in paſtures[q]. They have great, branching, nodding panicles. *Barren Brome* is not very tall; but the *Giant* and *Wood Bromes* are three feet in height. Their ſize, added to the character and air of the genus, mark them out ſo well, that you will not eaſily miſtake, when you ſee them.

You will get an idea of the *Oat graſſes* from the corn of that name, which having the parts of fructification larger than in the graſſes, gives you an advantage in the examination. *Bearded Oat graſs*, vulgarly called *Wild Oats*, is alſo well known as a dreadful weed among corn. *Yellow Oat graſs* is common in meadows and paſtures: it is a neat pretty graſs; and will diſcover itſelf to you by the fineneſs and yellowneſs of its panicle. Avena.

The characters of the above-mentioned ſpecies are theſe:

1. Two flowers in one calyx: the ſeeds ſmooth, and one of them bearded. *Cultivated Oats*.

[q] Bromus ſterilis, Curtis I. 9. giganteus Curt. 5. 7. & nemoralis.

2. Three

2. Three flowers in one calyx: hairy at the base; and all of them bearded. *Wild Oats*.
3. Panicle loose: three flowers in a short calyx; and all of them bearded. *Yellow Oat grass*[r].

Arundo. The woollyness of the flowers in the *Reed* will show you this genus as soon as it unfolds its panicle. It is a grass, though vulgarly not regarded as such, because it is not used for the same purposes with the grasses. That however makes no difference to us, whose province it is not to regard the uses to which plants are put, but their structure. If husbandmen will not admit Reed to be a grass, they take in other plants to their idea of grass which we exclude, such as Clover, Lucerne, Saintfoin, &c. The reason is, that they consider grass as an herb adapted to feed cattle: whereas naturalists define it to be an herb which has generally three stamens and two pistils; always an unbranched, knotted, hollow stem, and simple linear leaves.

Though you are perfectly acquainted with the Reed[s], it is perhaps rather by seeing it nodding its large panicles in the water at a distance; or else by the use which your gardener makes of the long light stems

[r] Avena sativa, fatua & flavescens *Linnæi*. Curtis, Lond. III., 5.

[s] Arundo phragmitis *Linnæi*. Moris, 8. 8. 1.

for

for hedges to guard his tender plants, than by its fructification. You will not therefore be displeased to be told that it is distinguished from the other species, which are six, by the looseness of its panicle, and by having five flowers growing together.

You are now arrived at the last division of corn and grasses, containing those whose fructification is always in a spike properly so called. Of these,

Secale or *Rie*, has two flowers included in the same calyx.

Triticum or *Wheat*, has several flowers in one calyx.

Hordeum or *Barley*, has a six-leaved involucre, containing three flowers; and the flowers simple.

Lolium or *Darnel*, has a one-leafed involucre, containing one flower only; but that flower compound.

Cynosurus or *Dog's-tail grass*, has a one-leafed lateral involucre, and a compound flower.

In *Rie*, the exterior valve or chaff of the corolla ends in a long beard or awn. The flowers are sessile, and there is frequently a third between these, which is less and pedunculate: the filaments hang out of the flower. Our cultivated species[t] is known by the rough hairs upon the chaff. Secale.

[t] Secale cereale *Linnæi*.

In

Hordeum. In *Barley* alſo the exterior valve of the corolla ends in a long awn. The flowers are ſeſſile. The filaments being ſhorter than the corolla do not hang out, and therefore *Barley* is not liable to be damaged by rain as Rie and Wheat.

There are four ſorts of Barley.

1. The *common*, diſtinguiſhed by its two rows of erect beards; all the flowers being perfect and bearded.

2. The *long-eared*, having the grains regularly ranged in a long double row, lying cloſe over each other; and flowers on the ſides, without piſtils or beards.—Theſe two ſpecies have the chaff very thin.

3. *Sprat Barley*, with ſhorter, broader ears, longer beards, the grains placed cloſer, and the ſtraw ſhorter and coarſer. This alſo has imperfect flowers on the ſides of the ear.

4 *Winter* or *Square Barley*, very diſtinct by having ſix rows of grains equally ranged, all furniſhed with awns, and perfect. The grain of this is large.

Beſides theſe ſpecies of corn, the genus contains ſeveral graſſes. *Wall Barley graſs* [u] is very common by way ſides, and under

1. Hordeum vulgare. 2. Hordeum zeocriton. 3. Hordeum diſtichon. 4. Hordeum hexaſtichon; called alſo *bear* and *big*.

[u] Hordeum murinum *Linnæi*. Curt. Lond. 5. 9. Fl. Dan. t. 629. Mor. hiſt. t. 6, f. 4.

walls:

walls: and *Meadow Barley graſs* [v], which is very like it, only that it has a longer ſtalk, and a ſhorter ſpike, is found in moiſt meadows. The common name of this laſt is *Rie-graſs*; and indeed it reſembles Rie more than Barley. I have ſeen it cultivated alone; but the ſort which is generally ſown, and vulgarly called Rie-graſs, is in reality *Ray-graſs*, which will be announced to you preſently. Theſe two ſorts, though apparently ſo alike, and thought to be but varieties by many, are however very diſtinguiſhable: the Wall Barley-graſs having the imperfect lateral flowers bearded, and the intermediate involucres ciliate; whereas the Meadow Barley-graſs has the ſame flowers beardleſs, and the involucres very narrow, like briſtles, and rough.

In *Wheat* the exterior valve of the corolla is ſometimes bearded, but not always. There are generally three or four flowers in the ſame calyx, and the middle one is frequently imperfect. The filaments hang out, but not ſo much as in Rie. Triticum.

1. *Common Wheat* has four flowers in one calyx, the chaffs are ſmooth, turgid, imbricate; ſometimes it has ſhort beards, but more often none: hence and from the colour, &c. are ſeveral varieties which huſbandmen notice, and we have nothing to do with.

[v] Hordeum pratenſe. Fl. dan. t. 630. Mor. hiſt. t. 2, f. 6.

1. Triticum hybernum.

2. *Summer* or *Spring Wheat*, has alſo four flowers together, and agrees with the former in the other characters, except that it is always bearded.

3. *Gray Wheat* has villous, turgid, imbricate obtuſe chaffs, containing four flowers. The ears are large, heavy, and nodding; the beards are very long, and drop off when the grain is full grown: the chaff being villous all over, gives the ear a gray appearance.

4. *Cone Wheat* has villous, turgid, imbricate chaffs; and the ear of a pyramidal form, ending in a ſlender point: the beards are long and rough.

5. *Polonian Wheat* has two flowers only in each calyx, naked, and having very long awns; with the teeth of the *rachis* or receptacle of the ſpike bearded. The ears are long and heavy.

6. *Spelt* has four flowers, but two only produce any grain; the outer ones are abortive, as the lower ones are in every ear: the outer chaff of the perfect flowers has a beard about an inch long. The flowers are more conical, and the grain is leſs than in wheat: the chaff alſo is adherent.

2. Triticum æſtivum.

3. Triticum turgidum: called alſo *Gray Pollard*, *Duck-bill*, and *Fuller's Wheat*.

4. Not noticed by Linnæus.

5. Triticum Polonicum.

6. Triticum Spelta. I do not know that this ſort is ever cultivated in England.

Few

Few plants are more univerſal than one graſs of this genus: it is known by the name of *Dogs-graſs*, and generally execrated by huſbandmen under the name of *Couch*, or *Quich*, which is but a corruption of *Quick*, the ancient term for *living*. It well deſerves this appellation, for it runs prodigiouſly at the root, and, like Hercules's hydra, the more you hack and cut it, the faſter it propagates itſelf. It is diſtinguiſhed from the ſeveral ſpecies of corn by the ſmallneſs of the ear and the grain, and alſo in the being perennial; whereas all ſorts of corn are annual: from the other graſſes of the ſame genus, by having many flowers, about five generally to one calyx, and thoſe not bearded, but very ſharp-pointed at the end [w]. There is another ſpecies, which has about four flowers in a calyx, and is bearded [x]. This grows in woods and hedges.

Before I quit this genus I muſt obſerve, as a ſingularity, that it is not known, with any degree of certainty, to what country we are originally indebted for the ſeveral ſpecies of corn, or whether they now grow wild in any. One ſays that Wheat came firſt from Africa; others, with more probability, that it travelled into Europe from

[w] Triticum repens *Linnæi*. Schreb. t. 26. Fl. dan. 748. Mor. hiſt. t. 1. f. 8. The number of flowers varies from 3 to 8. *Hudſon*.

[x] Triticum caninum *Linnæi*. Mor. hiſt. t. 1. f. 2.

the East. Linnæus affirms that Rie grows naturally in Crete [y]; and Spring Wheat, with Sprat Barley (*Hordeum distichon*), in Tartary: but upon what authority I know not. A late traveller also found barley and oats in Sicily growing like weeds among the bushes, but he does not pretend to determine whether they grew there originally wild, or whether they were stragglers from the fields where they had been cultivated [z].

Lolium. *Lolium* or *Darnel-grass* is an exception to the general character; for it has only one chaff or leaf to the calyx. The reason of this is, that the spicules are sessile, and in the same plane with the culm, which by this position is enabled to perform the office of the deficient leaf of the calyx in protecting the seed. This single chaff contains several flowers. Of the two common species

[y] It is said also to be wild in Siberia.

[z] Voyage en Sicile, &c. Lausanne, 1773. Diodorus Siculus, from the report of others, and Pliny, assert that grain grew in the Leontine fields, and other parts of Sicily, spontaneously; but this was only during the reign of Ceres. Aristotle also says (de Mirabil. Auscult.), that there is a wild Wheat in the neighbourhood of Mount Ætna. The passage in Homer's Odyssey is well known:

"The soil untill'd a ready harvest yields,
"With Wheat and Barley wave the golden fields."

Wheat, Barley, Vetches, Sesame, &c. are said, by Berosus, to be wild in Babylonia, between the Tigris and Euphrates.

in

in this genus one is *perennial*[a], the other *annual*[b]. The firſt is found naturally in meadows, paſtures, and by way-ſides. The diſtinctive marks of the ſpecies are, that the ſpicules in the firſt are longer than the calyx, and the flowers beardleſs: whereas in the ſecond, which is a weed among the corn, the ſpicules are only of equal length with the calyx, and the flowers have ſhort beards. Sometimes however it happens that the flowers of the perennial ſort have little beards, and thoſe of the annual none: but you may always know them, not only from their duration and place of growth, but becauſe the ſecond is larger in every reſpect; the ſtalk higher, the ſpike longer; the ſpicules alſo are much more remote, ſo that they do not touch each other, as they do in the firſt.

Cynoſurus, or *Dog's-tail graſs*, was the laſt-mentioned of this diviſion. The character of the genus is taken from a lateral leaf to each calyx, which Linnæus calls the receptacle, involucre or bracte: this Cynoſurus.

[a] Lolium perenne *Linnæi*. Schreb. t. 37. Fl. dan. 747. Mor. hiſt. t. 2. f. 2. Pl. 9. f. 1. This is the ſort which has been long cultivated in England under the name of *Rie-graſs*, which is a corruption of *Ray-graſs*; and that is derived from the French *Yvray*, a name given to the ſecond ſort, from its quality of affecting the nerves, ſomething like drunkenneſs: which makes it to be reputed a dangerous weed among Wheat.

[b] Lolium temulentum *Linnæi*. Schreb. t. 36. Fl. dan. 160.

gives the ſpike an air by which the genus is eaſily known from all others. There is an elegant ſpecies [c], very general in parks and on commons, and found alſo in other paſtures, which has theſe bractes pinnatifid, or toothed like a comb: the corolla does not open, but cloſely inveſts the ſeed, which therefore does not fall; the ſpicules have from three to five flowers, are all turned the ſame way, and do not ſit cloſe to the receptacle, or common ſtalk of the ſpike; one peduncle ſupports ſometimes two or three of theſe ſpicules. The ſtalk is very erect and ſlim, and the leaves are narrow and ſmooth.

There remain ſtill ſome graſſes which militate againſt the artificial ſyſtem, and are therefore not to be found in the third claſs of Linnæus's. But as we are not bound to follow him ſervilely, we will rather follow nature, who is a better guide.

Anthoxanthum. Earlier than moſt of the reſt flowers a graſs, called from thence *Vernal Graſs* [d] Linnæus has named it *Anthoxanthum*, from the yellowneſs of its ſpike. This will ſerve at preſent to introduce it to your acquaintance, until you have an opportunity next ſpring to examine the flowers more minutely. It has obtained the epithet of

[c] Cynoſurus criſtatus *Lin.* Creſted Dog's-tail. Schreb. t. 8. f. 1. Stillingfleet, t. 11. Curtis obſ. t. 6.

[d] Curtis, Lond. I. 4. and obſerv. t. 1. Stillingfleet, t. 1.

odoratum

odoratum from the ſweet odour which it communicates to hay. This genus ſtands alone in the ſecond order of the ſecond claſs. Each calyx ſuſtains but one flower; each valve of the corolla has an awn, one bent, and proceeding from the baſe, the other almoſt from the top: the two filaments are very long; and the two ſtyles are filiform: the chaff of the corolla adheres to the ſeed. There are three ſpecies of the genus: ours is diſtinguiſhed by the ſpike being of an oblong form; and the flowers growing on ſhort peduncles, and being longer than the beards.

There is alſo one ſpecies of graſs, called *Cinna*, in the ſecond order of the firſt claſs.

But in the firſt order of the twenty-third claſs [c] are ſeveral genera; of which the *Holcus* or *Soft graſs* is moſt likely to come under your obſervation. This, and all the others, have ſmaller imperfect flowers among the perfect ones; a circumſtance which conſtitutes them of that claſs. They have all bivalvular chaffs for calyx and corolla; three ſtamens, two piſtils, and one ſeed, together with the whole port or air of the plants we have been juſt conſidering: circumſtances which plainly denominate them graſſes. *Holcus* differs from its neighbours, in having two flowers incloſed in one calyx, which is beardleſs; whereas the Holcus.

[c] Polygamia Monœcia.

outer valve of the corolla generally has a beard. The imperfect flowers have neither corolla, pistil, nor seed; but only three stamens within the bivalvular chaff of the calyx. The two common wild species are thus distinguished: *Meadow Soft grass*[f] has villous chaffs: the perfect flowers are beardless; the imperfect have a bent awn. *Creeping Soft grass*[g] has smoothish chaffs: the perfect flowers are beardless, but the imperfect have a jointed awn. They are very much alike, but the calyx is more acute in this than in the former, or indeed than in any of the species. The first grows in pastures; the second in corn-fields and hedges.

Since it is not uncommon to find incomplete or imperfect flowers among those which are perfect, in many of the grasses, which are ranged by Linnæus in his third class; you will perhaps ask me why he has not either put them also in the twenty-third, or else ranged them all together in the third. To this question I cannot return you a better answer, than that the imperfect flowers seem not so constant and regular in the one as in the other; or perhaps are to be met with only in one species of the genus.

[f] Holcus lanatus *Lin.* Curtis, Lond. IV. 11. Schreber, t. 20. f. 1.

[g] Holcus mollis *Lin.* Curtis, Lond. V. 8. Schreber. t. 20. f. 2.

We

We have now run through the graſſes: there are many other plants very nearly allied to them; as *Schœnus* or *Bog ruſh*, *Cyperus*, *Scirpus*, *Club ruſh* or *Bulruſh*[h], all three very numerous genera, *Eriophorum* or *Cotton graſs*[i], &c. in the firſt order of the third claſs. *Cat's-tail*[k], *Bur-reed*[l], and all the *Carices* or *Sedges*[m], in the third order of the twenty-firſt. Theſe have the manner of growth, the leaves, the appearance of graſs; they have alſo three ſtamens: but the ſtalk is filled with a ſpongy ſubſtance, and the flower is deſtitute of petals. Finally the *Ruſhes* and ſome few others, in the firſt order of the ſixth claſs, have a ſix-leaved calyx, a hexapetalous corolla, or none, ſix ſtamens, and the ſeeds in a triangular capſule.

I have not told you all this while that *Sugar*[n] is a graſs of the firſt diviſion, which perhaps you did not expect. But if you are not tired, dear couſin, I am; ſo adieu for the preſent.

[h] Curt. Lond. 4. 4. S. maritimus.
[i] Curt. Lond. 4. 9, 10.
[k] Curt. Lond. 3. 61, 62.
[l] Curt. Lond. 5. 66, 67.
[m] Some of the ſpecies are figured in Curtis, Lond. 3. 63. & 4. 60, 61, 62.
[n] Saccharum officinarum. *Lin.* Sloan. jam. t. 66. Rumph. amb. 5. t. 44.

LETTER XIV.

OF OTHER PLANTS IN THE CLASS TRIANDRIA.

July the 1ſt, 1774.

YOU are not to ſuppoſe that, becauſe the laſt letter was engroſſed wholly by Graſſes, the third claſs therefore of the ſyſtem contains no other plants. In truth there are no fewer than ſeventy-ſix genera, and ſix hundred and eighteen ſpecies, in the three orders of this claſs taken together. You ſee however, that though the graſſes do not occupy the whole, they make a very large proportion of it.

There are ſome very beautiful genera in the firſt order of this claſs, particularly the *Ixia* and *Iris*, or *Fleur-de-lys* *. Theſe with *Crocus*, *Gladiolus*, *Antholyza*, and a few others not eaſily met with, agree in having a *Spathe* or ſheath inſtead of a calyx; a corolla of ſix petals, or at leaſt cut into ſix parts; generally three ſtigmas, or one that is trifid; and a triangular, trivalvular, trilocular capſule to incloſe the ſeeds: they have alſo long, narrow leaves, ſomething reſembling thoſe of graſs—Linnæus calls

* Corrupted into *Flower-de-luce*.

them

them *Enſiform*, or *ſword-ſhaped*[p]. Theſe plants are very nearly allied to the liliaceous tribe[q], and are indeed enrolled in it by the generality of authors who have aimed at framing a natural arrangement.

Take any ſpecies of *Iris*, either the *blue*[r] or *white*[s] ſorts, which you have ſo abundantly in the borders of your ſhrubberies and plantations; or elſe the *yellow*[t] one, common in wet places, and uſually called *flag*. In the firſt place you will obſerve, that whether the flowers are open or cloſed, each has its own ſheath, ſeparating it from the others. The corolla at firſt ſeems to conſiſt of ſix petals, but you will quickly ſee that the parts are all united at the baſe: the three outermoſt of theſe parts or petals are bent downwards, and thence are called *falls*; the three inner ones ſtand erect, and have the name of *ſtandards*. In the centre of them are three other petals, as they ſeem to be; but in reality they are the ſtigma thus divided into three parts; and under each diviſion you will detect a ſingle ſtamen lurking, with the filament bent along with the ſtigma, and terminated by a large oblong, flatted anther: Iris.

[p] Hence in his *Natural Orders* he has kept theſe together, with the addition of ſome others, under the title of *Enſatæ*.

[q] See Letter I.

[r] Iris Germanica *Linnæi*. Blackw. t. 69.

[s] Iris Florentina *Linnæi*. Mill. fig. t. 154.

[t] Iris pſeudacorus *Linnæi*. Curtis, Lond. III. 4.

for the germ you muſt ſearch below the flower, and there you will find it a green oblong body; which when the flower is faded and fallen, becomes in moſt ſpecies a three-cornered capſule, opening by three valves, and having the ſeeds ranged in three cells. We have not yet noticed a ſet of ſmall bodies forming a villous line along the middle of the reflex petals; but this you perceive is not common to all the ſpecies; your blue and white Iris having it, but not your yellow flag: it cannot therefore be a mark of the genus. However it may ſerve the purpoſe of ſubdividing it, or furniſhing a ſpecific character. When you have finiſhed with the fructification, you will remark that the leaves are very narrow in proportion to their length; and that they are not unaptly termed enſiform from the ſimilitude of their ſhape to that of a broad-ſword. If you can have the heart to pull one of theſe fine plants out of the ground, you will ſee that the roots are not fibrous, but oblong and fleſhy: I gueſs however that you will take my word till the autumn, when the gardener will be removing ſome of them, or at leaſt expoſing their roots, when he digs his borders.

You may diſtinguiſh the *blue* or *German*, the *white* or *Florentine*, and the *yellow* or *marſh Iris*, ſpecifically thus: The two firſt have the corollas bearded; the firſt and third have ſeveral flowers upon the ſtem; the

the ſecond has only one or two flowers, and the peduncles are not ſo long as in the firſt; the third has the corollas beardleſs, and the interior petals leſs than the diviſions of the ſtigma [u]. But why all this parade, ſay you, when we know them by their hues; blue, white, and yellow? Truſt not too much to colour, fair couſin. What if an Iris were to preſent itſelf with blue flowers, and only one or two on the ſtem, or without beards; or with the flowering ſtem ſhorter than the leaves, would ſuch be of the ſame ſpecies, merely becauſe the corolla is of a blue colour? No ſurely: and we pay more reſpect to theſe circumſtances than to colour, not becauſe we eſteem them more, but becauſe they are more certain and permanent.

The *Chalcedonian* Iris [v] has ſtems two feet and an half high, ſupporting one very large flower; the three ſtandards are very broad and thin, with black and white ſtripes; the three falls are of a darker colour: this is one of the bearded ſorts.

Among theſe handſome ſpecious plants, let us not forget the humble *Perſian* Iris [w], ſeldom riſing three inches from the ground, but beautiful in its colours, fragrant in its ſcent, and flowering at a time when few

[u] They are all three diſtinguiſhed from ſome other ſpecies by the flowering ſtalk ſtanding up ſuperior to the tips of the leaves.

[v] Iris ſuſiana *Linnæi.* Curt. Magaz. 91.

[w] Iris Perſica *Linnæi.*

beauties

beauties dare truſt themſelves to dubious ſkies and inclement air [x]. One or two flowers come out together: the ſtandards are of a pale ſky blue; the falls are of the ſame colour on the outſide, but the lip has a yellow ſtreak running through the middle, and on each ſide are many dark ſpots with one large deep purple ſpot at the bottom: they have no beard. The leaves are hollowed like the keel of a boat, and are about ſix inches long. You will be glad to entertain this pretty dwarf, when there is little elſe to amuſe you in this way beſides Crocuſes and Snowdrops.

I have ſent you this little noſegay of handſome flowers, to make you amends for all the dry chaff and hay with which I fatigued you in my laſt.

[x] February. This is figured in Curtis's Magazine, n. 1. And ſeveral other ſorts are figured in that elegant work:—as I. pumila t. 9.—variegata 16.—verſicolor 21.—ſibirica 50.—ſpuria 58.—ochroleuca 61.—ſuſiana 91.—By this aſſemblage we are much helped in diſtinguiſhing the ſpecies.

LETTER

LETTER XV.

OF THE CLASS TETRANDRIA.

July the 8th, 1774.

CONSCIOUS, dear coufin, that the nofegay of my laft was too fmall to employ you long, I have hafted to fend you the fourth clafs, which is rather more numerous than the third in the genera, of which it contains eighty-five; but far lefs fo in the fpecies, there being no more of thefe than three hundred and ninety.

You will have fome examples in this clafs of *aggregate* flowers, the general nature of which I explained to you before [y]; but you will be perfect miftrefs of it I am perfuaded, when you have confidered the ftructure of the *Teafel* and *Scabious*. Thefe and all others of this natural order have monopetalous corollas, fucceeded by one feed, to which they are fuperior. A number of thefe are included within one common calyx, as in the compound flowers, from which they differ, in having the ftamens four in number, and totally diftinct, with a calyx proper to each little flower; they might however eafily be confounded

[y] In Letter VI.

with compound flowers, if the general form and appearance only were attended to.

Dipſacus. The two genera of *Teaſel* and *Scabious* agree in having the common calyx polyphyllous, or conſiſting of many leaves. The firſt has chaffs between the flowers on the receptacle, or common baſe of them all; the form of which is conical. The ſecond has theſe chaffs in ſome ſpecies, but in others the receptacle is naked; the form of it is convex: it is remarkable for a double calyx to each little flower, beſides that which is common to the whole. The leaves of the calyx are very long in the Teaſel, and in ſeveral rows in the Scabious.

Such are their principal generic diſtinctions. Common Teaſel is ſeparated from its congeners, by its ſeſſile leaves, which are ſerrate or toothed about the edges. The conical head of the Teaſel is furniſhed with ſtiff beards, which in the wild ſort [z] are ſtraight, but in the cultivated hooked [a]. This difference did not ſeem to Linnæus conſiderable enough to make them ſpecifically diſtinct. Haller, Jacquin, and others, are of a different opinion; and it is now generally allowed that the cultivated Teaſel is of a ſpecies diſtinct from the wild one.

Scabioſa. Of Scabious there are no leſs than thirty-

[z] Dipſacus ſylveſtris. Curtis, Lond. III. 9. Ger. 1167. 2.

[a] Dipſacus fullonum *Linn.* Ger. 1167. 1. Mor. 7. 36. 1.

5 four

four ſpecies. The genus divides conveniently into ſuch as have the corollas of the little flowers divided into four, and ſuch as have them divided into five ſegments: of the firſt there are fourteen, of the ſecond twenty ſpecies. Of our three wild ſorts two are in the firſt diviſion, and one in the laſt. The common field Scabious [b] is a large, tall plant; the ſtalk is hairy: the lower leaves are ſometimes almoſt entire; ſometimes they, as well as the leaves upon the ſtem, are pinnatifid. The outer flowers are larger, and have the corolla deeper cut than the middle ones, and the outer ſegments are alſo largeſt: they are of a pale purple colour.

The other ſpecies with quadrifid corollas is called *Devil's-bit* [c], becauſe it has a ſhort tap root, which appears as if the end were bitten off. The ſtalks of this are not ſo high, nor are they branching as in the firſt: they generally ſend out two ſhort peduncles from the upper joint, oppoſite to one another, each terminated by one ſmall blue flower, as is the principal ſtalk by one larger; the little component flowers are not irregular as in the former. The leaves are ſimple and entire, (except ſome on the middle of the ſtem, which have a few teeth,) oblong and drawing to a point at each end. This ſpecies grows in paſtures and woods,

[b] Scabioſa arvenſis *Lin.* Curtis, Lond. IV. 13.
[c] Scabioſa ſucciſa *Lin.* Curtis, Lond. III. 10.

and flowers later than the firſt, which is common in corn fields, and not uncommon in paſtures.

Small Scabious [d], beſides having quinquefid corollas, is diſtinguiſhed from the two others by having the leaves next the ground ovate and notched about the edges, whilſt thoſe upon the ſtem are pinnate; towards the bottom the pinnas are broader, but in the upper ones very narrow: there are about eight pairs of theſe, and the terminating leaflet is large. The aggregate flower is produced ſingle, on a long peduncle, the outer little flowers larger, and very irregular, as in the firſt ſpecies, of a pale blue colour. It is common in paſtures, eſpecially where the ſoil is chalky.

Before you are got thus far, I am perſuaded your own mind has ſuggeſted to you that a plant with dark purple flowers, and a ſtrong ſweet odour, which your gardener ſows every year in the borders, is of this genus. The name of *Sweet Scabious* has not led you, who are not governed by mere names, to ſuppoſe this, but the evident ſimilitude in the ſtructure. An accurate examination of the flower will confirm your ſuſpicion; and you will find it to be one of thoſe which have quinquend irregular corollas: the receptacle of theſe is oblong; the common calyx conſiſts of

[d] Scabioſa columbaria *Lin.* Fl. dan. t. 314. Pl. 11. f. 1.

twelve

twelve linear folioles, of the length of the aggregate flower, and bent back: the leaves are finely cut [e]. The colour of the corolla varies from black to pale purple, red and variegated, and ſometimes the main flower is ſurrounded by a ſet of very ſmall ones on ſlender peduncles, as in the *Hen and Chicken Daiſy*; but all theſe are confeſſedly no other than ſeminal varieties: though now ſo common with us, this plant is originally from the Indies.

This claſs compriſes another natural order of plants, entitled *Stellated*, from the manner in which the leaves grow upon the ſtem, ſeveral together in ſets one above another, radiating like the points of a ſtar, as it is commonly repreſented. I muſt obſerve to you, that though in this caſe, and in many others, a claſs or order takes its name from an obvious or ſtriking circumſtance in its ſtructure, yet it does not follow that all plants which have that ſtructure are to be looked for there, or that this is the only or even principal reaſon of their being kept together. When a plant of this or that general appearance preſents itſelf, you may reaſonably preſume that it ranks in this or that order; but outward appearances muſt not carry you beyond preſumption, and it is the ſtructure of

[e] Scabioſa atropurpurea *Lin.* Ger. 724. 16.

the fructification that muſt determine you at laſt [f].

In the *Stellated* plants the ſtructure is this: the calyx is extremely minute, divided into four parts, and permanent: the corolla is monopetalous divided into four ſegments; the ſtamens are four in number; the germ is double, and below the flower; the ſtyle is bifid; the fruit is globoſe, and contains two ſeeds. The ſtalk is quadrangular.

All the genera of this order reſemble each other ſo much, that ſome authors have reduced them into one. *Madder* has a bell-ſhaped corolla, ſucceeded by two berries with one ſeed in each. *Sherardia* and *Woodroof* [g] have funnel-ſhaped corollas: the firſt has a little crown to the ſeeds, the ſecond

Galium. has them globoſe, without any crown. *Galium* has a ſalver-ſhaped corolla, and two roundiſh ſeeds. This laſt genus has twenty-ſix ſpecies, twenty of which have the fruit ſmooth; in the remaining ſix it is rough. The number of leaves in each ſtar or whorl, together with the ſhape of them, gives the principal ſpecific diſtinctions.

[f] See what was ſaid upon this ſubject with reſpect to the Elder in Letter V. I muſt add that uſe and practice is neceſſary to give the proper tact in natural objects as well as in works of art: the ſimilitudes and analogies that ignorant perſons find being uſually truly ridiculous.

[g] Aſperula odorata. Curtis, Lond. IV. 15.

White

White Galium, or *White Ladies Bedſtraw* has four leaves in a whorl towards the bottom of the ſtem, and ſix narrower ones higher up. *Great Ladies Bedſtraw*[i], has eight, a little notched about the edges, ovate in form, and terminating in a point or little hook. *Yellow Ladies Bedſtraw*[k] has alſo eight leaves, but they are very narrow, and furrowed; the flowering ſtalks are very ſhort, and the corollas are yellow. The firſt grows in moiſt meadows, and by river ſides; the ſecond in hedges, and on heaths among the buſhes; the third is very common in paſtures, on balks, and by way ſides. Theſe three all have ſmooth ſeeds. The common *Galium*[l], known by the name of *Gooſe-graſs* or *Cleavers*, every body knows to have rough ſeeds, by their ſticking to the clothes as we paſs near the hedges. The leaves alſo are rough, lance-ſhaped, and eight in number. The flowers of all the ſpecies, and indeed of the whole tribe, are very ſmall, but the plants are known at firſt ſight by their air.

The *Plantains* are alſo of the firſt order of this claſs *Tetrandria*: they are numerous, for there are twenty-four ſpecies of them. As a great number of ſmall flowers grow together in a ſpike or oblong head, you Plantago.

[h] Galium paluſtre *Lin.* Fl. dan. 423.

[i] Galium Mollugo *Lin.* Fl. dan. t. 455.

[k] Galium verum *Lin.* Curtis, Lond. n. 63. Mill. fig. t. 139. f. 1.

[l] Galium Aparine. Curtis, Lond. II. 9.

muſt ſeparate one of them to examine the parts of the fructification diſtinctly. You will then find that each of theſe ſmall flowers has a quadrifid calyx and corolla, with the border of the latter reflex: the filaments are remarkably long: and the ſeed-veſſel is a bilocular capſule, opening horizontally, and placed above the receptacle.

The *Great* [m] and *Ribwort* [n] Plantains are doubtleſs well known to you; the firſt ſo common by way ſides, and the ſecond in paſture grounds. The *Great Plantain* is diſtinguiſhed by its ovate, ſmooth leaves, and its round, naked, flowering-ſtalk [o] terminated by a long ſpike of flowers lying cloſe over each other [p]. *Hoary Plantain* [q] is nearly allied to this, but the leaves are longer, and white with hairs; the ſpike is cylindric, but ſhorter and thicker than in the firſt. *Ribwort Plantain* has the leaves lance-ſhaped; a ſhort, naked, ovate ſpike; the ſcape angular, and twiſted. This, and the other ſpecies have the leaves marked lengthwiſe, with very prominent ribs or nerves.

By ſubmitting to examine theſe plants, which you were already acquainted with, you will acquire a facility in diſcovering ſuch

[m] Plantago major *Lin.* Curtis, Lond. II. 11.

[n] Plantago lanceolata *Lin.* Curtis, Lond. II. 10. Pl. 11. f. 3.

[o] This Linnæus calls *ſcapus*, from its reſemblance to the ſhaft of a column.

[p] Imbricate.

[q] Plantago media *Lin.* Curtis, Lond. IV. 14.

ſuch as are ſtrangers to you; for you have too much ſenſe to deſpiſe them becauſe they are common, or deſtitute of beauty: in confidence of this, I have been ſtudious to ſelect rather ſuch plants as you may readily meet with, and are proper for examination, than thoſe that are moſt rare and valuable. If you were in the neighbourhood of a famous botanic garden, I might be nicer in my choice, and at the ſame time preſent you with greater variety, but perhaps after all, I might not be more uſeful, or you more amuſed: at leaſt I ſhall hope for the continuance of that indulgence a little longer with which you have hitherto honoured me [r].

But to return to our buſineſs; there is a plant of this fourth claſs and firſt order, which I muſt not omit preſenting to you, were it but for the name's ſake. *Ladies Mantle* has a calyx of one permanent leaf, divided into eight ſegments, four of which are larger, and four ſmaller; it has no corolla; and only one little ſeed to each flower. There are three ſpecies of *Ladies Mantle*. 1. The Common, 2. The Alpine, and Alchemilla.

[r] Students in Botany who live in or near London, or come occaſionally to the great city, will be happy to profit by Mr. Curtis's excellent Garden, at Brompton, where a conſiderable number of plants is arranged and named, ſo that he that runs may read.

1. Alchemilla vulgaris. *Lin.* Mor. hiſt. ſ. 2. t. 20. f. 1. Mill. fig. pl. 18.

2. Alchemilla alpina. *Lin.* Fl. dan. t. 49.

3. The five-leaved. The firſt is known by its ſimple, lobate leaves, nicely ſerrated about the edge, and divided into from eight to twelve greater parts: before the leaf expands it is folded or plaited at each of theſe diviſions, and hence the name. The flowers grow in bunches, are inconſiderable in point of ſize, and alſo of colour, for having no corolla they are only green, or what botaniſts call herbaceous. It is an humble, but an elegant plant, and grows in high paſtures, but not common.

Alpine Ladies Mantle is much more elegant than this, with its ſhining ſilky leaves, which are digitate, and indented at the end: the folioles or component leaves vary in number from five to nine. The third ſpecies is very uncommon: it is a ſmall plant, quite ſmooth, with digitate leaves, but each of its five folioles divided half way into ſeveral ſmaller ones.

The ſecond order of this claſs has a ſingular plant, *Cuſcuta* or *Dodder*. It is without leaves, has a ſtalk ſlender as a thread, which would trail along the ground did it not lay hold on ſome plant ſtronger than itſelf for ſupport; not content with ſupport, where it lays hold, there it draws its nouriſhment; and, at length, in gratitude for all this, ſtrangles its entertainer! I imagine this account will not beſpeak your af-

3. A. pentaphyllea *Lin.*

fection

fection for *Dodder*[s]. If you will be at the pains of disembarrassing a poor suffering bean from its entangling stalks, you will see that the flowers come out in sessile knots; that each of these has a calyx divided half way into four or five parts; that the corolla is of one petal divided into four or five segments at the edge: and that the seed-vessel is a bilocular capsule. This parasite, as Linnæus justly calls such plants, fastens itself about beans, nettles, clover, flax, heath, &c. and feeds upon them by means of innumerable teats or glands which it inserts into the pores of it s supporter's bark.

The *Pondweeds*, which are many, and sufficiently common, will serve for an instance of the third order. If your own fish-ponds are kept too clean to furnish these plants, you may probably procure them from some of your neighbours; or, if they were worth the carriage, I could send you abundance from our moat. You will know them by the leaves lying flat upon the water; and by the stem's pushing up a spike of inconsiderable flowers, that have no calyx, a corolla of four deciduous petals, four germs terminated by obtuse stigmas, with-

[s] Cuscuta Europæa *Lin.* Fl. dan. 199. The divisions of the calyx, and corolla, and the stamens, are five in the British species; ours therefore is C. Epithymum, and according to the strict laws of the artificial system, should appear in the next class. It is figured in Fl. dan. 42.

out

out the interpoſition of any ſtyle, and becoming in time four roundiſh ſeeds.

The *broad leaved*[t] ſpecies is one of the moſt common, and is known by its oblong ovate leaves. *Perfoliate* Pondweed[u] has heart-ſhaped leaves embracing the ſtalk, and grows in running waters. *Curled* Pondweed[v] has lance-ſhaped, waving leaves, notched about the edges, and ſtanding alternate upon the ſtem: this is found both in running and ſtagnant waters.

But of theſe enough—don't hazard getting wet, or catching cold, in ſearch of them. If any of theſe plants which I have hitherto recommended to your notice, elude your ſearch, or have paſſed their ſtated time of flowering before you find them, note them down for next year: ſo adieu, dear couſin.

[t] Potamogeton natans. *Lin.* Miller illuſtr. Ger. 821. 1.

[u] P. Perfoliatum. *Lin.* Fl. dan. 196. Ger. 822. 3.

[v] P. Criſpum *Lin.* Curtis, Lond. 5. 15. Ger. 824. 2.

LETTER

LETTER XVI.

THE FIRST ORDER OF THE FIFTH CLASS, PENTANDRIA MONOGYNIA.

March the 25th, 1775.

MY indiſpoſition of laſt autumn has given you ample leiſure, dear couſin, to make yourſelf miſtreſs of the general arrangement of plants, and of the firſt four claſſes in particular. Since it is your earneſt deſire, I have reſumed my former prate as early as poſſible, that nothing may eſcape us this ſeaſon. We have now a large claſs to encounter with, containing more than a tenth part of the vegetable world, for it has two hundred and ſixty-one genera, and one thouſand five hundred and five ſpecies. It includes, as you may ſuppoſe, ſeveral natural orders, and ſome ſpecies are even now ready for examination.

We will open the year, by your leave, with the *Primroſe*, which has its name from being one of the firſt flowers that blow. This, with ſome others that reſemble it, form a natural order, entitled, for the ſame reaſon, *Preciæ*[w]; and agreeing in having a monophyllous, quinquefid, permanent ca- Primula.

[w] Præcoces, early.

lyx;

lyx; a monopetalous, quinquefid corolla; and a capſule for a ſeed-veſſel, ſuperior or incloſed within the calyx. The characters of the genus are, an involucre under the flower, or knot of flowers; the corolla funnel-ſhaped or ſalver-ſhaped, with the tube cylindric, and open at the top; the ſtigma globoſe: the capſule unilocular. The ſpecies[x] is diſtinguiſhed by its pentagonal calyx, its cylindric oblong capſule, and the wrinkled ſurface, and indented edges of its leaves. The three principal varieties, if they are but varieties, are thus commodiouſly ſeparated. The *Primroſe*[y] has one flower on a naked ſtem, and the corolla ſalver-ſhaped. The *Ox-lip*[z] has ſeveral flowers on one naked ſtem, and the corolla ſalver-ſhaped. The *Cowſlip*[a] has many flowers on a naked ſtem, and the corolla funnel-ſhaped. The yellow of the two firſt is very pale; the corolla of the Primroſe is much the largeſt; that of the Ox-lip a middle ſize, between the two others: the ſimple unbranched flowering ſtem of the Primroſe is weak, and rather a peduncle than a ſtalk; the ſcape of the Ox-lip is ſometimes near a foot high, and ſtrong; that of the Cowſlip is generally lower and weaker. I do not know whether

[x] Comprehending Primroſe, Ox-lip, Cowſlip, and Polyanthus.

[y] Primula acaulis *Lin.* vulgaris *Hudſon.* Fl. dan. 194.

[z] Primula vulgaris β. *Hudſ.* Fl. dan. 434.

[a] Primula veris *Lin.* & *Hudſ.* Fl. dan. 433.

I dare to tell you that all the beautiful ſorts of *Polyanthus*, by you prized ſo much, are but an accidental variety of this ſpecies, which is certainly much diſpoſed to vary even in its wild ſtate. Thus the primroſe has ſometimes two flowers together, or changes to green, or to red, or doubles its corolla; the Ox-lip ſometimes has very few flowers, and they are nearly as large as a Primroſe; and the Cowſlip has frequently red flowers, then much reſembling a ſmall Polyanthus.

See now by how many ſteps you arrive at a knowledge of theſe plants. You firſt determine their claſs and order, by ſeeing that they have five ſtamens, and one piſtil; having ſtill an hundred and fifty-five genera to encounter, you next ſettle what ſubdiviſion of the order they range under; and finding that the corolla is monopetalous, inferior, and ſucceeded by a veſſel incloſing the ſeeds, you are reduced to ſeventy-three genera. Next you diſcover that they are of the natural order of *Preciæ*, which leaves you but ten genera to chooſe out of. You are now got within ſo ſmall a compaſs that it cannot be very difficult to aſcertain the genus, the ſpecies which are ten in number, and the ſubordinate varieties. I do not make all this parade, in order to enable you to diſcover a plant which you were perfectly acquainted with beforehand, but to ſhew

you

you how you are to proceed with a plant you do not know, from this inſtance of one which you do.

Or you may take it thus—You have a plant in flower, which for the preſent we will ſuppoſe you to be unacquainted with. You firſt examine the ſtamens and piſtils; and by the number of theſe you determine your plant to belong to the fifth claſs and the firſt order. You next conſult the ſubdiviſions of that order, and find it belonging to that which has monopetalous inferior corollas, with the ſeeds incloſed in a veſſel. Seeing farther that your plant has a monophyllous calyx cut into five ſegments, that the corolla is alſo divided in the ſame manner: this added to the foregoing circumſtances ſhows you that it ranges under the natural order of *Preciæ*. Here remarking an involucre under the flowers, the tube of the corolla cylindric, and open at top, and the capſule unilocular or one-celled, you are aſſured at length that your plant is of the genus *Primula*. But finding that the leaves, inſtead of being wrinkled, are perfectly ſmooth, fleſhy, and either entire, or ſharply notched about the edges, you are well aſſured that it is a diſtinct ſpecies; and upon inquiry diſcover it to be the *Auricula* [b], the elegant, the powdered Auricula, ſo much eſteemed by floriſts, and ſo various

[b] Primula Auricula *Lin.* Ger. 784, 5, 6.

in

in the ſize and colours of its corolla, when in a ſtate of cultivation.

All the other plants of this natural order are pretty, if not ſpecious. *Meadia*, perverſely altered by Linnæus to *Dodecatheon*[c], is an American plant, but flowers well and early in our climate. It has a rotate or wheel-ſhaped corolla with reflex petals: the ſtamens ſit upon the tube; and the capſule has one cell only, and is oblong. This is ſufficient for the complete detection of the plant, ſince there is only one known ſpecies. The leaves however are ſmooth; the flowering ſtems are naked, eight or nine inches high, and ſuſtain many flowers, each of which has a long ſlender peduncle, which is recurved ſo that the flower hangs down; the corolla is of a beautiful light purple. If you have not this plant already in your garden, procure it againſt next ſpring; you will be pleaſed with the ſtructure and appearance of it. Meadia.

Cyclamen reſembles Meadia in its wheel-ſhaped reflex corolla, but the tube is globular, and remarkably ſhort, with the neck prominent; the ſtigma, which was obtuſe in that, is acute in this. The ſeed-veſſel is roundiſh and fleſhy, incloſing ſeveral angular ſeeds: Linnæus calls it a berry covered with a capſular ſhell. There are ſeveral ſpecies or varieties of Cyclamen; for it is doubtful whether they are poſitively diſ- Cyclamen.

[c] Curtis's Magaz. 12. Mil. fig. pl. 174. Pl. 12. f. 2.

tinct

tinct or not. The most common[d] has heart-shaped angular leaves, marked with black in the middle. The flowers appear alone, before these, rising immediately from the root: when they fall, the peduncles twist up like a screw, inclosing the germ in the centre, and lie close to the ground among the leaves, which grow very thick together, and protect them all winter. The common colour of the corolla is red, but it varies to purple and white. There is one sort which has the leaves purple underneath; and another which has the veins only purple, and the upper side veined and marbled with white: the flowers white with a purple base. The Persian sort has leaves like the last in colour, but quite entire about the edges, the flowers large, pale purple with a bright red or purple base[e]. All these, and other differences, whether specific or not, make a most agreeable variety, and are very beautiful.

There are two wild plants of this natural order which I must recommend to your inspection for their beauty. They grow in the water, and therefore you must procure them by another hand.

Meny-anthes. *Marsh Trefoil*, *Buckbean* or *Bog-bean*[f] will discover itself to you immediately by

[d] Cyclamen Europæum *Lin.* C. coum is figured in Curt. Magaz. t. 4.—Persicum, in t. 44.

[e] Miller's fig. pl. 115.

[f] Menyanthes trifoliata *Lin.* Curtis, Lond. IV. 17.

the

the corolla being fringed all over; it is funnel-ſhaped, with a ſhort tube, and the border divided beyond the middle; the colour is white, but red on the outſide; the ſtigma bifid; and the ſeed-veſſel a capſule of one cell. The ſpecies is diſtinguiſhed by its ternate leaves; whence, and from its ſituation, it has the name of *Marſh-trefoil*; and becauſe each of the component leaves is of the ſize and ſhape of a bean-leaf, it is alſo called *Buckbean* or *Bogbean*. The flowers grow in a looſe ſpike at the top of the ſtem.

Water Violet[g] has a ſalver-ſhaped corolla not fringed, the tube longer than in the laſt, the colour white or faint purple, with a yellow eye: the ſtamens are placed upon the tube of the corolla; the ſtigma globoſe; and the ſeed-veſſel a capſule of one cell, as in the laſt. The leaves are wholly immerſed in the water, and finely pinnate; the flower-ſtem is naked, and riſes five or ſix inches above water; towards the top are two or three whorls of flowers, and it is terminated with a cluſter of them; the whole forming a kind of conical ſpike. Hottonia.

Another natural order of this claſs contains the plants entitled *Aſperifoliæ* or *rough-leaved*. Theſe are not ſo beautiful as the laſt; but you are by this time become too good a naturaliſt to be led away by gaudy colours or ſpecious appearances. Though roughneſs of the leaves and ſtem be a general

[g] Hottonia paluſtris *Lin.* Curtis, Lond. I. 11.

character of this order, yet it is more neceſſary that the following character ſhould be found in the fructification. The calyx is of one leaf divided into five ſegments, and permanent: the corolla is monopetalous, divided alſo into five ſegments, tubulous, and extending below the germs: the five ſtamens grow from the tube of the corolla: and there are four naked ſeeds to which the calyx ſerves as a capſule. We may remark farther, that the leaves are placed alternately, or without order on the ſtem; and that the ſpike of flowers, before they open, is reflex. With ſo ample a train of circumſtances to direct you, there cannot be much difficulty in knowing when you meet with one of this rough-leaved tribe of plants; eſpecially as they wear the ſame dreſs, and have a ſtrong family likeneſs.

Out of eighty-three ſpecies, which this order contains, you may perhaps know ſome of the following, and from them you will have an idea of the reſt. Heliotrope or Turnſole, Mouſe-ear Scorpion-graſs, Gromwell, Alkanet, Hound's-tongue, Pulmonaria, Comfrey, Cerinthe, Borage, Bugloſs, and Viper's Bugloſs. If you examine the corolla of theſe plants, you will obſerve that ſome of them have five ſcales in the tube of it, whilſt others have none; this circumſtance, together with the ſhape of the corolla, will furniſh the principal generic diſtinctions. Thus Gromwell, Pulmonaria, Cerinthe, and Viper's Buglofs, have the tube of the corolla naked; the reft have the five fcales. Heliotrope and Moufe-ear Scorpion-grafs have falver-fhaped flowers; Gromwell, Alkanet, Hound's-tongue, Pulmonaria, and Buglofs, have funnel-fhaped flowers; in Comfrey and Cerinthe the corolla is ventricofe, fwells or bulges out towards the top; Borage has a rotate corolla; and in Viper's Buglofs it is an irregular kind of bell-fhaped corolla. Heliotrope has the fcales; but the top of the tube is not clofed by them, as it is in the Moufe-ear Scorpion-grafs, Alkanet, Hound's-tongue, Comfrey, Borage. Hound's-tongue has flat feeds fixed to their ftyle by their inner fide only. Pulmonaria has a pentagonal or prifmatic calyx. Cerinthe has only two hard, fhining bilocular feeds. Buglofs has the tube of the corolla bent.

Common Turnfole [h] has the leaves ovate, entire, wrinkled, and covered with a nap; the lower fpikes of flowers are fingle, and the upper ones double. The colour of the corolla white, with a greenifh eye, and fometimes light red. This is an annual lant. Heliotropium.]

Peruvian Turnfole [i] has a fhrubby ftem; the leaves of a long ovate form, wrinkled and rough, on fhort petioles; the flowers are produced at the end of the branches in

[h] Heliotropium Europæum *Lin.* Jacq. auftr. 3. t. 207.
[i] Heliotropium Peruvianum *Lin.* Mill. fig. pl. 144.

ſhort ſpikes, growing on cluſters, the peduncles divide into two or three others, and theſe again into ſmaller ones, each ſuſtaining a ſpike of pale blue flowers, which have a peculiar odour.

Myoſotis. *Mouſe-ear Scorpion-graſs*[k] is common both in dry paſtures and heaths, and by the ſides of ditches and ſtreams; in the former it is hairy, in the latter ſmooth, with the flowers much larger, and extremely beautiful when ſeen ſufficiently near, of a moſt elegant blue with a yellow eye. Linnæus diſtinguiſhes this ſpecies by the ſmoothneſs of the ſeeds, and by the tips of the leaves being callous.

Litho-ſpermum. There are two ſorts of Gromwell wild. The *true* Gromwell[l], which name is a corruption from *Gray Millet*, is not very common; it affects dry ſoils, eſpecially chalk, and is found chiefly in woody places, or among buſhes. You will know it by its whitiſh, ſhining, oval, hard ſeeds; which latter quality gave occaſion to the Latin name, from the Greek, *Lithoſpermum*[m]. Or if it be not far enough advanced to ſhow the ſeeds, obſerve that it is a much larger and more branching plant than the next; the leaves are lance-ſhaped; the flowers are ſmall, and come out ſingle from the axils

[k] Myoſotis ſcorpioides *Lin.* Curtis, Lond. III. 13.
[l] Lithoſpermum officinale *Lin.* Mor. hiſt. ſ. 11. t. 31. f. 1. Ger. 609. 2.
[m] Stone-ſeed.

of

of the leaves on ſhort peduncles; the corolla is white or yellowiſh, with a greeniſh tube.

Corn Gromwell [n] is a common weed among corn, and differs from the former in its wrinkled, conical ſeeds; the leaves alſo are ovate, and ſharp-pointed; the flowers are chiefly on the top of the ſtem among the leaves; the corolla is white, with the tube ſwelling at top. Both ſpecies have the corollas ſcarcely extending beyond the ſegments of the calyx; and both have the roots tinged with red, whence the latter has the name of *Baſtard Alkanet*.

Hound's-tongue [o] is a large plant that grows common by hedges and way ſides; it has a ſtrong ſmell like that of mice. The corolla is of a dirty red, or the colour of blood that has ſtood ſome time. It is diſtinguiſhed from the other ſpecies by the ſtamens being ſhorter than the corolla; the leaves broad lance-ſhaped, nappy, and ſitting cloſe to the ſtem without petioles. Cynogloſſum.

Comfrey [p] is common by water ſides. The leaves are large, long, hairy, and ending in a point; from their baſe on each ſide runs a border down the ſtalk [q]. From the upper part of the ſtalk come out ſome ſide- Symphytum.

[n] Lithoſpermum arvenſe *Lin.* Fl. dan. 456. Mor. f. 7. Ger. 610. 4.

[o] Cynogloſſum officinale *Lin.* Curtis, Lond. IV. 16.

[p] Symphytum officinale *Linnæi.* Curtis, Lond. IV. 18.

[q] This is what Linnæus calls *decurrent*.

branches, with two ſmaller leaves, terminated by looſe bunches of nodding flowers; the corolla of a yellowiſh white, in ſome places purple.

Cerinthe. Of *Cerinthe* there are two ſpecies only, diſtinguiſhed by the larger ſort [r] having obtuſe, open corollas; the leſs [s] having ſharp, cloſe corollas. The leaves of the firſt are ſea-green ſpotted with white; it varies with prickly and ſmooth leaves, with yellow and purpliſh red corollas. It grows wild in Italy, the ſouth of France, Germany, and Switzerland. The ſecond has more ſlender ſtalks; the calyx large, the corolla ſmall and yellow. This is found naturally in the Alps. Both are not uncommon in gardens.

Borago. *Borage* [t] is an annual plant, which comes up in your kitchen garden, without the care of the gardener. The whole plant is rough; the leaves are large, and broad lance-ſhaped. The flowers came out in looſe, naked bunches, on long peduncles, at the end of the ſtalks: the calyx, with the corolla, ſpreads out quite flat: the colour of the corolla is a fine blue, which ſometimes fades to white, or changes to red.

Lycopſis. *Bugloſs* [u] is common among corn, and by

[r] Cerinthe major *Lin.* Mill. flg. 91.
[s] Cerinthe minor *Lin.* Jacq. auſtr. 2. t. 124.
[t] Borago officinalis *Lin.* Mor. hiſt. ſ. 11. t. 26. f. 1. Ger. 797. 1, 2.
[u] Lycopſis arvenſis *Lin.* Curt. Lond. 5. 17. Mor. t. 26. f. 8. Ger. 799. 3.

way

way ſides. A very rough plant, with blue corollas veined with white.

Viper's Bugloſs [v] is a much larger plant than this, with a large handſome ſpike of blue flowers. The ſtalk is very erect and ſpotted: the leaves lance-ſhaped, the lower ones petiolate, the upper ſeſſile. It is common among the corn in ſome countries; alſo in ſome paſtures, by way ſides, and on walls. Echium.

You will find ſome plants of this fifth claſs and firſt order which have a bell-ſhaped corolla of one petal. If they have a permanent calyx divided into five parts, and a capſule for a ſeed-veſſel, they belong to a natural order entitled *Campanaceæ* [w]. Three very large genera [x], beſides ſome others, belong to this order.

The genus *Convolvulus* [y] is diſtinguiſhed from all others by its large, ſpreading, plaited corolla, with the edge either marked with ten notches, or ſlightly quinquefid; two ſtigmas; and a capſule wrapped up in the calyx, generally bilocular, with two roundiſh ſeeds. Convolvulus.

From this genus I will ſelect two wild

[v] Echium vulgare *Lin.* Fl. dan. 445. Ger. 802. 2.

[w] Bell-flowers.

[x] Convolvulus, Ipomæa, and Campanula: the firſt has ſixty-four; the ſecond twenty-two; and the third ſixty-ſix ſpecies.

[y] So called from twining round any thing it comes near; this property however is not common to all the ſpecies.

 and

and two cultivated ſpecies, for your examination.

Small Bindweed[z], which is ſo common a weed among corn, has ſagittate leaves[a] acute both ways, and one flower upon a round long peduncle. The weak ſtalks trail on the ground, unleſs they meet with ſome other plant to ſupport them; the corolla is either white, or red, or variegated; and if the plant came from India it would be cultivated for the beauty of the flower: I do not however recommend you to grow fond of it, for it creeps intolerably at the root.

Great Bindweed[b] has ſagittate leaves as well as the laſt, but truncate or cut off behind; the flowers come out ſingle alſo, but on ſquare peduncles. This is a much larger, ſtronger plant than the other, riſing in hedges or among buſhes and ſhrubs, ten or twelve feet high: the corolla is very large, and always pure white; immediately under the calyx is a large heart-ſhaped involucre of two leaves. The former ſpecies has theſe two leaves, but they are very narrow, and in the middle of the peduncle.

Purple Bindweed[c], an annual ſpecies cultivated in flower gardens under the name of *Convolvulus major*, has heart-ſhaped undi-

[z] Convolvulus arvenſis *Lin.* Curtis, Lond. II. 13.

[a] Shaped like the head of an arrow.

[b] Convolvulus ſepium *Lin.* Curtis, Lond. I. 13. Pl. 12. f. 3.

[c] Convolvulus purpureus *Lin.* Ehret. pict. t. 7. f. 2. Curtis's Magaz. 113.

vided

vided leaves, the ſeed veſſels hanging down after the flower is gone, and the peduncles ſwelling. This, if ſupported, will climb to the height of ten or twelve feet. Though the moſt uſual colour of the corolla is purple, yet there are varieties white, red, and whitiſh blue.

Tricolor Bindweed[d], or, as it is vulgarly called, *Convolvulus minor*, has lance-ſhaped, ſmooth leaves, a weak falling ſtalk, that never climbs, and the flower coming out ſingly. The corolla is a beautiful blue with a white eye; but ſometimes all white or variegated. This is alſo annual. Its native country is Portugal. The former is wild both in Aſia and America.

This genus contains ſeveral remarkable plants; *as Scammony*[e], *Turpethum* or *Turbith*, and *Jalap*.

Ipomæa has rather a funnel-ſhaped than a campanulate corolla; a globoſe ſtigma, and a trilocular capſule[f]; but the plants that range under this genus being natives of the Weſt Indies, and conſequently requiring much heat to raiſe and preſerve them, may probably not come within your view; and therefore I ſhall not enlarge upon them.

In *Campanula* you will of courſe expect to find a campanulate or bell-ſhaped corolla; but it is worth your obſervation that Campanula.

[d] Convolvulus tricolor *Lin.* Curtis's Mag. 27.
[e] Conv. Scammonia. *Lin.* Mill. fig. 102.
[f] See Mill. fig. 214.

the

the bottom of it is clofed with five valves, concealing the receptacle, and that the ftamens take their rife from thefe valves. The ftigma is trifid, and the feed veffel is a capfule, below the flower, having three or five cells, and at the top of each a hole, through which the feeds are fcattered when ripe. You fee by this time how curious and how various the ftructure of the parts of fructification is. By thus examining them fingly, and comparing them one with another, you will in time grow an eminent botanift, and acquire a facility in determining the genus, fpecies, analogy, and connexion of vegetables.

There is a little *Bell-flower* that grows frequent in dry paftures, and on almoft every heath and common, with is nodding blue corolla anfwering well to its name. The botanifts have confpired to call it *round-leaved* Bell-flower[g]; for what reafon perhaps you will wonder, fince you will difcover no leaves upon the ftem but what are linear, or very long, narrow lance-fhaped: if however you take a young plant, or at leaft one in full vigour, and fearch among the grafs clofe to the ground, you will fee thefe leaves, which are not fo properly round as heart[h] or kidney-fhaped[i]. This fort flowers towards the latter end of the fummer, and all the autumn, till froft puts an

[g] Campanula rotundifolia *Linnæi.* Curtis, Lond. IV. 21. [h] Haller. [i] Linnæus.

end

end to it; and frequently has a white corolla. *Rampion*[k], which was formerly cultivated for its roots to eat in ſallads, is now ſo much neglected, that your kitchen garden perhaps may not furniſh it; and in its wild ſtate it is by no means common. This has upright ſtalks, two feet high; the leaves undulating, thoſe next the root ſhort, lance-ſhaped, inclined to oval: towards the upper part of the ſtem, and cloſe to it, ſmall flowers are produced, with a blue or white corolla.

Peach-leaved Bell-flower[l] is abundant in your flower borders, both blue and white; but ſince your gardener has obtained the double ſorts, he has probably deſpiſed the ſingle ones ſo much as to have deſtroyed them, and at the ſame time to have deprived you of the power of determining the genus: you will however know this to be a Campanula by its air; and you will determine the ſpecies by the leaves, which are ovate near the root, and on the ſtalk are very narrow lance-ſhaped approaching to linear, ſlightly ſerrated about the edge, ſit cloſe to the ſtem, and are remote from each other.

I remember your hall chimney uſed to be adorned in ſummer with the *pyramidal* or *ſteeple* Bell-flower[m], ſtrutting out like a fan,

[k] Campanula Rapunculus *Linnæi.*
[l] Campanula Perſicifolia *Linnæi.*
[m] Campanula pyramidalis *Linnæi.*

by

by means of a frame of little ſticks. This has ſmooth, heart-ſhaped leaves, ſerrated about the edge; thoſe on the ſtem lance-ſhaped: the ſtems are ſimple and ruſh-like: the flowers come out in ſeſſile umbels from the ſide of the ſtem. Such are Linnæus's ſpecific characters.

There is the *Giant Throatwort*[n], wild, but not common, in buſhy places and hedges: known by its ſtrong, round, ſingle ſtalks; its long ovate leaves, inclined to lance-ſhaped, ſlightly ſerrated or toothed like a ſaw on their edges: towards the upper part of the ſtalk the flowers come out ſingly upon ſhort peduncles. Pray remark, that after theſe are faded, the ſeed-veſſels turn downwards till the ſeeds are ripe, and then riſe up again.

Great Bell-flower[o], vulgarly called *Canterbury Bells*, is much more common in the like places. This has ſtiff, hairy, angular ſtalks, putting out a few ſhort ſide-branches. The leaves are like thoſe of nettles, hairy, and deeply ſerrated on their edges: towards the upper part of the ſtalks the flowers come out on ſhort trifid peduncles, and have hairy calyxes.

Small Canterbury Bells[p] is common in

[n] Campanula latifolia *Lin.* Fl. dan. 85. Ger. 448. 3.

[o] Campanula Trachelium *Lin.* Mor. hiſt. ſ. 5. t. 3. f. 28. Ger. 448. 1.

[p] Campanula glomerata *Linnæi.* Mor. t. 4. f. 40 & 43. Ger. 449. 4.

paſtures,

paſtures, eſpecially in a chalky ſoil. In dry places it is very ſmall, and in a moiſt ſoil will grow to the height of two feet. The ſtalk is hairy, angulate, and unbranched; the lower leaves are broad, and pedunculate; thoſe on the ſtalk long, narrow, ſitting cloſe to the ſtalk, and even embracing it: towards the top of the ſtalk, from the axils of the leaves, two or three flowers come out together, and a larger bunch terminates it: the flowers are ſeſſile.

Venus's Looking-glaſs[q] is a Campanula, with a weak, low, and very branching ſtalk; the leaves oblong, and a little notched; the flowers ſolitary, and the ſeed-veſſels of a priſmatic form. *Corn-bell-flower*[r] very much reſembles this; but the ſtalk is ſtiff, and branches little; the leaves are more deeply notched, and waving; the flowers come out in parcels, and the calyx is longer than the corolla. This is a common weed among corn. Theſe two have ſcarcely bell-ſhaped corollas, any more than another plant of this Campanulate order, entitled *Greek Valerian* or *Jacob's Ladder*[s], which has the corolla rather rotate, with the tube ſhorter than the calyx, but cloſed with five valves, into which the ſtamens are inſerted, as in

Polemonium.

[q] Campanula ſpeculum *Lin.* Curtis Magaz. 102.
[r] Campanula hybrida *Lin.* Mor. t. 2. f. 22. Ger. 439. 2.
[s] Polemonium cæruleum *Lin.* Fl. dan. 255. Ger. 1076. 5.

Campanula:

Campanula: the ſtigma alſo is trifid, as in that, and the ſeed-veſſel a trilocular or three-celled capſule, but incloſed within the flower. The circumſtances that diſtinguiſh this from the other two ſpecies are, that the leaves are pinnate, the flowers erect, and the calyx full as long as the tube of the corolla ; in which you ſee it recedes a little from one character of the genus. It is blue, and cut into five roundiſh ſegments. I ſcarcely need caution you not to be miſled by names, which being uſually given by ignorant perſons, are very fanciful or erroneous. Thus here, you may as well ſuppoſe *Polemonium* to have an affinity with a ladder as with valerian : indeed the ſame circumſtance of the pinnate leaves probably gave occaſion to both names.

I am almoſt afraid to preſent you with a ſet of plants, which from their lurid, duſky, diſmal, gloomy, appearance, are kept together under the title of *Luridæ.* They have alſo moſt of them a diſagreeable ſmell, which, with their forbidding look, will deter our young couſin from examining them, ſhe not being yet ſufficiently tinctured with enthuſiaſm to go on in ſpite of ſuch circumſtances. Indeed I would not wiſh her to be too buſy with ſome of theſe *inſane roots that take the reaſon priſoner*, and which I can never collect and examine myſelf, without their affecting my head. You will conſider that nature has kindly given us notice

notice in general of approaching danger, by means of our ſenſes; and accordingly ſome of theſe *Lurid* plants are highly poiſonous; moſt of them are ſo in ſome degree, though ſoil and climate may mitigate the poiſon, and even render them wholeſome. I will ſelect ſome of the leaſt diſagreeable in ſmell and appearance; or, if they be otherwiſe, will announce it to you. Beſides the circumſtances of five ſtamens and one piſtil, theſe plants agree in a permanent calyx divided more or leſs deeply into five ſegments; a monopetalous corolla, divided alſo into five ſegments, tubulous, irregular; the ſeed-veſſel bilocular, and either a capſule or a berry, incloſed within the flower.

Of *Verbaſcum*, or *Mullein*, there are ſeveral ſpecies wild, one very common, and another not uncommon. Their general characters are, that the corolla is rotate, and ſlightly irregular; the ſtamens unequal in length, bending down, and generally clothed at bottom with a coloured fringe; the ſtigma obtuſe, and the capſule bivalve, and opening at top. Verbaſcum.

The common ſpecies is the *Great* or *Hoary Mullein* [t], which grows moſtly under banks or hedges. It is a biennial plant; the firſt year forming its root, and a ſet of large, broad leaves, extremely woolly on both ſides, and ſpreading on the ground,

[t] Verbaſcum Thapſus *Linnæi.* Fl. dan. 631. Mor. hiſt. ſ. 5. t. 9. f. 1. Ger. 733. 1.

with

with ſcarcely any petioles: the ſecond year it ſends up a ſingle ſtem, ſometimes five feet in height, with decurrent leaves on it, woolly as the radical ones; and on the top a cloſe ſpike of yellow flowers, which have an odour not diſagreeable.

The other which I hinted at is the *Black Mullein*[u], growing in ſimilar places, abundantly in ſome, but by no means ſo extenſively. It has not ſo high a ſtem; the ſhape of the lower leaves is that of a heart much lengthened out, and they are petiolate; the leaves on the ſtem ovate, ſharp-pointed and ſeſſile; all of them are pale green on the upper, and hoary on the under ſurface; and are indented about the edges. The ſtalk is terminated by a long ſpike of yellow flowers, formed by ſhort cluſters or ſpicules on the ſides of the principal ſtalk. The corolla is yellow, with the filaments fringed or bearded with purple. It has the name of black, I preſume, merely becauſe it is not white, like the other.

Datura. *Datura*, *Stramonium*, or *Thorn Apple*, has the calyx tubulous, ſwelling in the middle, five-cornered, and deciduous; the corolla funnel-ſhaped, ſpreading out gradually very wide from a long cylindric tube, into a pentangular border with five plaits: the capſule is quadrivalvular, or opens into four parts. The flowers of theſe are large, and rather

[u] Verbaſcum nigrum *Lin.* Mor. hiſt. ſ. 5. t. 9. f. 5.

ſpecious,

ſpecious, and the capſules are remarkable for their ſize.

The *common* Thorn Apple[v] has ſmooth leaves, irregularly angular, and ſmelling diſagreeably; the flowers come out from the firſt diviſions, and near the extremities of the branches; the corolla is white, and each angle of it ends in a long point; the capſule is ovate, covered with ſtrong thorns, and grows erect.

Another ſort[w], cultivated ſometimes in flower gardens, has purple flowers; it has alſo purple ſtalks, which are ſtouter and taller than thoſe of the laſt; the leaves are alſo larger, and more angular and notched; the capſule is larger, but much like that of the common ſort. One of them, having the capſule armed with very ſtrong ſpines, has the epithet of *fierce*[x].

Henbane[y] is a very common plant, and has often done miſchief to ſuch as will not ſuffer their appetites to be corrected by their ſenſes. You will agree with me that the ſmell is ſufficient to deter any perſon from eating it. I cannot however diſpenſe with your examining the flower, which is really beautiful on a near view. The corolla is funnel-ſhaped, and obtuſe; of a pale yel- Hyoſcyamus.

[v] Datura Stramonium *Lin.* Curtis, Lond. n. 61. Fl. dan. 436. Ger. 348. 2.

[w] Datura Tatula *Lin.*

[x] Datura ferox. *Lin.* Mor. t. 2. f. 4.

[y] Hyoſcyamus niger *Lin.* Ger. 353. 1.

lowiſh colour, beautifully veined with purple. The ſtamens are of different lengths and bent; and the capſule is involved in the calyx, of an oval form, and covered with a hemiſpherical lid, which, by falling off, announces that the ſeeds are ripe.

The common wild ſpecies is diſtinguiſhed from the others by its ſinuate leaves, embracing the ſtalk, and by the flowers ſitting cloſe to it. The whole plant is covered with long hairs, from which exudes a clammy, fetid juice: the leaves are very large, and remarkably ſoft; and the flowers come out in a very long ſpike, rather on one ſide. It grows on banks, dunghills, and way-ſides about villages, and is a biennial plant. There are other ſorts, but neither wild nor much cultivated.

Nicotiana. You who have ſuch an averſion from tobacco in all the ways of uſing it, will not be diſpleaſed at finding it in this lurid order. Notwithſtanding it is ſo generally taken, the oil of it is the ſtrongeſt of the vegetable poiſons. It is a plant however neither unornamental for your garden, nor dangerous, nor even diſagreeable to examine. The eſſential generic characters are, that the corolla is funnel-ſhaped, the border plaited; the ſtamens a little inclined; the ſtigma notched; the capſule ovate, marked with a furrow on each ſide, bivalvular, and opening from the top.

Common

Common or *broad-leaved* Tobacco [z] is diſtinguiſhed by its broad lanceolate leaves, which are about ten inches long, and three and an half broad, ſmooth, ending in acute points, and ſitting cloſe to the ſtalks; the corollas are of a pink purple, and end in five acute points. There is a ſort like this, or perhaps a variety of it, called *Oroonoko* Tobacco, which is a larger plant, the leaves more than a foot and half long, and a foot broad; very rough and glutinous; the baſe embracing the ſtem: the corollas are of a pale purple.

Another ſpecies, called *Engliſh* Tobacco [a], might eaſily be miſtaken for a Henbane, if you did not remark the regular form of the corolla, and the want of a lid to the capſule. It is a lower plant than the others; the leaves are ovate, entire, and on ſhort petioles. The flowers come out in looſe bunches on the top of the ſtalks; the corolla has a ſhort tube, ſpreading out into five obtuſe ſegments, of a greeniſh yellow colour. Though this has the epithet of Engliſh, you are not to ſuppoſe it to be an European plant, for it is a native of America, as well as all the other ſpecies, which are at leaſt ſeven in number.

How the ſame plant ſhould come to have the gentle appellation of *Bella-donna*, and Atropa.

[z] Nicotiana Tabacum *Linnæi*. Mill. fig. 185. 1. Pl. 12. f. 1.

[a] Nicotiana ruſtica *Linnæi*. Blackw. t. 437.

the tremendous name of *Atropa*[b], ſeems ſtrange, till we know that it was uſed as a waſh among the Italian ladies, to take off pimples and other excreſcences from the ſkin; and are told of its dreadful effects as a poiſon. Linnæus has joined them, making *Atropa* the generic, and *Bella-donna* the ſpecific or trivial title. The principal characters which he gives of the genus are theſe—the corolla is bell-ſhaped; the filaments grow from the baſe of it, are cloſe at bottom, but at top diverge from each other, and are arched; the ſeed-veſſel is a globoſe berry, ſitting on the calyx, which is large.

Our ſort, for there are ſix ſpecies of the genus, is a great branching plant, with ovate, entire leaves, and large flowers coming out among the leaves ſingly, on long peduncles; the corolla is of a duſky brown colour on the outſide, and of a dull purple within; the ſtalks have a tinge of the ſame colour, as have alſo the leaves towards autumn. The berry is round, of a ſhining black when ripe, and not unlike a black cherry in ſize and colour; it contains a purple juice of a mawkiſh ſweetneſs, and has frequently enticed children to taſte it at their peril. I have known however the ſame poiſonous effects follow from eating the young ſhoots

[b] From *Atropos*, the name of one of the furies. Figured by Miller, pl. 62. Fl. dan. 758. Ger. 340. Blackw. 564. Curtis, Lond. 5. 16.

of

of the ſpring boiled, as from the crude berries of autumn. *Deadly Nightſhade* is rarely cultivated, and not common wild; it ſkulks in gloomy lanes, and uncultivated places, but is too frequent near villages in ſome countries.

You have heard of the *Mandrake's Groan*, and " of ſhrieks, like Mandrakes torn out " of the earth:" ſuperſtition having endued this plant with a ſort of animal life, fatal to whoever preſumed to deſtroy it by digging up the root. It was famous, as Opium now, for procuring ſleep; whence Cleopatra ſays,

——" Give me to drink *Mandragora*,
" That I might ſleep out this great gap of time
" My Anthony is away."

And the vile Iago boaſts that

——" Not Poppy, nor *Mandragora*,
" Nor all the drowſy ſyrups of the world,
" Shall ever med'cine thee to that ſweet ſleep
" Which thou hadſt yeſterday."

Since Mandrake groans and ſhrieks when injured, it muſt needs have a human form; and accordingly ſuch have been carried about for ſale, notwithſtanding the danger that attends the procuring it; but this is cunningly avoided by tying a dog to the root, and thus making the blind fury of the poor Mandrake fall upon the innocent dog inſtead of the aggreſſor. Theſe pretended

Mandrakes are ſaid to be roots of Angelica or Bryony, either cut into form, or compelled to go through earthen moulds put into the ground for this purpoſe: they were uſed in magical incantations; and though theſe are now pretty much out of faſhion, yet I have had them very gravely offered me for ſale. Linnæus formerly made this a diſtinct genus from the laſt, but on ſecond thoughts he has made it a ſpecies of Atropa [c], diſtinguiſhing it from the others, by its having no ſtems except the ſcapes which ſupport a ſingle flower. The root is like that of a parſnep, ſometimes forked; next the ground there is a circle of large, broad leaves; the *ſcapes* or naked ſtalks that ſupport the flowers are but about three inches long; the corollas are five cornered, and of a greeniſh white or purpliſh colour; the berry is as large as a nutmeg, and of a yellowiſh green. The root and leaves are ſtinking, and the whole plant is poiſonous, though, in ſmall doſes, it is uſed medicinally.

Phyſalis. Another genus of this ſame natural order is *Phyſalis*; the characters of it are theſe—the corolla is wheel-ſhaped; the filaments and anthers are convergent or bend towards each other; and the ſeed-veſſel is a berry incloſed within the calyx, which grows to a large inflated, coloured bladder. *Winter-*

[c] Atropa Mandragora. Mill. fig. pl. 173. Blackw. 364.

Cherry,

Cherry [d], of which you have ſuch abundance under your ſhrubs, is a ſpecies of this genus. The diſtinguiſhing marks are, that the leaves are double or conjugate, that is, come out in pairs, are entire about the edges, or but very ſlightly indented, and ſharp pointed; the ſtalk is herbaceous, and a little branching at bottom. The roots creep ſo far as to be troubleſome; the ſtalks are only about a foot high; the leaves are of various ſhapes, and have long petioles: the flowers are produced ſingly from the axils of the ſtalks on ſlender peduncles; and have a white corolla, which, with the calyx, leaves, and ſtalks, is hairy. This plant, which is ſo humble and inconſiderable all the ſummer, attracts your notice in autumn, by its great inflated calyx turning red, and diſcloſing the round red berry within it, about the ſize of a ſmall cherry.

But the principal genus of this natural Solanum. order is the *Nightſhade*, or *Solanum*, whence ſome authors have entitled theſe plants *Solanaceæ*. There are no leſs than forty-ſix ſpecies of Solanum; out of which I ſhall ſelect, as uſual, both ſome wild and cultivated ſorts, ſuch eſpecially as are either moſt important, or moſt likely to be within your reach.

You will eaſily know the genus by its wheel-ſhaped corolla; by its large anthers cloſed in the middle of the corolla, and

[d] Phyſalis Alkekengi. Blackw. 161.

ſeeming to form but one body; and by its bilocular berry.

Some of the ſpecies have prickly ſtalks and leaves; others are unarmed: hence a commodious partition of the genus into two ſubdiviſions.

A ſhrubby, tall ſort, from the Madeiras, without any ſpines or prickles, has long been an inhabitant of the greenhouſe, which it adorns with its ſplendid red berries all the winter: the gardeners know it by the name of *Amomum Plinii*; and it is often called *Winter Cherry* [e]; ſuch is the dearth of diſtinctive names, and ſuch the confuſion ariſing from the want of a regular language, like that which Linnæus firſt introduced into Botany. The leaves are lance-ſhaped, and have a waving edge [f]: the flowers grow in ſmall umbels, cloſe to the branches; the corolla is white; and the berries are as large as a ſmall cherry; generally red, but ſometimes yellow.

Another ſhrubby ſort, without ſpines, is the *Woody Nightſhade*, or *Bitter-ſweet* [g], which grows commonly wild in moiſt hedges. This has a climbing, flexuous ſtalk: the lower leaves lance-ſhaped, the upper ones ſometimes trifid: the flowers are in bunches, or branched cymes, coming out from the axils of the leaves; the corolla

[e] Solanum Pſeudocapſicum *Lin.*

[f] Linnæus calls them repand.

[g] Solanum Dulcamara *Lin.* Curtis, Lond. I. 14.

revolute,

revolute, purple, marked with two ſhining green ſpots at the bottom of each ſegment; and the berries red.

Garden Nightſhade [h] is alſo unarmed, but not ſhrubby. It is an herb, an annual. The leaves are on long petioles, and being of a ſoft texture, are inclined to hang down. They are either of an ovate or rhomboid form, with long points, angulate and notched about the edges: the flowers grow on a kind of nodding umbel; the corolla is white, and the berry is black. It is a common weed on dunghills, in gardens, and other richly cultivated places. It varies with yellow and red berries; and in the form of the leaves.

Potatoe [i] is of this genus, as you will be convinced, if you compare the ſtructure of the flower with that of the other ſpecies. Linnæus characteriſes it by theſe diſtinctions—that the ſtalk is herbaceous and unarmed, the leaves pinnate and quite entire, the peduncles ſubdivided: the corollas are either purple or white, and the berry is large.

Tomatos or *Love-apple* [k] is another ſpecies of Nightſhade, which is alſo admitted to the table, and eaten with impunity, in ſpite of the ill neighbourhood in which it is found.

[h] Solanum nigrum *Lin.* Curtis, Lond. II. 14.

[i] Solanum tuberoſum *Lin.* The Engliſh name is evidently a corruption of the Indian *Batatas.*

[k] Solanum Lycoperſicum *Lin.* Blackw. 133.

found. This has an unarmed, herbaceous ſtem, which is very hairy; the leaves alſo are pinnate, but cut; and the flowers are borne on ſimple unbranched bunches; the corolla is yellow, and the fruit or berry is large, flatted, and deeply furrowed.

Melongena or *Mad Apple*[1] is alſo of this genus; it is cultivated as a curioſity for the largeneſs and ſhape of its fruit; and when this is white, it has the name of *Egg plant*; and indeed it then perfectly reſembles a hen's egg in ſize, ſhape, and colour. The ſtem of this is herbaceous, and without prickles; the leaves ovate and nappy; the peduncles pendulous, and growing thicker towards the top, and the calyxes unarmed. The corollas are purple, and the fruit varies much in colour. The three laſt ſpecies recede a little from the character of the order; for the Potatoe and Tomatos have many cells to the fruit, and this has but one.

The prickly ſorts of *Solanum* are natives of hot countries, and moſt of them are brought to us from the Spaniſh Weſt Indies: they will not therefore commonly fall under your obſervation.

Capſicum, or *Guinea Pepper*, is alſo of this lurid order; its beauty and uſe lies in the fruit, which Linnæus calls a dry or juiceleſs berry, and others a capſule or pod.

[1] Solanum Melongena *Lin.* Pluk. phyt. t. 226. f. 2.

This

This circumſtance, together with the rotate form of the corolla, and the anthers being connivent or converging, make up the eſſential characters of the genus. Linnæus has only five ſpecies, one annual[m], with an herbaceous ſtem, the reſt perennial with woody ſtems[n]. Others make many more ſpecies from the different form of the fruit; which indeed varies much both in ſhape and colour, and intermixt with the white flowers and green leaves, makes a pleaſing variety: but Linnæus does not allow the form of the fruit in this genus to be permanent enough to conſtitute ſpecific differences. They are all very hot, and hence have the names of *Bell Pepper*, *Hen Pepper*, *Barberry Pepper*, and *Bird Pepper*. The *Bell Pepper*, which has large, ſwelling, wrinkled fruit, with a fleſhy tender ſkin, of a red colour when ripe, is the only ſort fit for pickling. *Cayan Pepper* is made from the laſt, whoſe fruit is ſmall, oval, and of a bright red, and much more pungent than the reſt. Moſt ſorts of Capſicum come from both Eaſt and Weſt Indies. Though they are uſed in hot countries ſo univerſally with their food, yet the ripe fruits thrown on the fire will emit ſtrong noiſome vapours, which occaſion violent ſneezing, coughing, and often vomiting, in thoſe who are near; and mixt in ſnuff will have

[m] Capſicum annuum. Blackw. 129.

[n] Capſicum baccatum, ſinenſe, groſſum & fruteſcens.

the ſame effects to a violent and dangerous degree: ſo that theſe plants, though not ſtrictly poiſonous, are however worthy a place in the lurid tribe.

Lonicera. In this firſt order of the fifth claſs are to be found ſeveral well known ſhrubs; among which the *Honey-ſuckle* is eminent. Of theſe the *Italian*[o], and *Wild*[p] ſpecies are the principal. They are diſtinguiſhed by the firſt having the upper pairs of leaves connate, or ſo joined as to form but one, and the ſtalk running through the middle of them: whereas in the wild honey-ſuckle they are all diſtinct. The Dutch or German Honey-ſuckle of the gardens is ſuppoſed to be a variety only of this, though it is much ſtronger, and not ſo apt to climb. The Woodbind has indeed very ſlender trailing branches, twining round the boughs of trees, and climbing to the very tops of them.

Trumpet Honey-ſuckle[q] is a North American; it agrees with the Italian in having the upper leaves connate; with the Woodbind in its ſlender trailing branches: but differs from both in the whorls of flowers being naked or void of leaves, and the corollas being almoſt regular; the leaves alſo are

[o] Lonicera Caprifolium *Linnæi.* Hort. angl. t. 5. Pl. 12. f. 4.

[p] Lonicera Periclymenum *Lin.* Woodbind. Curtis, Lond. I. 15.

[q] Lonicera ſempervirens *Lin.* Riv. mon. 116.

are evergreen, and the corollas are bright ſcarlet on the outſide, and yellow within.

There are other ſpecies, which you will find among the ſhrubs, differing in appearance, and receding ſomething in character from Honey-ſuckles properly ſo called. Theſe have always two flowers only coming out together; whereas in the former the flowers go in whorls or heads many together. *Fly* Honey-ſuckle[r] has the two berries that ſucceed the two neighbouring flowers diſtinct; the leaves are entire and hoary; and the corollas are white. *Red-berried upright* Honey-ſuckle[s] has the two berries joined together; the leaves lance-ſhaped and ſmooth; the corollas are red on the outſide, but pale within. This is not ſo tall growing a plant as the other.

The five recited ſpecies agree in having a monopetalous irregular corolla, except that in the Trumpet Honey-ſuckle it is almoſt regular; in the genuine Honey-ſuckles the tube is remarkably long. The ſeed-veſſel in all is a berry growing below the flower, and incloſing ſeveral ſeeds; though the laſt has only two.

The numerous genus of *Rhamnus*, containing twenty-ſeven ſpecies, is alſo of the firſt order in the claſs *Pentandria*: theſe are either thorny, prickly, or unarmed. *Buck-* Rhamnus.

[r] Lonicera Xyloſteum *Lin.* Mill. fig. 167. 1.

[s] Lonicera alpigena *Lin.* Mill. fig. 167. 2.

thorn

thorn is one of the firſt; having thorns terminating the branches, the ſtem erect, the leaves ovate, and the calyx cut into four ſegments: the berries have four ſeeds in them, and if you wet them and rub them on white paper, they will ſtain it of a green colour. I mention theſe two circumſtances, becauſe they who gather the berries for ſale are apt to mix others with them: and I know you will be intereſted in them, when I inform you, that the fine green colour [u], which you uſe in your miniature painting, is made from theſe berries. If you ſhould have the curioſity to ſearch the hedges for them, in order to make this paint yourſelf, you muſt not be ſurpriſed if you do not find them on every Buckthorn ſhrub; for all the flowers are incomplete, ſome plants having them with ſtamens, others with a piſtil only; and the former of theſe are never ſucceeded by fruit.

Berry-bearing Alder [v] is one of the unarmed ſpecies. It grows in woods, is a black looking ſhrub, with bunches of inconſiderable herbaceous flowers, with a quinquefid corolla, ſucceeded by black berries containing four ſeeds: the leaves are ovate, ſmooth, and quite entire.

[t] Rhamnus catharticus *Lin.* Fl. dan. 850. Duham. 50. Ger. 1337.

[u] Verd de veſſie.

[v] Rhamnus Frangula *Lin.* Fl. dan. 278. Duham. 100. Ger. 1470.

Another

Another of the unarmed diviſion is the *Alaternus*[w], formerly ſo ſhorn and beclipped in hedges, and covering of walls; but now ſeen chiefly among other evergreens, taking its natural form. The leaves are extremely ſhining, generally notched or ſerrate about the edges; the flowers have a triſid ſtigma, and are incomplete, like thoſe of the Buckthorn: the corolla is quinquefid, and the berry has three ſeeds. There are ſeveral varieties of *Alaternus*, differing in the ſhape of the leaves, and depth of the ſerratures; they are alſo ſometimes blotched or variegated. This ſhrub is frequently confounded with *Philyrea*, from which it may be known at all times by the poſition of the leaves, which is alternate in this, and oppoſite in that: when the two ſhrubs are in flower, you perceive other more eſſential diſtinctions.

Paliurus, or *Chriſt's-Thorn*[x], is one of the prickly diviſion. It has double prickles, the under ones reflex; and is another inſtance of irregularity in this genus, the germ being trilocular, ſurrounded by a membranaceous rim, and crowned by three ſtyles. It has a pliant weak ſtem requiring ſome ſupport; the flowers grow in cluſters, and are of a greeniſh yellow colour the corollas are quinquefid. Being very common in Paleſtine, it is ſuppoſed to be the thorn with which our Saviour was crowned.

[w] Rhamnus Alaternus *Lin.* [x] Rhamnus Paliurus *Lin.*

The

The common characters of all theſe is, that there is only a calyx or corolla, with five ſmall ſcales, one at the baſe of each diviſion, bending towards one another, and defending the ſtamens; the ſeed-veſſel a roundiſh berry, divided within into fewer parts than the corolla or calyx.

Currants and Gooſeberries[y], the Ivy[z] and the Vine[a], are alſo of this order *Monogynia*; but being ſo well known to you and every body, I will not dwell on them, having already run out this letter to ſo great a length.

Coffea. Some other trees and ſhrubs are leſs known, becauſe they are the growth of hotter climes. Such is the coffee[b], originally of Arabia, though now common in both the Indies. It is known by its ſalver-ſhaped corolla, with the ſtamens growing upon the tube of it; and by its ſeed-veſſel, which is a berry below the flower, containing two ſeeds, covered with an *aril*, or detached coat. This tree does not grow above ſixteen or eighteen feet high; the leaves are large, of a lucid green, lance-ſhaped, and waving about the edges. The flowers are produced in cluſters, cloſe to the branches; the corollas are quinquefid, of a pure white colour, and a very grateful odour. It is an evergreen, and at all times makes a beautiful appearance.

[y] Ribes *Linnæi.* [z] Hedera Helix *Lin.*

[a] Vitis vinifera *Lin.*

[b] Coffea Arabica *Linnæi.* Blackw. 337. Dougl. et Ellis monogr.

Ceſtrum

Cestrum or *Bastard Jasmine* is a shrub of the West Indies, and therefore requires a stove to keep it alive in these northern countries. It has a funnel-shaped corolla; the filaments have a little process in the middle; and the seed-vessel is an unilocular berry, containing several seeds. One species [c] has clusters of herbaceous flowers on short peduncles, smelling sweetly in the night. And another [d], with leaves of a lively green, and great consistence, has clusters of white flowers, sitting close to the stalk, smelling sweet in the day time. Cestrum.

Diosma is a genus of shrubs from the Cape of Good Hope. These are of another phalanx, having five petals to the corolla, which is inferior, or incloses the seed-vessel. The germ also is crowned with five nectaries, and becomes three or five united capsules, containing each one seed, with an elastic *Aril* involving it. The flowers are small, but elegant; white, and of an agreeable spicy odour. Diosma.

Other foreign trees and shrubs of this class and order are, the *Iron-wood tree* [e], the *Phylicas*, the *Mango-tree* [f], and some others: but since it is not probable that you will meet with these, I have not troubled you with their characters, or any account of them.

[c] Cestrum nocturnum *Lin.* Dill. elth. t. 153. f. 185.
[d] Cestrum diurnum *Lin.* Dill. elth. t. 154. f. 186.
[e] Sideroxylon. [f] Mangifera Indica *Lin.*

Phlox. There remain ſome ſpecious plants to be noticed, which are commonly cultivated in flower gardens for their beauty. Such are all the ſpecies of *Lychnidea*[g]: which you will know by their ſalver-ſhaped corolla, with a bent tube; their filaments of unequal length; their trifid ſtigma; their priſmatic calyx; their three-celled capſule, with one ſeed in each cell. They are perennial plants; the corollas of moſt of the ſpecies are large, and of a purple colour; and the leaves are lance-ſhaped. They are the produce of North America.

Upon the firſt diſcovery of the New World, as America was vauntingly called, every thing found there was repreſented as wonderful. Strange ſtories were related of the plants and animals they met with, and thoſe which were ſent to Europe had pompous names given them. One of theſe is the *Marvel of Peru*, the only wonder of which is the variety of colours in the flower. It appertains to this claſs and order, and has the following generic marks—the corolla is funnel-ſhaped, the ſtigma globoſe; and there is a globoſe nectary incloſing the germ, which afterwards hardens to a kind of nut. There are three ſpecies: firſt, the common Marvel of Peru[h], which has ſo much variety of colour in the flowers of the ſame plant; theſe are produced plentifully

Mirabilis.

[g] Phlox *Linnæi*. See Mill. fig. 205

[h] Mirabilis Jalapa *Lin.* Blackw. t. 404.

at the ends of the branches, and in hot weather do not open till towards evening; but when it is cool covered weather, continue open the greatest part of the day. Secondly, that whose root was supposed, though erroneously, to yield the Jalap[i]; the stalks of this are swollen at the joints, the leaves are smaller and the flowers sit singly, close in the axils of the leaves: they are not variable, but all of a purplish red, and not much more than half the size of the others: the fruit also is very rough. In the West Indian islands, where it is very common, they call it *four o clock flower*. Thirdly, the *long-flowered* Marvel of Peru[k], whose corollas are white, and have remarkably long tubes; they have a musky odour, and keep close shut all the day, expanding as the sun declines: they grow in bunches like the first sort, and the seeds are rough like the second: this differs from both the others in having weak stalks that require some support; and these, with the leaves, are hairy and viscous. This species is from Mexico, and has not been long known.

The *Crested Amaranth* belongs also to this place; it is commonly called *Cock's comb*, from the form in which the head of flowers grows. It ranges in the division of incomplete, inferior flowers: and the generic characters are—that the exterior calyx con- Celosia.

[i] Mirabilis dichotoma *Lin.* Mart. cent. t. 1.
[k] Mirabilis longiflora *Lin.*

 sists

fifts of three dry, coloured leaves, within which is a corolla or fecond calyx, confifting of five ftiff, fharp-pointed leaves: that there is a fmall rim furrounding the germ, from which the filaments take their rife; and that the feed-veffel is a round capfule, opening horizontally, and containing three feeds.

There are many fpecies; but that which is fo much efteemed for the variety of form and colours in its fine creft of flowers, is diftinguifhed by oblong ovate leaves; round, ftriated peduncles; and oblong fpikes[l]. The colours are red, purple, yellow, white, and variegated; and fome are like a fine plume of fcarlet feathers. You muft not however confound thefe plants with the *Amaranth* or *Prince's Feather*, which you will find in a place far diftant from this.

One natural order more fhall, if you pleafe, conclude your labours, and my prate, for the prefent. It has its name[m] from this circumftance; the divifions of the corolla are turned or bent in the fame direction with the apparent motion of the fun. But befides this fingularity, the flowers of this order have a one-leafed calyx divided into five fegments; a corolla of one petal; and a fruit confifting of two veffels, containing many feeds. In moft of the genera thefe

[l] Celofia criftata *Lin.*

[m] Contortæ *Lin.*

fruits are *follicles*[n]. The corollas in the greater part are funnel-ſhaped; and are furniſhed with a remarkable *nectary*.

The common Periwincle, which covers the ground and creeps about the bottoms of the hedges, in many parts of your plantations, may ſerve you very well for an example of this order. It has a ſalver-ſhaped corolla, ſucceeded by two erect follicles, which contain ſeeds that are called naked or ſimple, to diſtinguiſh them from thoſe of ſome other genera, which are winged. You will obſerve alſo that the tube of the corolla forms a pentagon, at top; nor will it eſcape you, that there are two large ſtigmas, one over the other.

Linnæus will not allow that the little running ſort[o], and the upright one with larger flowers[p], are diſtinct ſpecies. Without entering into any controverſy on a matter not eaſy to ſettle, you know them aſunder not only by their ſize, but by the ſtalks of the firſt lying on the ground, and the leaves being narrower, and ſharp-pointed towards either end, that is lance-ſhaped, and on very ſhort petioles; whereas the ſtalks of the ſecond are upright, and will climb a little, and the leaves are hollow at

[n] This is a dry ſeed-veſſel, of one cell and one valve; the ſeeds lie looſe in a down, and the ſhell opens on one ſide to let them eſcape.

[o] Vinca minor *Lin.* Curtis, Lond. III. 16.

[p] Vinca major *Lin.* Curtis, Lond. IV. 19. Pl. 12. f. 5.

the baſe, and ovate, ſharper pointed at the end, and on longer petioles.

There is a third ſort, called *Upright Periwincle*[q], for which we are obliged to the Iſland of Madagaſcar, and of courſe it requires the protection of a ſtove, in our colder climates. It has a ſtiff, upright, branching ſtalk, woody at bottom; the leaves are of an oblong ovate ſhape, ſmooth and ſucculent, and ſitting pretty cloſe to the branches; from the axils of theſe come out the flowers, on very ſhort peduncles, generally ſingle, but ſometimes two together: the tube of the corolla is long and ſlender, the brim very flat, the upper ſurface of a bright crimſon or peach colour; the under of a pale fleſh colour: and there is a conſtant ſucceſſion of theſe beautiful flowers from February to October: the corolla is ſometimes white.

Nerium. The *Oleander*[r] is one of the moſt beautiful plants of this tribe. The genus has two erect follicles, like the laſt; but the ſeeds incloſed in them are downy: there is a ſhort crown alſo terminating the tube of the corolla cut into narrow ſegments, and the diviſions of the corolla are oblique to the tube. This ſhrub grows to the height of eight or ten feet; the branches come out by threes from the main ſtem; and the leaves alſo come out by threes from the branches, on very ſhort petioles, point up-

[q] Vinca roſea *Lin.* Mill. fig. 186.

[r] Nerium Oleander *Lin.* Figured in Miller's illuſtr.

wards,

wards, are very ſtiff, and end in ſharp points. The flowers come out in bunches at the ends of the branches; the corolla is of a bright purple, varying to crimſon or white. It grows wild in ſeveral countries about the Mediterranean Sea, but with us is generally kept in tubs, not being hardy enough to ſuſtain the ſeverity of all our winters.

But the moſt admired of this tribe is the *Cape Jaſmine* [s], which was firſt diſcovered near the Cape of Good Hope by the ſuperior fragrancy of its flowers. The diviſions of the calyx are uniform and vertical, and the ſeed-veſſel is a two or four-celled berry, below the flower. The branches come out by pairs; and the leaves are oppoſite, cloſe to the branches, of a ſhining green, and thick conſiſtence: the flowers are produced at the ends of the branches; the corolla is of one petal only, but cut into many ſegments, of which it has ſometimes three or four rows, and then it is as large and as double as a roſe: the anthers are inſerted on the tube without filaments. The colour of the corolla is white, changing as it decays to a buff-colour; and the odour is that of Orange flowers or Narciſſus. Gardenia.

There is another plant of this order of twiſted corollas, called alſo a *Jaſmine*, with the addition of *Red*, but of a very different genus from the Jaſmines properly ſo called. *Plumeria* or *Red Jaſmine* has two reflex Plumeria.

[s] Gardenia florida *Lin.* Mill. fig. 180.

follicles, with the ſeeds flat, winged, and imbricate. There are four or five known ſpecies, all natives of the Spaniſh Weſt Indies, except one, which comes from Senegal. The ſort moſt known[t] has oblong ovate leaves, with two glands upon the petioles: it grows to the height of eighteen or twenty feet; the ſtalks abound with a milky juice, and towards the top put out a few thick ſucculent branches; at the ends of which come out the flowers in cluſters, ſhaped like thoſe of the Oleander; of a pale red colour, and having an agreeable odour. Theſe being never ſucceeded by the fruit in our northern climes, you will not be able to diſcern the generic character.

Cinchona. The famous *Jeſuits' Bark* is from a tree of this claſs and order[u], approaching in its characters to the natural tribe of *Contortæ:* to which alſo belong ſome plants of the ſecond order of this fifth claſs, becauſe they have two piſtils: ſuch are the *Periplocas*, the *Cynanchums*, and the numerous genus of

Aſclepias. *Aſclepias*, containing twenty-ſeven ſpecies. Of this laſt, you have the common *Swallow-wort*, or *Tame poiſon*[v], whoſe root is ſuppoſed to be a powerful antidote to poiſons: it has a ſhort upright ſtalk, ovate leaves bearded at the baſe, white flowers growing

[t] Plumeria rubra *Lin.* Cateſb. car. 2. 92. Ehret. t. 10.

[u] Cinchona officinalis *Lin.*

[v] Aſclepias Vincetoxicum *Lin.* Fl. dan. 849.

in proliferous umbels [w], and each of them ſucceeded by two long, jointed follicles, incloſing ſeveral compreſſed ſeeds, crowned with a ſoft white down. This is a native of the ſouthern countries of Europe, and is very hardy. Other ſpecies are much larger, growing to the height of ſix or ſeven feet. Some creep very much at the root, and become troubleſome in a garden. Others coming from the Cape, or the warm parts of America, require care and heat to preſerve them. Some have white, others purple, orange, or red corollas. Some have the leaves oppoſite; others have them alternate; in ſome again they are flat, whilſt others have their edges rolled back. Many of the ſorts are very handſome. They all agree in the following circumſtances, which therefore form the generic character—that the ſegments of the corolla are bent back; that five ovate, hollow nectaries, ending at bottom in a ſharp ſpur, involve the ſtamens and piſtils; and that each flower is ſucceeded by two follicles, incloſing many downy ſeeds.

Stapelia is ſo remarkable a plant of this tribe, that I muſt not omit mentioning it. This has a very large wheel-ſhaped corolla, divided beyond the middle into five ſegments, which are broad, flat, and ſharp-pointed. The nectary is a double ſtar, one of them ſurrounding, the other covering Stapelia

[w] That is, the large umbels have ſmaller ones iſſuing from them.

the

the ſtamens and piſtils. Two follicles, incloſing many flat, downy ſeeds, follow each flower.

There are three known ſpecies, all growing naturally at the Cape of Good Hope, and all having ſucculent branches, as thick at leaſt as a man's finger. The three ſorts are diſtinguiſhed by the indentures on the ſides of theſe leafleſs branches; which in the firſt [x] ſpread open horizontally, ending in acute points; in the ſecond [y] have their points erect; and in the third [z] obtuſe.

In the firſt ſpecies the flowers come out ſingly on a ſhort peduncle from the ſide of the branches towards the bottom: the corolla is greeniſh on the outſide, but yellow within, having a purple circle round the nectaries, and the whole is finely ſpotted with purple, like a frog's belly. The branches of the ſecond ſort are much larger, and ſtand more erect; they have four longitudinal furrows, and the indentures are on the ridges between them. The flowers are much bigger than thoſe of the laſt, of a thicker ſubſtance, and covered with fine purpliſh hairs: the ground of it is a greeniſh yellow, ſtreaked and chequered with purpliſh lines.

But the great ſingularity of theſe plants is that the flower when fully open has a fetid

[x] Stapelia variegata *Lin.* Bradl. ſucc. 3. t. 22. Curtis Mag. 26.

[y] Stapelia hirſuta *Lin.* Mill. fig. 258.

[z] Stapelia mammillaris *Lin.* Burm. afr. t. 11.

ſmell

ſmell ſo perfectly reſembling that of carrion, that the common fleſh-fly depoſits her eggs in it, which frequently are hatched into little worms, but never proceed any farther, or become flies. A rare inſtance this of an animal miſtaking its inſtinct.

Having by this time ſufficiently fatigued you, I leave you, dear couſin, to meditate on this irregularity in the operations of nature, and once more heartily bid you adieu.

LETTER

LETTER XVII.

ON THE OTHER ORDERS OF THE FIFTH CLASS, PENTANDRIA DIGYNIA, &c.

May the 1ſt, 1774.

I AM not ſurpriſed, dear couſin, at your being ſolicitous to know what the nectary is, which I mentioned ſeveral times in my laſt. But I am not diſpoſed at preſent to ſatisfy your curioſity any farther, than to inform you, that it is an appendage to the corolla, and that there is a juice in it, probably of uſe to the plant, certainly ſerving for the food of bees, and numberleſs other inſects. It is a perfect Proteus, and puts on a far greater variety of forms than the ſon of Neptune. Another time I may perhaps enter more deeply into this matter; but at preſent we will go ſtraight on our way.

You will have great pleaſure when I inform you, that the ſecond order of the fifth claſs [a] is almoſt wholly made up of the *Umbellate* tribe of plants [b], which you are already ſo well acquainted with: there are however ſome, which the circumſtances of having five ſtamens and two piſtils bring into the ſame diviſion of the arbitrary ſyſ-

[a] Pentandria Digynia *Lin.* [b] See Letter V.

tem,

tem, though they are not naturally related to them. A few of theſe we will examine, before we enter into a detail of the *Umbellate* tribe.

Many of them have incomplete flowers, or are deficient in the corolla; and may be found among the *Oleraceous* plants in the natural orders of Linnæus, by other authors called *Apetalous*.

Such are all the *Gooſefoots*, of which there are no leſs than twenty ſpecies, moſt of them growing common on dunghills, and in waſte places, and having no beauty to attract your notice. They are known by their five-leaved, five-cornered calyx, incloſing one round, flattiſh ſeed, ſhaped like a lens. One of the moſt reſpectable ſpecies is the *Engliſh Mercury* or *Allgood*[c], growing frequently in waſte places, and by walls and way-ſides; and cultivated in ſome places as a ſubſtitute to Spinach. The leaves of this are triangular, quite entire, waving, and having the under ſurface covered with a kind of meal; the flowers grow in compound ſpikes, which are deſtitute of leaves, and ſpring from the axils. Chenopodium.

Beet is very nearly allied to theſe in its characters; but it is diſtinguiſhed by having a kidney-ſhaped ſeed, wrapped up in the ſubſtance of the calyx. In its wild ſtate, on the ſea-coaſt, and in ſalt marſhes[d], it Beta.

[c] Chenopodium Bonus Henricus *Lin.* Curtis, Lond. III. 17. Ger. 32.

[d] Beta maritima *Lin.*

it has two flowers coming out together, the ſtalks are weak, and lie moſtly on the ground, the leaves are triangular and oblique or vertical; the diviſions of the calyx are equal and not toothed at bottom, and it flowers the firſt year of its riſing from ſeed. The garden ſort[e] has many flowers coming out together, the ſtalks erect, the leaves oblong lance-ſhaped, thick and ſucculent; the diviſions of the calyx are toothed at the baſe, and it does not flower till the ſecond year.

It ſometimes has pale green leaves, and ſmall roots; ſometimes dark red or purple leaves, with large purple roots ſhaped like a carrot; but theſe are not generally ſuppoſed to be diſtinct ſpecies.

Salſola. The *Glaſſworts* are alſo of this *Oleraceous* tribe. They are diſtinguiſhed by having a large ſeed, ſpiral like a ſcrew, covered with a kind of capſule which is wrapped up in the calyx. There is one ſort that grows wild in the ſalt marſhes[f], which has a herbaceous ſtalk that lies on the ground; awl-ſhaped, rough-leaves terminating in ſpines; the calyxes edged, and ſitting cloſe in the axils, and a trifid ſtyle.

Another ſort which grows wild in warmer countries[g], has alſo herbaceous

[e] Beta vulgaris *Lin.*

[f] Salſola Kali *Lin.* Fl. dan. 818. Mor. hiſt. 3. 5. t. 33. f. 11.

[g] Salſola Soda *Lin.* Jacqu. hort. t. 68.

ſpreading

ſpreading ſtems; but it is a much larger plant than the other, and the leaves have no ſpines. Theſe or any of the ſorts yield the cauſtic alkaline ſalt, which is ſo neceſſary in that moſt elegant and uſeful manufacture of glaſs; but this is the ſort generally uſed.

The *Globe Amaranth* [h] is of this claſs and order. Its fine round head is compoſed of many flowers, which have a large, boat-ſhaped, flat, coloured calyx, of two leaves; a corolla divided into five rude, villous ſegments; a cylindric nectary, divided into five parts at top; a ſtyle cut half way into two; and a capſule opening horizontally, and containing one ſeed. India is its native country: the ſtalk is erect and annual; the leaves are lance-ſhaped, as are the branches and peduncles, which are long and naked, except that a pair of ſhort leaves grows cloſe under each head of flowers, which always comes out ſingle. The calyx and corolla being dry and chaffy, will retain their colour ſeveral years, and hence their name of *Amaranth* or *incorruptible*. Bright purple is the uſual colour, but ſometimes the heads are brilliant white, or ſilver-coloured. The name muſt not lead you to ſuppoſe this, any more than the *creſted Amaranth*, to be of the ſame kind with the true *Amaranth* [i]. When you are told that

Gomphrena.

Ulmus.

[h] Gomphrena globoſa *Lin.* Mill. fig. pl. 21.
[i] See Letter XXVIII.

the

the *Elm* is of the ſame claſs and order, and alſo one of the incomplete tribe, as having no corolla, you will probably reflect that an artificial ſyſtem is very diffeıent from a natural arrangement: and in this you are not miſtaken; but then you muſt conſider, that an artificial ſyſtem is the only one that can enable you to find out the genera and ſpecies of plants, which is the art I propoſe to inſtruct you in. Few perſons know that the Elm has any flower, becauſe it is inconſiderable in ſize and appearance, and comes out in an early inclement ſeaſon: however this tree in reality abounds in flowers, before the leaves make their appearance. They have no corolla, but a quinquefid calyx: the flower quickly paſſes, and is ſucceeded by one ſeed covered and ſurrounded by a flat membrane. The different ſorts, known by the names of *Rough Witch Elm*, *Smooth-leaved Witch Elm*, *Witch Hazel*, *Engliſh Elm*, *Dutch Elm*, *Upright Elm*, &c. are ſuppoſed to be varieties of one ſpecies [k]; and all have doubly-ſerrated leaves, unequal at the baſe.

Gentiana. The *Gentians* are alſo of this claſs an order, and of that ſubdiviſion which h s monopetalous inferior corollas. They are diſtinguiſhed from the other genera of this ſubdiviſion by the capſule, which is oblong, round, and ſharp-pointed; has one

[k] Ulmus campeſtris *Lin.* Duham. t. 108. Hunter's vel. ſilva, p. 114.

cell.

cell, opens by two valves; and has two receptacles on the inside, each adhering lengthwise to one of the valves. The form of the fruit is constant; whereas the figure and number of parts in the flower vary in the different species, which are numerous[l]. Great part of the skill and sagacity of the botanist consists in seizing those parts which are constant in all the species, for the generic characters, and in this consists the great merit of Linnæus; writers before him having either taken all parts indiscriminately, or else the same part invariably for this purpose.

The species have either four or five petals, and the latter have either funnel-shaped corollas, or else approaching to bell-shaped; hence a threefold division of the genus.

The principal of the genus is the *Great Yellow Gentian*[m], which has a single stalk, three feet high, covered with leaves that are large, ovate, marked underneath with nerves meeting at the tip; the lower ones petiolate, the upper sessile. There is but one flower to a peduncle, but they grow round the stalk in whorls: the calyx resembles a double spathe: the corolla is rotate, cut into five segments[n]; the colour yellow irregularly dotted. The root is very

[l] Thirty-nine.
[m] Gentiana lutea *Lin.* Mill. fig. 139. 2.
[n] Varying sometimes as far as eight.

large, and remarkably bitter; it communicates the bitterneſs ſo much to the whole plant, that it remains always untouched by the cattle in the mountainous paſtures of Germany and Switzerland, where it grows naturally.

The *Leſſer Centaury* [o] is of this genus, and is diſtinguiſhed by its dichotomous ſtalk, and its funnel-ſhaped corollas divided into five ſegments; they are of a bright purple colour, but often fade to white. This plant is annual, and varies much in height according to the ſoil, from three or four inches to a foot. This is extremely bitter as well as the other.

There are ſeveral beautiful little Gentians, with flowers of the fineſt blue that can be imagined, growing wild in the Alps. One of them is frequently cultivated in gardens, under the name of *Gentianella* [p], and is ſingular for having its fine bell-ſhaped azure flowers larger than the whole plant beſides.

Chlora. *Yellow Centaury* [q] is alſo naturally of this genus; but has been removed to the eighth claſs; firſt with the title of *Blackſtonia*, and now under that of *Chlora*.

But methinks you are languiſhing to be

[o] Gentiana Centaurium *Lin.* Chironia Centaurium Curtis, Lond. IV. 22.

[p] Gentiana Acaulis *Lin.* Jacquin auſtr. 2. t. 135. Curt. Magaz. 52.

[q] Chlora perfoliata *Lin.* See Letter XIX.

on

on ground you are better acquainted with. And indeed you are already ſo well verſed in the nature of the umbellate tribe, that I am perſuaded you will find little difficulty in determining the genera and ſpecies. Many of them are very generally known, either for their uſe in medicine or the kitchen, or elſe for their poiſonous qualities. Moſt of thoſe which grow on dry ſoils have roots that have an aromatic pungent ſmell and taſte; whilſt thoſe which grow in moiſt places or in the water, as many of them do, are in a greater or leſs degree poiſonous.

You have long ſince been able to diſtinguiſh true Parſley and Chervil from Fool's-Parſley [r]. There is another wild plant that grows upon banks and by way-ſides, called *Hemlock-Chervil* [s], which has been miſtaken for *Garden-Chervil* [t], and has produced bad effects, when put into ſoups: it is not however ſo dangerous, becauſe it does not grow wild in gardens, and we muſt go out of our way to poiſon ourſelves: on another account however it is more dangerous, becauſe it is not only of the ſame diviſion, as having partial involucres only, but alſo of the ſame genus; and therefore liable to be miſtaken for the true Chervil, even when in flower, which Fool's-Parſley Scandix.

[r] See Letter V.

[s] Scandix Anthriſcus *Lin.* Curtis, Lond. I. 19.

[t] Scandix Cerefolium *Lin.* Jacquin auſtr. 4. t. 390. Compare Pl. 13. f. 2. & Pl. 5. f. 3.

 cannot

cannot be. They have both a radiate corolla, petals notched at the end, the flowers in the middle often incomplete and producing no ſeed, and the fruits of an oblong ſhape. However, notwithſtanding all this ſimilitude of character, they are eaſily to be diſtinguiſhed both in and out of flower. *Hemlock-Chervil* is a much lower plant; the ſtalks are ſmooth indeed, and the leaves finely cut, but they are hairy, the diviſions much ſmaller and cloſely placed, and the green much deeper than in *Garden Chervil*; the corollas alſo are uniform, the ſeeds ovate, and very rough. *Garden Chervil* is a tall, genteel, ſmooth plant; the umbels come out on the ſides of the branches, and ſit cloſe to them; and the ſeeds are long, narrow and ſhining. After all, I am perſuaded that when you have an opportunity of comparing theſe two plants together, as you eaſily may, the gardener furniſhing you with one, and the other being ſo common in a wild ſtate, you will wonder that any perſon ſhould ever have confounded them. Here you ſee we have an inſtance of an umbellate plant, growing on dry land, that is poiſonous; you are not therefore to conclude that all theſe are wholeſome, any more than that every water ſpecies is poiſonous.

Sium. We have another inſtance of fatal confuſion, not in two plants of this tribe, but in one of this, with another of a different claſs;

namely, of the *Creeping Water Parſnep*[u], with *Water Creſs*[v], which belongs to the cruciform flowers. You are ſo well miſtreſs of both tribes, that it is impoſſible you ſhould miſtake them when in flower; but this is not the time when Water-Creſſes are eaten, and this plant is ſo different in its flowering ſtate, that I am perſuaded an eater of it would think himſelf impoſed upon, if he were then ſhown it for Water-Creſſes. When they are both young they are really not unlike; and ſince they frequently grow together, the one may ſometimes be gathered for the other; though I muſt confeſs that I have not met with the miſtake more than twice, and that only in a ſingle piece among a conſiderable quantity: however, the leaves of Water Parſnep are of a light green; the ſmall leaves compoſing the whole winged or pinnate leaf are longer and narrower, ſerrated on the edges, and pointed at the end; whereas thoſe of Water-Creſſes have a tincture of brown upon them, the leaflets are roundiſh, and particularly the odd one at the end is very large and blunt, and they are none of them regularly ſerrated, but have only a few indentures on their edges.

[u] Sium nodiflorum *Lin.* Fl. dan. t. 247. Mor. hiſt. ſ. 9. t. 5. f. 3.

[v] Siſymbrium Naſturtium *Lin.* Fl. dan. t. 690. Mor. hiſt. ſ. 3. t. 4. f. 8. Ger. 257. 5. Compare Pl. 13. f. 1. with Pl. 21.

The characters by which you will know the Water-Parsnep when in flower are these—it has both an universal and partial involucre, the flowers are all fertile, the petals are heart-shaped, and the seeds are ovate and streaked. This species is distinguished from the others by its pinnate leaves, and the umbels of flowers sitting close to the stem, in the axils.

Conium. Another poisonous herb of great fame is the *Hemlock* [w]. A tall plant, three feet high and more, easily known by its purple-spotted stalk. It has both involucres, the universal of three, four, five, or seven broadish reflexed leaves; the partial of three or four broad leaves only, on one side of the umbel; both very short. The flowers are all fertile; irregular without, regular within: the petals heart-shaped. The fruit is almost spherical, marked with five notched ridges. The common species is distinguished by its smooth streaked seeds. The leaves are large, abundant, of a dark green but shining, triply pinnate, with the last divisions obtusely indented; it has many umbels of white flowers, with numerous spreading rays. It grows wild on ditch banks, in shady lanes, about dunghills and church-yards: and is a biennial plant.

The waters afford other poisonous herbs,

[w] Conium maculatum *Lin.* Curtis, Lond, 1. 17. Ger. 1061.

as *Water-Hemlock* [x], *Long-leaved Water-Hemlock* [y], *Hemlock Water Dropwort* [z], and *Common Water Dropwort* [a]: but let us quit theſe ill-omened plants, and proceed to others more innocent, and more within your reach.

Two umbellate plants you will be ſure to find under every hedge, called *Wild Chervil* [b] and *Rough Chervil* [c]: they are both of the ſame genus, but of a different genus from Garden Chervil. They have partial, but no univerſal involucres; theſe are of five leaves, concave and bent back; ſome flowers in the middle drop without leaving ſeeds; the petals are bent in and heart-ſhaped; and the fruit is oblong and ſmooth. The firſt, vulgarly called *Cow-weed* or *Cow-parſley*, has a ſmooth ſtreaked ſtalk, and the joints ſwelling but a little. The ſecond has a rough ſtalk, and the joints more tumid. The firſt is remarkably leafy, and the leaves very large, and generally ſmooth, except the nerves. The ſecond has hairy Chærophyllum.

[x] Phellandrium aquaticum *Lin.* Mor. hiſt. ſ. 9. t. 7 f. 7. Ger. 1063. 2.

[y] Cicuta viroſa *Lin.* Fl. dan. 208. Mor. hiſt. ſ. 9. t. 5. f. 4. Ger. 256. 4.

[z] Oenanthe crocata *Lin.* Philoſ. Tranſact. for 1747. Ger. 1059. 4.

[a] Oenanthe fiſtuloſa *Lin.* Fl. dan. 846. Mor. hiſt ſ. 9. t. 7. f. 8. Ger. 1060.

[b] Chærophyllum ſylveſtre *Lin.* Curtis, Lond. IV. 25. Mor. hiſt. t. 11. f. 5.

[c] Chærophyllum temulum *Lin.* Curt. Lond. n. 61. Mor. hiſt. t. 10. f. 7. Ger. 1038. 2.

leaves, not ſo large, nor ſo much divided; the umbels uſually nod, and the ſeeds are deeply ſtreaked. Both ſometimes have a leaf at the origin of the univerſal umbel: both have a ſtrong ſmell, and approach in their qualities to the forementioned plants, but not enough to denominate them poiſonous.

Some of this tribe are ſo generally uſed in food, that they are univerſally known, and therefore it ſeems impertinent to ſay any thing to you about them; and yet you may have eaten the roots of Carrots and Parſneps, the ſtalks of Angelica, Celeri and Finochia, the leaves of Parſley, Fennel, and Sampire, the ſeeds of Coriander and Carraways, without knowing one of the plants when they they are preſented to you. However, when you meet with any of theſe in flower, you aſcribe them immediately to the umbellate tribe. Carrot, Sampire, and Angelica range among thoſe which have both involucres; Coriander has a partial involucre only; and the reſt have neither one nor the other.

Daucus. *Carrot*[d] has a large winged involucre: ſome flowers in the middle drop without ſeed, and the fruit is ſtiff with briſtles. The outer flowers are very irregular: and the whole umbel, as it approaches a ſtate of maturity, takes a hollow form, very like a bird's neſt.

[d] Daucus Carota *Lin.* In the cultivated ſort all the flowers are fertile. Fl. dan. 723. Mor. umb. t. 2. Ger. 1028.

The

The leaves are rough and hairy. The garden Carrot differs little from the wild one, but in the ſize and tenderneſs of the root.

Sampire[e] has the umbel not flat, or hollow like the laſt, but hemiſpherical, the flowers all alike and fertile, the petals flat, the fruit ovate, flatted. The ſtalks are ſucculent, the leaves pinnate, compoſed of three or five diviſions, each of which has three or five ſmall, thick, lance-ſhaped leaves; the corollas are yellow. This herb ſtrikes its roots deep into the crevices of the rocks, and hangs down; growing chiefly in places difficult of acceſs, the herb-gatherers are tempted to ſubſtitute another plant[f], which they obtain without trouble on the beach, but which has none of the warm, aromatic quality of the Sampire. Thoſe who live on the Eaſt coaſt muſt wonder what is meant by calling the occupation of a Sampire-gatherer, *dangerous trade*, when they obtain it walking at their eaſe on the flat ſandy ſhore. But theirs is a roundiſh, jointed, taſtleſs ſtalk, with a tough ſtring running through the middle of it[g], inſtead of a flat leaf, with a pungent taſte. This Marſh Sampire ranges in the firſt order of the firſt claſs, and is burnt to make kelp for the glaſs-works. Crithmum.

[e] Crithmum maritimum *Lin.* Jacqu. hort. 2. 187. Ger. 533. 1.

[f] Inula crithmoides *Lin.* Golden Sampire.

[g] Salicornia europæa *Lin.* Marſh Sampire, called alſo jointed Glaſſwort or Saltwort. Fl. dan. 303. Blackw. 598.

Here

Here you ſee what confuſion of names we have again, and how difficult it muſt be to obtain the plant you want, without knowing ſomething more of it than the name. It is generally true of objects much in requeſt, that where people have them not, they ſubſtitute others, to which they give the ſame title, whether they have the ſame qualities or no; by which, if they do not injure themſelves or their neighbours, they at leaſt miſlead the incautious and unexperienced naturaliſt.

Angelica. *Angelica* has large globoſe umbels, all the flowers in them are regular and fertile, the petals are inflex, or bent upwards at the end; the fruit is roundiſh, cornered, or furrowed, and terminated with two reflex ſtyles.

The cultivated[h] and wild[i] Angelica are allowed on all hands to be diſtinct ſpecies. They have both pinnate leaves; but the firſt has the odd lobe at the end divided generally into three parts; the ſecond has all the leaflets equal, lance-ſhaped, and ſerrated about the edges. The firſt is a much larger plant in all reſpects, the leaflets broader, rather ovate than lance-ſhaped, and the corollas greeniſh: the ſecond has a thinner and leſs ſucculent ſtem, ſcarcely

[h] Angelica Archangelica *Lin.* Fl. dan. t. 206. Ger. 999. 1.

[i] Angelica ſylveſtris *Lin.* Mor. hiſt. ſ. 9. t. 3. f. 2. Ger. 999. 2.

any

any univerſal involucre, and the corollas tinged with red.

Coriander [k] has no proper univerſal involucre, though there be ſometimes one leaf, as in the Wild Angelica; the partial one conſiſts of three leaves, and is ſhort. The flowers in the middle produce no ſeed; the petals are bent inwards, and heart-ſhaped; the outer ones large. The fruit is ſpherical, as you know. The calyx of each little flower is more evident in this than in the other umbellate plants. The diviſions of the leaves next the ground are broad; thoſe of the upper ones narrow: they and the whole plant are ſmooth, and have a ſtrong rank ſmell, like bugs. Coriandrum.

Parſnep [l] has all the flowers fertile and regular, the petals entire, and bent inwards; the fruit oblong, flatted and ſurrounded with a membrane. The leaves are ſimply pinnate. The garden Parſnep differs not ſpecifically from the wild, which has hairy leaves, whereas thoſe of the firſt are ſmooth; but ſmoothneſs is a common effect of culture. The cultivated plant is alſo of courſe much larger, and the roots ſucculent and eſculent: both have yellow corollas. Paſtinaca.

Fennel [m] has likewiſe all the flowers fertile and regular; and the petals entire and Anethum.

[k] Coriandrum ſativum *Lin.* Blackw. 176. Ger. 1012.

[l] Paſtinaca ſativa *Lin.* Ger. 1025.

[m] Anethum Fœniculum *Lin.* Mill. Illuſtr. Moriſ, ſ. 9. t. 2. f. 1. Ger. 1032.

bent inwards, as in the laſt: the fruit is nearly ovate, flatted, and ſtreaked. *Dill*[n], which is alſo of this genus, has the fruit ſurrounded with a membrane, and more flatted than that of Fennel. *Sweet Fennel* is but a variety of the common ſort, though the lobes of the leaves are longer, more ſlender, and not ſo denſe as in that; the ſeeds are longer and much ſweeter. *Finochia* is probably another variety, though a much humbler plant, ſwelling much in breadth and thickneſs juſt above the ground. The leaves of all theſe are very finely cut.

Carum. *Carraway*[o] has no proper involucre, but a ſingle leaf at the origin of the univerſal umbel; the middle flowers fall without ſeed; the petals are keeled, bent inwards, and notched at the end; the ſeeds are of an oblong ovate form, and ſtreaked.

Apium. *Parſley*[p] and *Smallage*, or *Celeri*[q], are of the ſame genus. They have a ſort of involucre, generally of one leaf; all the flowers fertile; the petals equal, and bent inwards; the fruit ſmall, ovate, and ſtreaked. They have both winged leaves, with the leaflets linear on the ſtalk in Parſley, wedge-ſhaped in Smallage, of which Celeri is only an improvement from warmer countries. Our wild Smallage however, which is common

[n] Anethum graveolens *Lin.* Ger. 1033.
[o] Carum Carui *Lin.* Mor. umb. t. 8. Ger. 1034.
[p] Apium Petroſelinum *Lin.* Pl. 5. f. 1. Ger. 1013.
[q] Apium graveolens *Linnæi.* Fl. dan. 790. Moris, t. 9. f. 8. Ger. 1014.

by

by ditches and brooks, cannot be rendered esculent by culture.

Earth-nut or *Pig-nut*, [r] whose roots are like a small potatoe and eatable, has both involucres, the lesser ones narrow as a hair; the flowers in a close umbel, all fertile; the corollas regular, with heart-shaped petals; and the fruit ovate. It grows, not uncommonly, wild on dry pastures. **Bunium.**

Ferula [s], in the dry stalk of which Prometheus brought fire from heaven, has both involucres; all the flowers fertile, the petals heart-shaped; the fruit oval, flat, and marked with three streaks on each side. It is so lofty and large a plant as to have acquired the name of *Fennel-giant*; the lower leaves spread two feet, and are subdivided into very long, narrow, simple leaflets; the stalk is hollow, jointed, and will grow ten or twelve feet high: when these are dry they have a light dry pith, which readily takes fire; and the people of Sicily use it as tinder. It is a species of Ferula that produces the *Assa fœtida* [t]. **Ferula.**

Cow-Parsnep [u] is a very large plant, though not so gigantic as the last. It has two involucres, but as they are very apt to drop off, you may easily be deceived in that respect. **Heracleum.**

[r] Bunium Bulbocastanum *Lin.* Curtis, Lond. IV. 24. Ger. 1064. 1, 2. There is a smaller and a greater sort.

[s] Ferula communis *Lin.* Ger. 1056.

[t] Ferula Assa fœtida *Lin.* Kœmpf. amœn. t. 536.

[u] Heracleum Sphondylium *Lin.* Mor. hist. s. 9. t. 16. f. 1. Ger. 1009.

ſpect. The corolla is very irregular, bent in and notched. The fruit is ovate, notched, flatted, ſtreaked, and with a membrane round the edge. In moſt of the ſpecies, the middle flowers fall ſeedleſs; but in our common one all the flowers are fertile: the leaves are winged, and the lobes pinnatifid. This plant grows common in meadows and paſtures.

Scandix. *Shepherd's-needle* or *Venus's-comb*[v] is remarkable for long proceſſes or beaks terminating the ſeeds, and giving it the appearance of Geranium, when in fruit. It is of the ſame genus with Chervil, and is a common weed among corn. But of theſe umbellate plants enough.

Of the third order of this fifth claſs we have ſeveral trees and ſhrubs; as the Varniſh-trees and Sumach, Wayfaring trees and Lauruſtinus, Caſſines, Elder, Bladder-nut, &c. The firſt are known by their inferior flowers, their five-leaved calyx, their corolla of five petals, and their berry with one ſeed in it.

Rhus. *Virginian Sumach*[w] is common among your ſhrubs, and known to you by the young branches being covered with a velvet-like down, reſembling both in colour and texture a ſtag's horn when firſt budding; the branches are crooked and deformed; the

[v] Scandix Pecten *Lin.* Curt. Lond. 5. 21. Mor. hiſt. ſ. 9. t. 11. f. 1. Ger. 1040. 1. Pl. 13. f. 3.

[w] Rhus typhinum *Lin.* Duhamel.

leaves

leaves are winged, with ſix or ſeven pair of lance-ſhaped lobes, ſharply ſerrated, and nappy beneath. The flowers are produced in cloſe tufts at the ends of the branches, and are followed by ſeeds incloſed in purple, woolly, ſucculent covers, which give them their autumnal hue, when the leaves fade firſt to purple and then to feuillemort colour.

Wayfaring-tree[x], *Marſh-elder*[y], and *Lauruſtinus*[z], are all of one genus; having ſuperior flowers, a five-leaved calyx, a corolla divided into five ſegments, and a berry incloſing one ſeed. Viburnum.

The firſt has heart-ſhaped leaves very much veined, ſerrated about the edges, and white underneath. The ſecond has lobed leaves, with glands upon the petioles; the flowers round the outſide of the *cyme* are barren, with the corollas much larger than the others. The *Gelder Roſe* is a remarkable variety of this, with the flowers growing in a ball, and every one of them barren. The third has the leaves ovate, and entire, with the veins underneath villous: this is an evergreen.

The fourth order is a very ſmall one, comprising only two genera; of which *Par-* Parnaſſia.

[x] Viburnum Lantana *Lin.* Duhamel, t. 103. Ger. 1490.

[y] Viburnum Opulus *Lin.* Fl. dan. 661. Duham. t. 16. Ger. 1424. 1.

[z] Viburnum Tinus *Lin.* Curt. Magaz. 38.

naſſia

naſſia[a] is one. This grows wild in wet meadows, and on the borders of marſhes, but not very common. It is eaſily known by its calyx divided into five parts; its corolla of five petals; five heart-ſhaped nectaries, furniſhed with hairs, upon the top of which are little balls; a large ovate germ, without any ſtyle; but four ſtigmas; and a capſule of one cell and four valves. It has a ſingle ſtalk, with one heart-ſhaped leaf on it, embracing the ſtalk, and one flower only; the corolla is white.

Statice. Of the fifth order, *Pentagynia*, is *Thrift*, *Flax*, &c. *Thrift*[b] has the calyx of one leaf, entire, plaited and dry, like chaff[c]; a corolla of five petals; and one ſeed crowned with the calyx. Theſe are the characters of the genus, which has twenty-two ſpecies. Common Thrift has a threefold involucre or common calyx, and the flowers growing in a round head, upon the top of a naked ſtalk; the leaves, which form a cloſe tuft near the ground, are linear. The corollas are red, of different ſhades, from pale fleſh colour to bright ſcarlet; varieties occaſioned by ſoil and ſituation; for this plant is found both on ſalt marſhes and mountains. Thrift was much uſed formerly for edging the borders in flower gardens, but it is now almoſt entirely out of date.

Linum. *Flax* has alſo a corolla of five petals; but the calyx is five-leaved, and the capſule

[a] Mill. illuſtr. Fl. dan. 584. Ger. 840. 1.
[b] Statice Armeria *Lin.* Ger. 602. [c] Scarioſe

opens

opens by five valves, having ten cells within, in each of which is one ſeed. There are no leſs than twenty-two ſpecies of Flax: that whoſe uſe is ſo extenſive[d] is diſtinguiſhed from the reſt by the calyx and capſule being pointed, the petals being notched, the leaves lance-ſhaped, and alternate upon the ſtem, and the ſtalk unbranched. On the top of this are four or five flowers, with beautiful blue corollas, very apt to fall off. It is an annual plant, about a foot and half high, in the fields. In the garden it will grow ſix inches higher, and branch a little where it ſtands detached.

Both the uſe and beauty of Flax will intereſt you; ſo I leave you with this impreſſion, and bid you once more adieu.

[d] Linum uſitatiſſimum *Lin.* Curt. Lond. 5. 22. Mor. hiſt. ſ. 5. t. 26. f. 1. Ger. 556.

LETTER XVIII.

THE CLASS HEXANDRIA.

May the 15th, 1775.

WE are returned, dear cousin, to the point from which we first set out [c]; the liliaceous tribe of plants being included in the first order of the sixth class, in the System of Linnæus. These superb and beautiful flowers have gained so much on the esteem of the curious in Europe, that they have spared neither trouble in fetching them from the farthest parts of the East, nor expense in cultivating them at home. Hence they are so generally known, that persons not at all versed in Botany readily find them to be of the same family. You certainly are at no loss to determine their general relation and analogy, from the hints which were thrown out in the first letter, and the experience you have since acquired. It remains therefore only to be acquainted with their generic and specific characters; to which end I shall present you with some that may be most within your reach: were I to set every liliaceous plant before you, the beauty of which merits your attention, I

[c] See Letter I.

should

ſhould almoſt exhauſt the tribe. Two cautions you are to obſerve: firſt, that the whole liliaceous tribe is not confined to the claſs *Hexandria*[f], though the far greater part of it is; ſecondly, that other plants, few indeed in number[g], are to be found in the ſame order.

You remember that the Lily had no calyx; you are not however to ſuppoſe that the whole tribe is deſtitute of this important part of the flower. It is a circumſtance that occaſions a threefold ſubdiviſion of the order, into ſuch as have a calyx; ſuch as have a ſpathe or ſheath, covering the corolla whilſt a bud, but torn and forſaken by the corolla when it is expanded; and laſtly, ſuch as have the corolla quite naked.

You would not perhaps have ſuſpected at firſt ſight that the *Ananas* or *Pine-Apple* is of this tribe. It is almoſt the only genus capable of miſleading you. The flower has a trifid, ſuperior calyx, a corolla of three petals, a ſcale faſtened to the baſe of each petal; the fruit is a ſort of berry. The ſpecies[h] is diſtinguiſhed by its long, narrow, pointed leaves, like thoſe of Aloes, ſerrated on the edges, and ſet with tender Bromelia.

[f] See Letter XIV.

[g] Eighteen genera out of 65. The whole claſs has eighty-one genera and four hundred and ſeventy-three ſpecies.

[h] Bromelia Ananas *Linnæi*. Comm. hort. 1. t. 57. Trew Ehret. t. 2.

ſpines; and by the fruit being terminated with a buſh of leaves, commonly called the crown, which being planted takes root, and produces another fruit. There are differences in the fruit, proper to be remarked by thoſe who cultivate this luxury; but they are no more than varieties of the ſame ſpecies, and therefore do not concern us as botaniſts.

Tradeſcantia. *Tradeſcantia*, or *Virginian Spiderwort*[i], is another of the liliaceous tribe furniſhed with a calyx, which in this is three-leaved; the corolla alſo has three petals, and the capſule has three cells. It is remarkable for having the filaments fringed with purple jointed hairs. The ſpecies common in gardens is diſtinguiſhed from ſeven others, by its ſmooth, erect ſtalk, and by the flowers growing in cluſters at the top of it. Theſe are of a fine purple, and blow in ſucceſſion moſt part of the ſummer, though each flower continues open but a day. From the number of parts in the fructification, and its enſiform leaves, this plant will range in the ſame natural order with *Iris* and its congeners[k].

Galanthus. Of thoſe which have a *ſpathe* or *ſheath* inſtead of a calyx, there is the modeſt, the humble, the early *Snow-drop*[l]; that comes

[i] Tradeſcantia Virginica *Lin.* Mor. hiſt. ſ. 15. t. 2. f. 4. Curt. Mag. 105. Pl. 14. f. 1.

[k] Called *Enſatæ* by Linnæus. See Letter XIV.

[l] Galanthus nivalis *Lin.* Jacq. auſtr. 4. 313. Ger. 147. Park. parad. 107.

one

one of the firſt of the year to ſalute us, and, no leſs white than the ſnow itſelf, is frequently covered by it. This is diſtinguiſhed by its ſuperior corolla of ſix petals, of which the three inner ones are ſhorter by half than the others, and notched at the end. Theſe are ſuppoſed to be the nectary, More needs not to be ſaid of a flower ſo univerſally known.

Narciſſus is another of this diviſion. Narciſſus. There are many ſpecies, all united by theſe characters: a ſuperior corolla of ſix equal petals, and a funnel-ſhaped nectary, of one piece, within which are the ſtamens. The moſt known ſpecies are the *common white Narciſſus* [m], the *Daffodil* [n], the *Polyanthus Narciſſus* [o], and the *Jonquil* [p]. The firſt and ſecond, in a natural ſtate, have only one flower burſting from the ſame ſheath; the third and fourth have ſeveral: the firſt has the nectary or cup in the middle of the flower, wheel-ſhaped, very ſhort, chaffy, and a little notched at the edge: the ſecond has a large, erect, curled, bell-ſhaped cup [q], ſometimes as long as the ovate petals of

[m] Narciſſus poeticus *Lin.* Ger. 124. 7. Park. parad. 75. 1.

[n] Narciſſus Pſeudonarciſſus *Lin.* Ger. 133. 2.

[o] Narciſſus Tazetta *Lin.* Pl. 14. f. 2. of this work.

[p] Narciſſus Jonquilla *Lin.* Curtis, Bot. Mag. 15.

[q] Milton has made poetical uſe of this cup:——

"And Daffodillies fill their cups with tears
"To ſtrew the laureate hearſe where Lycid lies."

of the corolla: the third has a bell-ſhaped, plaited cup, truncate at the end, and one third of the length of the petals; this has flat leaves, whereas the fourth has them ſubulate, long, and narrow like a ruſh; this alſo has a ſhort bell-ſhaped cup. The eſteem in which theſe flowers have been always held, is the occaſion that a great number of beautiful varieties have been produced from the plain ſimple parents. The Dutch catalogues have no leſs than thirty varieties of *Polyanthus Narciſſus*: and in the other three the cup is entirely changed into petals by culture. The petals of the firſt are white, and the cup yellow: the petals of the ſecond are naturally pale brimſtone, and the cup yellow: the petals of the third are either white or yellow, with orange-coloured cups: and the fourth is all yellow.

Amaryllis. There is no genus of plants in the whole round of vegetable nature more ſuperb in its flowers than the beautiful *Amaryllis*: known by its ſuperior, bell-ſhaped corolla of ſix petals; its ſtamens of unequal length; and its trifid ſtigma. Beſides ſeveral other ſpecies, either leſs obvious, or leſs beautiful [r], you will find here the *Jacobea*

Shakſpeare informs us of the early appearance of this flower:——

——" The Daffodil
" That comes before the ſwallow dares, and takes
" The winds of March."

[r] A. vittata. Curt. Magaz. 129.—A. criſpa, figured by John Miller in his 8th Plate.

Lily,

Lily [s], which produces but one, or at moſt two, of its large, deep-red flowers, from the ſame ſheath; the three under petals are larger than the others, and with the ſtamens and piſtil are bent downwards: the whole flower ſtands nodding on one ſide of the ſtalk, and makes a moſt beautiful appearance, eſpecially in the ſun, when it appears to be powdered with gold duſt.

The *Mexican Lily* [t] has ſeveral flowers, generally from two to four, burſting from the ſame ſpathe; the corolla is bell-ſhaped and regular, the three outer petals are reverſed or reflex at the tip, the three inner ones are ciliate at the baſe; the ſtamens and piſtil are bent downwards. The flowers are large, of a bright copper colour, inclining to red; and the ſtyle is red, which is unuſual: the baſe of the corolla is of a whitiſh green.

The *Guernſey Lily* [u] has alſo many flowers in the ſame ſheath, the corollas revolute, or rolled back, and the ſtamen and piſtil upright. The corollas are of the richeſt red colour, powdered with gold. This fine flower is ſuppoſed to have come originally from Japan; and to have been

[s] Amaryllis formoſiſſima *Lin.* Mill. fig. pl. 23. Curt. Magaz. 47.

[t] Amaryllis Reginæ *Lin.* Mill. pl. 224. J. Mill. illuſtr.

[u] Amaryllis ſarnienſis *Lin.* Douglas monogr. Ehret. t. 9. f. 3.

left by a wrecked veſſel on the coaſt of the iſland of Guernſey; where, being protected among the ſand by the ſea reed, it ſprung up to the great ſurpriſe of the inhabitants.

Tulipa. The Tulip and ſome others which I ſhall now preſent to you, agree with the Lily in having naked, unprotected corollas [v]. The *Tulip* [w], unbounded in the variety of colour, in the cultivated ſtate of its gaudy flowers, has an inferior bell-ſhaped corolla of ſix petals, and no ſtyle, but only a triangular ſtigma, ſitting cloſe to a long, priſmatic germ. The ſpecies is diſtinguiſhed by its ſhort lance-ſhaped leaves, and its upright flowers, from the *Italian Tulip* [x], whoſe flowers nod a little, have longer and narrower lance-ſhaped leaves, yellow corollas never varying in colour, ending in acute points, and having a ſweet ſcent. The common colour of the Eaſtern Tulip, in a ſtate of nature, is red. This, when broken into ſtripes by culture, has obtained the imaginary value of a hundred ducats for a ſingle root, among the Dutch floriſts.

Convallaria. How different is the ſweet, the elegantly-

[v] Linnæus has ſplit the liliaceous tribe, in his natural orders, into the *Enſatæ* before mentioned; the *Spathaceæ* juſt gone through; and the *Coronariæ* into which we now enter. Some alſo of his *Sarmentaceæ* belong to this tribe.

[w] Tulipa Geſneriana *Lin.* Ger. 138. 3. 4. & 139—146.

[x] Tulipa ſylveſtris *Lin.* Fl. dan. 375. Ger. 138. 1, 2.

modeſt *Lily of the valley* [y], from the flaunting beauty of the Tulip! The pure, bell-ſhaped corolla, is divided at top into ſix ſegments, which are bent back a little: and the ſeed-veſſel is not a capſule, as in moſt of this claſs, but a berry, divided however into three cells, in each of which is lodged one ſeed: this berry, before it ripens, is ſpotted. I doubt not but that you have often ſearched for it in vain, becauſe this plant ſeldom produces its fruit: the reaſon is, that it runs very much at the root, and increaſes ſo much that way as almoſt entirely to forget the other. I have ſeen large tracts covered with it, in the remote receſſes of woods, without a ſingle berry; and the way to obtain them is to impriſon the plant within the narrow circuit of a pot, when, by preventing it from running at the root, it will take to increaſing by the red berry. This ſpecies is diſtinguiſhed from *Solomon's-ſeal*, and others of the genus, by the flowers growing on a *ſcape* or naked ſtalk; it has only two leaves, which take their riſe immediately from the root.

The *Hyacinth* is one of the moſt favoured plants of the floriſts. In the natural ſtate, wherein you ſeldom ſee it, the corolla is ſingle, and cut into ſix ſegments; and there Hyacinthus.

[y] Convallaria majalis *Lin.* Curt. Lond. 5. 24. Fl. dan. 854. Ger. 410. This is one of the *Sarmentaceæ* in the natural orders.

are

are three pores or glands, at the top of the germ, exuding honey. The ſpecies from whence all the fine varieties take their riſe[z], has the corollas funnel-ſhaped, divided half way into ſix ſegments, and ſwelling out at bottom. This muſt not be confounded with the *Wild Hyacinth* or *Blue-bells* of the European woods[a], which has longer, narrower flowers, not ſwelling at bottom, but rolled back at their tips; the bunch of flowers is alſo longer, and the top of it bends downwards. This is frequently found with white corollas.

Aloe. *Aloe* is a remarkable, beautiful, and numerous genus, diſtinguiſhed by its erect corollas, with a ſpreading mouth, divided into ſix ſegments, and exuding a nectareous juice at bottom: the filaments are inſerted into the receptacle. Linnæus reduces them to ten ſpecies, but there are many very diſtinct varieties, if not ſpecies, under each. They have all thick ſucculent leaves, and the ſpecies may be ſeparated either by the forms of theſe, or by the forms and manner of growth of the flowers.

Agave. If you ſhould hear of the *Great American Aloe*[b] flowering any where in your neigh-

[z] Hyacinthus orientalis *Lin.* Mill. fig. pl. 148. Ger. 112—115.

[a] Hyacinthus non ſcriptus *Lin.* Curtis, Lond. II. 18. Ger. 111.

[b] Agave Americana *Lin.*

bourhood,

bourhood, you will find that it differs from the Aloes properly ſo called, by the corolla being ſuperior, or ſitting on the top of the germ, and the filaments being longer than the corolla. In the firſt circumſtance this differs from almoſt all the liliaceous tribe, which have the germ incloſed within the corolla. I ſhould advertiſe you, that you muſt mount a ladder or ſcaffold to examine the flowers, for they grow on a ſtem that is ſometimes twenty feet in height. You know it is a vulgar error that this plant flowers once only in a hundred years; the truth is, that in its own country it flowers in a few years from its birth; but in our cold inhoſpitable climes, it takes many years to produce its vaſt ſtem and numerous flowers, but the term of its life with us is uncertain; after having flowered, it produces a number of off-ſets, and dies. This is not the caſe in the Aloes properly ſo called, and in them the flowering ſtem is produced from the ſide of the heart or central leaves, whereas in this it iſſues from the very centre, where you obſerve that the leaves lie very cloſe over each other before they expand.

Of plants not liliaceous, belonging to this firſt order of the ſixth claſs, there is one ſhrub, the *Barberry* [c]; and ſeveral plants deficient in the corolla, as the *Calamus Aro-*

[c] Berberis vulgaris. Mill. fig. pl. 63. Ger. 1325.

maticus or *Sweet Rush*[d], the *Rattan*[e], and all the species of *Rush*[f].

Oryza. The *Rice*[g] is almost the only plant to be found in the second order of this class. It has the exact form and structure of the Grasses, differing from them only in the number of stamens.

Rumex. In the third order is the *Dock*, a numerous and prolific genus, containing thirty-one species. It is known by the calyx of three leaves, the corolla of three converging petals, and one triangular seed. These plants will not attract you by their beauty. Their flowers are more numerous than considerable. *Bloody Dock*[h] has the valves of the flowers quite entire, one of them bearing a seed, and the leaves are lance-shaped and hollowed next the petiole. *Curled Dock*[i] has the valves entire and graniferous; the leaves lance-shaped, waving about the edges, and sharp-pointed at the end. *Fiddle-Dock*[k] has the valves notched about the edges, one of them usually graniferous, and the leaves next the ground shaped like the

[d] Acorus Calamus *Lin.* Blackw. 466. Mor. hist. s. 8. t. 13. f. 4. Ger. 62.

[e] Calamus Rotang *Lin.* Rheed. malab. 12. t. 64, 65.

[f] Juncus *Lin.* See Letter XIII. at the end.

[g] Oryza sativa *Lin.*, Catesb. carol. 1. 14. Mill. illustr.

[h] Rumex sanguineus *Lin.* Blackw. 492. Ger. 390.

[i] Rumex crispus *Lin.* Curtis, Lond. II. 20.

[k] Rumex pulcher *Lin.* Mor. hist. s. 5. t. 27. f. 13.

body

body of a violin. The great *Water Dock*[l] has the valves entire and graniferous; the leaves lance-ſhaped and ſharp-pointed: the common *Blunt Dock*[m] has the valves notched and graniferous; the leaves oblong, hollowed at the baſe, near which they are notched, and obtuſe at the end. Common *Sharp Dock*[n] has the valves oblong, entire, very ſmall, the outer one graniferous; the leaves oblong and hollowed at the baſe, but drawn out into a long point. Two common ſpecies differ in one remarkable circumſtance from all the reſt; for they have the ſtaminiferous and piſtilliferous flowers on ſeparate plants, and therefore ſtrictly belong to the twenty-ſecond claſs; but they are evidently, as you will confeſs upon examination, of the ſame natural genus with the Docks. Theſe are the *Common*[o] and *Sheep's Sorrel*[p], the firſt growing in meadows and paſtures, the ſecond on dry ſandy grounds; the firſt with oblong, arrow-head leaves; the ſecond with leaves ſhaped like the head of a halberd. Thus you have the means of diſtinguiſhing eight ſpecies of Dock.

[l] Rumex Hydrolapathum *Hudſ.* Pet. 2. 1.

[m] Rumex obtuſus *Lin.* Curtis, Lond. III. 22. Ger. 388. 3.

[n] Rumex acutus *Lin.* Pet. 2. 3. Mor. 5. 27. 3.

[o] Rumex Acetoſa *Lin.* Mor. hiſt. ſ. 5. t. 28. f. 1. Ger. 396. 1. Blackw. 230.

[p] Rumex Acetoſella *Lin.* Moris, t. 28. f. 11, 12. Ger. 397. 3. Blackw. 307. Curt. Lond. 5. 29.

Meadow-

Colchicum.

Meadow-Saffron [q] is alſo of this order, and clearly of the liliaceous tribe; its reſemblance to Crocus or Saffron is obvious. Like that it has a *ſpathe* for a calyx; a corolla divided into ſix parts, with the tube extending down to the bulb; and a trilobate capſule, of three valves and three cells. So that were it not that the one has three ſtamens with one ſtyle, and the other ſix ſtamens with three ſtyles, they would be of the ſame genus. Meadow-Saffron has flat, lance-ſhaped, erect leaves, and flowers of a light purple; the firſt coming out in the ſpring, the latter in the autumn.

Aliſma.

Of the laſt order of this ſixth claſs are the *Water Plantains*, eaſily known by the calyx of three leaves, the corolla of three petals, ſucceeded by ſeveral compreſſed capſules, each containing one ſeed. *Great Water Plantain* [r] is common enough in wet places, and on the banks of rivers and brooks: it is diſtinguiſhed from its fellows by its ovate ſharp-pointed leaves, and its obtuſely triangular fruits. This is one of the plants in which you cannot err; if the differences of all were as ſtrongly marked, your trouble would be diminiſhed, but then your genius and ſagacity, dear couſin, would not have ſo much room for exerciſe.

[q] Colchicum autumnale *Lin.* Ger. 157. Blackw. 566.

[r] Aliſma Plantago *Lin.* Curt. Lond. 5. 27. Fl. dan. 561. Mill. illuſtr. Ger. 417. 1.—A. Damaſonium. Curt. Lond. 5. 28. Ger. 417. 2.

LETTER

LETTER XIX.

THE CLASSES HEPTANDRIA, OCTANDRIA, ENNEANDRIA, AND DECANDRIA.

June the 1ſt, 1775.

NATURE ſeems to have no delight in the number ſeven; the ſeventh being the ſmalleſt of all the claſſes: containing no more than ſeven genera, and ten ſpecies. Of theſe I ſhall ſelect only one for your obſervation, which ſhall be the *Horſe-Cheſnut*[s]. It is of the firſt order, and theſe are the principal characters of the genus—a ſmall calyx, of one leaf, ſlightly divided at top into five ſegments, and ſwelling at the baſe; a corolla of five petals, inſerted into the calyx, and unequally coloured, a capſule of three cells, in one or two of which only is a ſeed. Linnæus ſays that though no more than one ſeed generally comes to perfection, yet there are two in the young capſule. But ſurely the third cell is not made for nothing; and therefore I ſhould ſuſpect that in Aſia, the native clime of this fine tree, the capſule contains three nuts. The form of the *Horſe-Cheſnut* is grand, the pyramids of flowers beautiful, and making, with the large digitate leaves, a fine whole. Æſculus.

[s] Æſculus Hippocaſtanum *Lin.* Mill. illuſtr. Hunt. Evel. ſilva, p. 159.

THE

THE CLASS OCTANDRIA.

The eighth claſs has forty-four genera, and two hundred and ſeventy-three ſpecies.

Tropæolum. *Indian Naſturtium* or *Indian Creſs*[t] is one of theſe; the calyx is inferior, of one leaf cut into five ſegments, and terminated by a ſpur; the corolla has five unequal petals, and is ſucceeded by three dry berries, in each of which is one ſeed. The greater ſpecies[u] is moſt common in the gardens, and is known by the leaves being divided at the edge into five lobes, and being peltate, or having the petiole faſtened to the middle of the leaf's ſurface: the petals are blunt at the end in this; whereas in the ſmaller ſort[v] the petals are ſharp-pointed. The corollas of both are large, and of a fine orange colour.

Oenothera. *Tree Primroſe*, a Virginian plant, now ſo common in the European gardens, has a calyx of one leaf, cut into four ſegments, a corolla of four petals, and a cylindric capſule of four cells, containing naked ſeeds. The broad-leaved ſort[w], which is moſt common, has flat, lance-ſhaped leaves, and a hairy ſtalk: the corolla is of a fine yellow, ſhut uſually during the day, but expanding in the

[t] Tropæolum *Lin.*
[u] Tropæolum majus *Lin.* Curtis Magaz. 23.
[v] Tropæolum minus *Lin.* Curtis Mag. 98.
[w] Oenothera biennis *Lin.* Fl. dan. 446. Mill. illuſtr.

evening; whence ſome call it *Nightly Primroſe*.

Our European *Willow-herbs* are nearly allied to this, differing only in having a calyx of four leaves, and downy ſeeds. There is one ſort common in old gardens called *French Willow*[x], with narrow lance-ſhaped leaves inclining to linear, irregularly ſet upon the ſtalk; irregular flowers, and ſtamens bent down. The hairy ſort[y] growing common in wet places, by ditches, hedges, and ſtreams, and vulgarly known by the names of *Codlins and Cream*, or *Gooſeberry Fool*, from the ſmell of the leaves when ſlightly bruiſed, has lance-ſhaped leaves, ſerrate about the edges, running down the ſtalk, the lower ones oppoſite: the ſtamens of this and of all our common ſpecies are upright, and the petals bifid. Four of the filaments are ſhort, and the other four riſe to the top of the tube of the corolla, each four forming a regular ſquare. I do not know whether it is generally ſo, but this year I could ſcarcely find any but what had been gnawn by inſects; ſo that if I had not known the plant well, I ſhould have been puzzled to determine even the claſs. The flowers are large, ſpecious, and of a purple colour. Epilobium.

The heath genus contains no leſs than ſe- Erica.

[x] Epilobium anguſtifolium *Lin.* Curtis, Lond. II. 24. Ger. 477. 7.

[y] Epilobium hirſutum *Lin.* ramoſum *Hudſ.* Curtis, Lond. II. 21. Ger. 476. 6.

venty-four ſpecies of lowly ſhrubs, which are by no means deſtitute of beauty, though the commonneſs of one ſpecies renders it contemptible[z]. They all agree in theſe characters—a calyx of four leaves, incloſing the germ, a corolla of one petal, cut into four ſegments; the filaments inſerted into the receptacle; the anthers bifid; and a capſule of four cells.

Common Heath[a], which is ſo general a plant, that vaſt tracts of land take their name from it, is diſtinguiſhed by the anthers being terminated with an awn, and lying within the flower, the ſtyle appearing behind it, the corollas bell-ſhaped, and not quite regular, the calyxes double, the leaves oppoſite and ſhaped like the head of an arrow. *Fine-leaved Heath*[b] has creſted anthers lying within the corolla; the ſtyle hardly iſſues from it; the ſtigma is capitate; the flowers grow many cloſe together; the corollas are ovate and of a bluiſh colour; the leaves are produced in threes; and the bark is aſh-coloured. *Croſs-leaved Heath*[c] has the anthers as in the firſt; the ſtyle lies within the corolla; the flowers grow in a head; the corollas are ovate; and the leaves

[z] E'en the wild heath diſplays its purple dies.

[a] Erica vulgaris *Lin.* Curtis, Lond. V. 30. Fl. dan. 677. Ger. 1380. 1.

[b] Erica cinerea *Lin.* Curtis, Lond. II. 25. Ger. 1382. 7.

[c] Erica Tetralix *Lin.* Curtis, Lond. I. 21. Fl. dan. 81.

are

are produced in fours: this grows in the wet and boggy parts of heaths, and is a handſome ſpecies. The foreign ſorts, moſtly from the Cape of Good Hope, are eminently beautiful, but not being commonly met with, I ſhall not trouble you with them.

Mezereon, which you value for viſiting Daphne. you at a time when you have very few viſitors, and alſo for its pleaſant odour, is of this claſs, and of the firſt order, as well as all the foregoing. It has no calyx, but a monopetalous, funnel-ſhaped corolla, incloſing the ſtamens, and the border cut into four ſegments: the fruit is a roundiſh berry containing one ſeed. This ſpecies [d] is diſtinguiſhed from the reſt of the *Daphne* genus by its ſeſſile flowers, growing by threes from the ſame joint; and by its lance-ſhaped deciduous leaves. The corollas are peach-coloured, deeper red, or white, and the berries of the two firſt are red, of the laſt yellow.

There is a ſort [e] not uncommonly wild in woods, and ſhady hedges, which is an evergreen, and has the flowers coming out by fives, from the axils; the corollas are of a yellowiſh green, and the leaves are lance-ſhaped. This is rather a diſmal plant in reſpect of its ſituation, time of flowering,

[d] Daphne Mezereum *Lin.* Fl. dan. t. 268. Ger. 1402. 2.

[e] Daphne Laureola *Lin.* Spurge Laurel. Ger. 1404. Blackw. 62.

and colour of the corollas; nor has it the ſame agreeable ſcent with the Mezereon: it is not however without its value as an evergreen, and flouriſhing under the deep ſhade of trees. Both ſpecies are very hot and cauſtic in their nature; notwithſtanding which birds are greedy of the berries.

Chlora. *Yellow perfoliate Gentian* [f] is now removed from the other Gentians, to the ſecond order of this claſs, becauſe the number eight prevails in the ſtamens, calyx, and corolla: in other circumſtances it agrees with the genus in which it formerly ranged. It is found in paſtures, on a chalky ſoil, and is eaſily known by its yellow corollas, and upright ſmooth perfoliate ſtalks.

Polygonum. The third order has a large genus containing twenty-ſeven ſpecies, among which, beſides other common plants, are *Biſtort*, *Knot-graſs*, *Buck-wheat*, and *Black Bindweed.*

Biſtort [g] has a ſingle, undivided ſtalk, terminated by one ſpike of flowers; and lance-ſhaped leaves, generally hollowed at the baſe, running along the petiole, or forming a membrane along each ſide of it, and waved. The root is large for the ſize of the plant, and turns and twiſts in the ground.

[f] Chlora perfoliata *Lin.* Ger. 547. 2.

[g] Polygonum Biſtorta *Lin.* Curtis, Lond. I. 22. and Mill. fig. pl. 66. Ger. 399. 1.

Knot-

Knot-graſs [h] is a very common weed in places that are trod. The little flowers are produced from the axils of the ſtalks, which are herbaceous, and trail upon the ground; the leaves are lance-ſhaped, and, being of different ſize and breadth in different ſoils, have given occaſion to the forming diſtinctions, which are but varieties.

Buck-wheat [i], which makes a pretty appearance when cultivated, has arrow-ſhaped leaves hollowed at the baſe, the ſtalk upright, though weak, ſmooth and unarmed, and the angles of the ſeeds equal.

Black Bindweed [k] is not very unlike this; but the leaves are heart-ſhaped, the ſtalk angular and twining, and the flowers obtuſe. The anthers alſo are purple; and the baſe of the petioles is perforated beneath with a pore. This is not an unfrequent weed among corn.

All the ſpecies agree in having no calyx; a corolla divided into five ſegments, that might eaſily be taken for a calyx; and one naked, angular ſeed.

THE CLASS ENNEANDRIA.

The ninth claſs has not ſo many genera as the ſeventh, but it has many more ſpe-

[h] Polygonum aviculare *Lin.* Curtis, I. 27. Ger. 565.

[i] Polygonum Fagopyrum *Lin.* Ger. 89.

[k] Polygonum Convolvulus *Lin.* Curtis, Lond. IV. 29.

cies[l], and among them ſeveral very remarkable ones; as the Bay, Cinnamon, Caſſia, Camphor, Benzoin and Saſſafras, all comprehended under one genus[m]; Acajou or Cashew Nut, and Rhubarb. The Bay genus has the following character: no calyx, but a corolla reſembling a calyx, and divided into ſix parts in moſt of the ſpecies; a nectary of three glands, each terminated by two briſtles, ſurrounding the germ; the filaments in three rows, with two round glands near the baſe of the three that form the inner row; the fruit an oval *drupe* or plum, incloſing a nut.

Laurus.

The true Bay[n] is known by its lance-ſhaped, veiny evergreen leaves; the corolla recedes from the general character in being quadrifid, or cut into four ſegments. It varies alſo in the number of ſtamens from eight to fourteen; and it recedes from the claſs in having incomplete flowers on ſeparate plants. Linnæus however has kept it here becauſe it has the eſſential characters of this genus, particularly the glands on the inner filaments. You will ſcarcely have the good fortune to meet with the other ſpecies, at leaſt in flower.

Anacardium.

Acajou or *Caſhew*[o] we know chiefly by the

[l] Twenty-eight: and only ſix genera.

[m] Laurus.

[n] Laurus nobilis. Laurel is known only to modern times, and ranges in the claſs *Icoſandria* under *Prunus*. Alexandrian Laurel is a Ruſcus in Claſs XXII.

[o] Anacardium occidentale *Lin.*

the nut, which grows at the end of a flefhy body as large as an orange, and full of an acid juice; this Linnæus calls the receptacle. Between the two fhells is a thick, black inflammable oil, with which you may mark your linen, for it will not wafh out. It alfo makes the fineft black varnifh. I need not caution you againft putting this nut into your mouth to crack it. The oil is very cauftic, and will raife blifters in the tongue. If it fhould ever be your fortune to fee this tree in flower, you will obferve that the calyx is five-leaved; that the corolla confifts of five reflex petals; and that there are ten filaments, whence Linnæus firft put it into the tenth clafs; but one of thefe being conftantly without an anther, he afterwards removed it to the ninth. More recent obfervations however have afcertained that the Anacardium has perfect and ftaminiferous flowers on diftinct individuals: it belongs therefore to the fecond order of the twenty-third clafs, *Polygamia Diœcia*.

Thefe are of the firft order, *Rhubarb* is Rheum. of the fecond, *Trigynia*; there being no plants known of this clafs with two piftils. The characters of this genus are, a flower without a calyx; a corolla of one petal, divided into fix fegments; and one large triangular feed, much like that of the Docks[p]. No lefs than four fpecies have

[p] They are both placed in the fame natural order, namely the fifth divifion of the *Oleraceæ*.

been ſent over and cultivated at different times under a notion of their being the true Tartarian Rhubarb. Of theſe the *Rhapontick*[q] has migrated from the apothecary's ſhop into the kitchen, the petioles of the leaves being much eſteemed for making tarts. The leaves are ſmooth, of a roundiſh heart-ſhape, with the petioles thick, reddiſh, a little channelled on their lower part, but flat at the top: the flower ſtems are red, grow from two to three feet high, and are terminated by thick, cloſe, obtuſe ſpikes of white flowers, coming out in June. This grows wild near the Pontic, Euxine or Black Sea.

There is a good teſtimony for the three others being the true Rhubarb; and I think it not improbable but that they may all be cultivated in Tartary for their roots. One of theſe[r] has longer leaves than the Rhapontic, running more to a point, much waved on their edges, a little hairy on their upper ſide, and they appear much earlier; the petioles are not ſo much channelled on their under ſide, and are plain on the upper; they are alſo neither ſo red nor ſo thick: the flower ſtem is of a pale browniſh colour, about four feet high, dividing into ſeveral looſe panicles of white flowers, which appear in May.

Another[s] has very ſmooth, ſhining,

[q] Rheum Rhaponticum *Lin.*
[r] Rheum Rhabarbarum *Lin.*
[s] Rheum compactum *Lin.* Mill. fig. pl. 218.

heart-ſhaped leaves, not running out ſo much to a point as the ſecond, but more than the firſt; they are very broad towards the baſe, and a little waved and indented on their edges: the petioles have ſcarcely any channels, and are flat on their upper ſide; they are pale green, and almoſt as large as thoſe of the firſt ſort. The flower-ſtem is pale green, five or ſix feet high, the upper part dividing into ſmall branches, each ſuſtaining a panicle of white flowers ſtanding erect, and appearing the latter end of May.

A fourth ſort, called *Palmated Rhubarb* [t], differs greatly from the others, and is known immediately by its palmated and very ſharp-pointed leaves. The flower-ſtem is red, and ſix or ſeven feet high: the flowers are in looſe panicles. Whatſoever may be the caſe with the other ſpecies, there is the moſt undoubted evidence of this being the true Tartarian Rhubarb.

There is one wild plant of this claſs, which is of the third order, having ſix ſtyles. It grows in the water, and having handſome roſe-coloured flowers, with long narrow leaves, is called *Flowering Ruſh* [u]; the flowers are produced at the end of a Butomus.

[t] Rheum palmatum *Lin.* Mill. illuſtr. Philoſ. Tranſ. 1765.

[u] Butomus umbellatus *Lin.* Curtis, Lond. I. 29. Fl. dan. 604. Mill. illuſtr. Mor. ſ. 12. t. 5. f. penult. Ger. 29.

naked

naked ſtalk, in an umbel. They have no calyx, but a three-leaved involucre, a corolla of ſix petals, and ſix capſules of one valve, gaping on the ſide towards the centre of the umbel, and containing many ſeeds.

THE CLASS DECANDRIA.

The tenth is a much more conſiderable claſs, having ninety-five genera, and five hundred and thirty-ſix ſpecies. The firſt order being very numerous, Linnæus has made a commodious ſubdiviſion of it into ſuch as have corollas of many petals, of one petal, or none; and the firſt of them he has ſubdivided again into ſuch as have irregular and ſuch as have equal corollas. Moſt of thoſe with irregular polypetalous flowers are very nearly allied to the papilionaceous tribe, with which you are already acquainted. Of theſe the moſt known are the *Judas-tree*, *Locuſt-tree*, *Flower-fence*, *Braſiletto*, all the numerous ſpecies of *Caſſia*, *Balſam of Tolu-tree*, and *Nickar-tree*; moſtly the produce of South America and the Weſt Indies. *White Dittany* or *Fraxinella* [u] is alſo of this ſubdiviſion, but not of the papilionaceous tribe.

Dictamnus. This elegant flower is known by its five-leaved calyx; its corolla of five ſpreading petals; the filaments ſet with glandulous

[v] Dictamnus albus *Lin.* Mill. fig. pl. 123. & Pl. 16. f. 2. of this work.

points:

points: it is ſucceeded by five connected capſules, containing two ſeeds covered with a common *aril.*

There is only one ſpecies of *Fraxinella*, varying in the colour of the flowers, which are either pale red ſtriped with purple, or elſe white. It has pinnate leaves, ſomewhat reſembling thoſe of the Aſh. The whole plant emits an odour of lemon peel, but when bruiſed has a balſamic ſcent.

Among the plants with regular or equal polypetalous corollas, you will find *Logwood*, *Melia* or the *Bead-tree*; *Guaiacum*, *Rue*, and *Dionæa Muſcipula*, ſo curious for that ſenſitive quality of the leaves, by which it entraps inſects that light upon them.

Rue is diſtinguiſhed by theſe generic Ruta. characters—a calyx divided into five parts; concave petals; ten honied pores at the baſe of the germ, which is raiſed on a receptacle punched with the ſame number of pores; and laſtly, a capſule cut half way into five parts, conſiſting of five cells within, and containing many ſeeds. If I do not give you a caution reſpecting the common Rue [w] of the gardens, you may probably be puzzled in examining its flowers; for there is only one flower on a branch which will anſwer to the generic characters; in all the reſt you are to ſubtract one fifth from every part of the fructification. This circumſtance

[w] Ruta graveolens *Lin.* Mor. hiſt. ſ. 5. t. 14. f. 3.

cumſtance is not peculiar to Rue, but is found in ſeveral other plants , and has been made an objection by ſome to the Linnæan ſyſtem. The illuſtrious author has extricated himſelf from the difficulty by forming his character upon the principal or primary flower, as he calls it, and announcing the anomaly. There are other plants, which in all the reſt, *add* a fifth to the number of parts in the primary flower [y].

Garden Rue is ſpecifically diſtinguiſhed, partly by this circumſtance, of having the ſide flowers quadrifid, and partly by the leaves being decompounded. There are ſome differences in this ſpecies: common garden Rue has the component lobes of the leaves wedge-ſhaped, and the ſtamens longer than the corolla; another, alſo frequently cultivated, has narrower lobes, the flowers in longer, looſer bunches, and the ſtamens equal in length with the petals, the ſeed-veſſel is alſo ſmaller; a third has the lobes of a linear ſhape.

Andromedas, *Rhododendrons*, *Kalmias*, *Arbutus*, and a few others, have regular monopetalous corollas. The characters of the laſt are a very ſmall calyx divided into five

[x] As in Cinchona, Myrſine, Euonymus europæus; Theſium alpinum, Herniaria fruticoſa, Gentianæ 23—27. Linum Radiola, &c.

[y] Such as *Adoxa Moſcha tellina*. Curtis, Lond. II. 26. and ſome others.

parts

parts: an ovate corolla pellucid at the baſe: and the fruit a berry, with the ſeeds lodged in five cells.

Strawberry-tree[z] is known by its woody ſtem, its ſmooth leaves ſerrate about the edges, and the cells of the berries having ſeveral ſeeds. Some of the other ſpecies have weak procumbent ſtems[a]; and ſome have only a ſimple ſeed to each cell[b]. You are well acquainted with the Arbutus, by the ornament which it affords to your plantations in the latter months, with its lucid leaves thick covering the plant; and its bunches of flowers of this year, accompanied by the red round berries of the laſt. Arbutus.

But let not the firſt order of the tenth claſs occupy too much of your time, ſince there are four other orders contained in it. In the ſecond you have all the *Saxifrages*, forty-two in number; agreeing in a calyx divided into five parts; a corolla of five petals; a capſule of one cell, filled with many ſmall ſeeds, and terminated by two beaks formed of the permanent ſtyles. Of theſe, *Pyramidal Saxifrage*[c] is eſteemed for adorning halls and chimnies with its beautiful pyramids of white flowers; which it will do for a long time. There are ſeveral va- Saxifraga.

[z] Arbutus Unedo *Lin.* Mill. fig. pl. 48. Ger. 1496.
[a] Arb. acadienſis, alpina & uva urſi.
[b] A. alpina & uva urſi.
[c] Saxifraga Cotyledon *Lin.* Mill. fig. 243. Fl. dan. 241.

rieties

rieties of it, but they have all ſtiff tongue-ſhaped leaves, with a cartilaginous ſerrate border, and collected into ſeveral rows cloſe to the ground. From the midſt of theſe iſſues the ſtalk, ſuſtaining the panicles of flowers.

Another ſpecies [d] was alſo formerly much ſhown out at windows and balconies in ſmoky towns, and hence, with its being really beautiful, had the names of *London Pride* and *None-ſo-pretty*, at a time when few plants were generally known. This has oblong or roundiſh leaves, deeply notched on the edges, ſpringing from broad, flat, furrowed petioles, near two inches long. They ſurround the flowering ſtalk, which itſelf is deſtitute of leaves, of a red colour, ſtiff, ſlender, and hairy. The corollas are white dotted with red.

Common White Saxifrage [e] flowers early and in great quantities among the graſs. The bottom leaves are kidney-ſhaped, hairy, and on pretty long petioles: the ſtalks are hairy, and in good ground a foot high, branching out from the bottom, and furniſhed with a few ſmall leaves, in ſhape like the others, but ſitting cloſe to the ſtem: the flowers terminate the ſtalk in ſmall cluſters; the corollas are white, and large for the ſize of the plant: if any doubt remains concerning it, pull it up, and you

[d] Saxifraga umbroſa *Lin.* Mill. fig. 141. f. 2.

[e] Saxifraga granulata *Lin.* Mill. illuſtr. Curtis, Lond. I. 30. Ger. 841. 1.

will

will find that the roots are like grains of corn, and of a reddiſh colour. In poor ground this plant is very ſmall, and has only two or three flowers, ſometimes but one, on a ſimple, unbranched ſtem.

Theſe, with moſt of the other ſpecies, have upright ſtems, but there are three which have weak trailing ſtalks. Of theſe there is one which has much reſemblance to a moſs, when it is out of flower; and, from the manner of its growth in a thick tuft, it has acquired the Engliſh name of *Ladies' Cuſhion*[f]. The leaves are linear, ſome entire and others trifid: the little flower ſtems are three or four inches high, ſlender, erect, and almoſt naked, terminated by ſmall flowers of a dirty white.

The genus *Dianthus*, of this ſecond order, is numerous, as well as the laſt, comprising twenty-two ſpecies, which agree in having a cylindric calyx of one leaf, ſurrounded at the baſe by four ſcales; a corolla of five petals; and a cylindric, unilocular capſule, for a ſeed-veſſel. Many of the ſpecies are beautiful, as *Sweet William*[g], the noble *Carnation*[h], the *Pink*[i], with all its numerous varieties, the *China Pink*[k] Dianthus.

[f] Saxifraga hypnoides *Lin.* Fl. dan. 348. Mor. hiſt. ſ. 12. t. 9. f. 26.

[g] Dianthus barbatus *Lin.*

[h] Dianthus Caryophyllus *Lin.* Mill. fig. 121.

[i] Dianthus plumarius *Lin.*

[k] Dianthus chinenſis *Lin.* Mill. fig. pl. 81. f. 2. Curtis Mag. 25.

diſtinct from the former: ſeveral alſo of the ſorts, which are wild in many parts of Europe, though adorned with leſs ſplendid flowers, and more modeſt in their pretenſions, are not however without their beauty. The *Carnation* is acknowledged, on all hands, for a worthy leader of one of the fineſt natural orders, entitled from the Latin name of this fragrant flower *Caryophylleous* plants. When we conſider the ſize of the flower, the beauty of its colours, the arrangement of its parts, and above all the ſingularly rich and ſpicy odour that it exhales, we cannot withhold that tribute of admiration which will ever be given it, unleſs by obtruding itſelf too frequently on the eye, its real beauties become at length diſregarded.

The leading feature, in diſtinguiſhing the ſpecies of this genus, is the *infloreſcence* or manner of flowering. *Sweet William* and ſome others have aggregate flowers; *Carnation*, *Pink*, *China Pink*, &c. have many flowers on the ſame ſtalk, not however in herds, but ſolitary or ſeparate; ſome few have one flower only on a ſtem; and two or three have ſhrubby ſtalks. The other circumſtances that diſcriminate the ſpecies are, that the ſcales at the baſe of the calyx in the *Sweet William* are of an ovate-ſubulate form, and as long as the tube of the corolla; in the *Carnation* and *Pink* they are

are ſubovate and very ſhort; in the *China Pink* they are ſubulate, as long as the tube, and hang looſe. The *Sweet William* has alſo lance-ſhaped leaves. *Carnation* and *China Pink* have the petals notched. The *Pink* has the corollas pubeſcent at the baſe, and the petals deeply cut. For ornament and beauty you will gather theſe flowers from your parterre; but as a botaniſt you will take them from a wall, or a dry untilled ſoil, where their ſimplicity and the clearneſs of their natural characters will make you full amends for the want of ſplendour. You would not always chooſe to be among full-dreſſed people at a ball, or in a drawing room; but ſometimes to take a rural walk, and entertain yourſelf with plain country manners.

Arenaria, &c.

In the third order, beſides ſome others, there are four genera containing many ſpecies which have a good deal of ſimilitude. They are however thus well diſtinguiſhed. *Arenaria* and *Stellaria* have a capſule of one cell; *Cucubalus* and *Silene*, a capſule of three cells: of the two former the firſt has the petals entire, the ſecond has them bifid: of the two latter, in both of which the petals are bifid, the ſecond has a crown compoſed of a ſet of minute petals in the centre; whereas the firſt has nothing of this, or is naked. *Arenaria* and *Stellaria* have alſo a five-leaved calyx; in *Cucubalus* it is much

 inflated,

inflated, and in *Silene* it is ſwelling. All four have five petals in the corolla.

Cucubalus. *Spatling Poppy*[l] is not an uncommon weed among corn and in meadows. You will know it by the almoſt round and much inflated calyx, beautifully veined, ſo as to have the appearance of a fine network thrown over it, and quite ſmooth: the corollas are not entirely naked, and are pure white.

Sedum. *Sedums* or *Stone-crops* are found in the fourth order (*Pentagynia*). They are known by the general prevalence of the number five in all parts of the flower: a calyx cut into five ſegments, a corolla of five petals, five nectariferous ſcales at the baſe of the germ, and five capſules: not to mention the twice five ſtamens, and five ſtyles, which form the characters of the claſs and order. Many of them are not uncommon in a wild ſtate, particularly a ſmall trailing ſort with yellow flowers growing in a trifid cyme; and ovate, blunt, ſmooth leaves, imbricate and alternately adhering to the ſtalk[m]: other ſpecies have white, and ſome red corollas. They grow chiefly on walls, or in very dry ſoils.

[l] Cucubalus Behen *Lin.* Fl. dan. 857. Mor. hiſt. ſ. 5. t. 20. f. 1. Ger. 678. 2. Blackw. 268.

[m] Sedum acre *Lin.* Wall-pepper. Curtis, Lond. I. 32. Ger. 517. album 31. Ger. 512. 2.

Cockle [n], which is ſo common a weed among corn, has a membranaceous, one-leafed calyx; a corolla of five obtuſe, undivided petals, and an oblong capſule of one cell. The ſpecies is diſtinguiſhed by the roughneſs of the plant, the length of the ſegments of the calyx, and by the petals being entire and naked. Agroſtemma.

Of *Lychnis* there are ſeveral ſpecies agreeing in theſe common characters. An oblong, ſmooth calyx of one leaf; a corolla of five petals ſlightly bifid; and a one-celled capſule of five valves. Lychnis.

Scarlet Lychnis [o], commonly cultivated in gardens, has the flowers growing in bunches, ſo that the whole forms nearly a flat ſurface at top; the colour of the corolla is a very high ſcarlet.

Catchfly [p], ſo called from the clammy juice exuding from the ſtalks under each pair of leaves, glutinous enough to entangle ſmall flies, is known by the petals being almoſt entire; the colour of them is red: the leaves are long, narrow, and graſs-like, eſpecially the lower ones. The flowers of this and the foregoing are uſually double in the gardens, and therefore uſeleſs to you in your botanical reſearches.

There is a ſort of *Lychnis* commonly wild

[n] Agroſtemma Githago. *Lin.* Curtis, Lond. III. 27. Ger. 1087. Fl. dan. 576.

[o] Lychnis chalcedonica *Lin.*

[p] Lychnis Viſcaria *Lin.*

by water-ſides and in moiſt meadows, called *Ragged-Robin*, *Meadow-Pinks*, *Wild-Williams*, or *Cuckow-flower*[q], which has red jagged petals, generally cut into four parts; and roundiſh capſules, the mouth of which has five teeth turning back. There is alſo another no leſs common in paſtures, called *White Lychnis*, or *White Campion*[r], which differs eſſentially from its congeners in having the piſtils ſeparate from the ſtamens, and on diſtinct plants. I leave you, dear couſin, with this irregularity, and wait a day of leiſure to purſue our botanical career.

[q] Lychnis flos cuculi *Lin.* Curtis, Lond. I. 33. Ger. 600. 1.

[r] Lychnis dioica *Lin.* Fl. dan. 792. Mor. 5. 21. 21. Ger. 469. 1. with red flowers.

LETTER

LETTER XX.

THE CLASS DODECANDRIA.

June the 10th, 1775.

NOTHING difficult has hitherto occurred, dear coufin, in your determination of the claffes, the number of the ftamens alone having fufficed for that purpofe. But no plant being yet difcovered with eleven ftamens, among thofe which have them diftinct[s], the eleventh clafs fhould be expected to contain thofe plants which have twelve; but here the number is found to be by no means conftant, and Linnæus is obliged to take into his clafs *Dodecandria*, all fuch plants as have from twelve to nineteen ftamens inclufive. Nor is the eleventh clafs, with all this latitude, an eafy one for a novice to determine; the number of ftamens in fome cafes being fewer than twelve, in others more than nineteen, or elfe coming out in parcels at different periods. It is not very numerous, containing but thirty-three genera and one hundred and fixty-four fpecies.

Of the firft order, the moft known or the moft remarkable are *Afarum*, or *Afarabacca*,

[s] *Brownea*, which has naturally eleven ftamens, is of the fixteenth clafs, *Monadelphia*.

the *Mangoſteen*, *Winter's Bark*, *Purſlain*, *Looſeſtrife*.

Aſarum. *Aſarabacca* has a calyx cut half way into three ſegments, and ſitting on the top of the ſtyle: no corolla: and a leathery capſule, of ſix cells within, and crowned at top. There are three ſpecies—the *Canadian*, the *Virginian*, and the *European*[t], which laſt is diſtinguiſhed by two kidney-ſhaped leaves, ending bluntly.

Portulaca. *Purſlain* has a bifid calyx incloſing the germ: a corolla of five petals: and a capſule of one cell, in which the receptacle is looſe; in ſome ſpecies it opens horizontally[u], in others it is trivalvular: the number of ſtamens varies in the different ſpecies. The *Purſlain*, cultivated for ſallads[v], is a native of the hot parts of America; it is known by its wedge-ſhaped leaves, and the flowers ſitting cloſe to the ſtalk; and it is one of thoſe which have the capſule opening horizontally.

Lythrum. *Looſeſtrife* has the calyx cut at the edge into twelve portions; and incloſing the germ: the corolla of ſix petals, inſerted into the calyx: the capſule bilocular, and containing many ſeeds. *Purple Looſeſtrife*[w] is a handſome plant, adorning the banks of

[t] Aſarum europæum *Lin.* Fl. dan. 633. Mill. fig. t. 53.

[u] Capſula circumſciſſa.

[v] Portulaca oleracea *Lin.* Blackw. t. 287.

[w] Lythrum Salicaria *Lin.* Curtis, Lond. III. 28. Ger. 476. 5.

rivers,

rivers, ponds, and ditches, with its fine ſpikes of purple flowers; the leaves grow in pairs, and are lance-ſhaped, with a hollowed baſe; ſometimes three leaves come out together from the ſame point, and the ſtalk is hexangular; but this is only an accidental variety. Our ſpecies anſwers to the character of the claſs in having twelve ſtamens; but there are ſome which have but ten, nay even only ſix ſtamens.

In the ſecond order are only two genera —*Heliocarpus*, an American plant, little known; and *Agrimony*, an European, and ſufficiently common. This has a ſmall quinquefid calyx, ſitting on the top of the germ, fortified with another: a corolla of five petals, inſerted into the calyx, and one or two roundiſh ſeeds in the bottom of the calyx. The number of ſtamens is very uncertain in this genus; ſome ſpecies having twelve, others ten, others ſeven. *Common Agrimony*[a], which is found in woods and by hedge ſides, has interruptedly-pinnate leaves on the ſtalk, with the leaflet at the end petiolate; the ſeeds are fortified with briſtles. The outer calyx grows faſt to the inner; and the ſtamens vary in number from twelve to twenty. Agrimonia.

The third order has alſo only two genera, but they are numerous; *Reſeda* having twelve and *Euphorbia* no leſs than ſixty-nine ſpecies.

[a] Agrimonia Eupatoria *Lin.* Curt. Lond. 5. 32. Fl. dan. 588. Mill. illuſtr. Ger. 712.

No genera are more difficult to determine than theſe; the number and form of the parts varying in the different ſpecies. The eſſential character of the firſt conſiſts in the trifid petals, one of them melliferous at the baſe; and in a capſule of one cell, always open: the calyx alſo is of one leaf, cut into ſeveral narrow ſegments, two of which gape more than the others on account of the melliferous petal; the ſtamens are from eleven to fifteen in number.

Reſeda. *Dyer's-weed* or *Weld* [y] grows common in barren paſtures, dry banks, and on walls; it is alſo cultivated for the uſe of the dyers [z]. The leaves are lance-ſhaped, and entire, except that they have one indentation on each ſide at the baſe; and the calyx is cut into four ſegments. The corolla alſo has three petals; the upper one melliferous, and divided half way into ſix parts; the oppoſite lateral petals are trifid; and ſometimes two ſmall entire petals are added below. *Dyer's-weed* is a biennial plant, producing the firſt year a circle of leaves cloſe to the ground; and the next a ſtalk terminated by a long looſe ſpike of yellowiſh flowers.

Sweet Reſeda, or *Mignionette* [a], has oblong

[y] Reſeda Luteola *Lin.* Fl. dan. 864. Ger. 494.

[z] This is thought to be the plant with which the ancient Britons dyed their bodies.

[a] Reſeda odorata *Lin.* Mill. fig. 217. Curt. Magaz. 29.

leaves,

leaves, ſome of which are entire, and others trifid; the calyx of the flower is large, equalling the corolla in ſize. The flowers are produced in looſe ſpikes, on long peduncles; are of an herbaceous colour, and much eſteemed for their agreeable odour, like that of freſh Raſpberries.

Euphorbia.

Euphorbia has a corolla of four, and ſometimes of five petals, glandulous in moſt ſpecies, in ſome ſhaped like a creſcent, or indented about the edges, in a few thin as a fine membrane; commonly placed as it were on the outſide of the calyx, which is of one leaf, divided at the edge into four, or in ſome into five parts, and ventricoſe or ſwelling out. The ſtamens are twelve or more, iſſuing forth at different periods. The ſeed-veſſel is a capſule of three diſtinct cells united, with one roundiſh ſeed in each cell, and on the outſide ſmooth, rough or warted in the different ſpecies. This genus being ſo numerous, ſome ſubordinate diſtinctions are neceſſary: and accordingly Linnæus has divided it into ſeven ſections. The firſt contains the *Euphorbiæ* properly ſo called; or ſuch as have a ſhrubby, angular, ſpiny ſtem, generally void of leaves. The ſecond contains the ſhrubby ſpecies without ſpines. In all the other ſections the ſtems are *dichotomous*, or divide always by pairs, and the flowers are borne in a kind of umbel; which, in the third ſection, is commonly *bifid*; in the fourth, *trifid*; in the

the fifth, *quadrifid*; in the ſixth, *quinquefid*; and in the ſeventh, *multifid*.

Several ſpecies of the firſt ſection yield indifferently that acrid milky juice, which when inſpiſſated is ſent us under the title of *Euphorbium*. The flowers are of little beauty, and theſe plants have been noticed rather for the ſingularity of their form, and the ſtriking difference of their ſtructure, from the plants of Europe, than for any charms that they poſſeſs. The ſpecies ſuppoſed to be that from whence the ancients had the drug[b], is known by a triangular, jointed ſtalk: the ſpecies from which it is ſaid we now have it[c], has a quadrangular ſtem, and double ſpines: and the ſpecies which Linnæus ſuppoſes ought to be uſed[d], is multangular with double ſpines.

Meduſa's-head[e] is of the ſecond ſection. The ſtalks are cloſely covered with tubercles, lying over each other, and from the ſides of theſe ſpring many branches, which are frequently ſo entwined as to give the idea of a parcel of ſerpents. The ends of the branches have narrow ſucculent leaves readily dropping off, and a ſet of white flowers.

The plants of the other ſections are com-

[b] Euphorbia antiquorum *Lin.* Comm. hort. 1. t. 12.
[c] Euphorbia canarienſis *Lin.* Comm. hort. 2. t. 104.
[d] Euphorbia officinarum *Lin.* Comm. hor. 1. t. 11.
[e] Euphorbia Caput Meduſæ *Lin.* Comm. hort. 1. t. 17.

monly

monly known by the name of *Spurge*, and are moſt of them wild in the different parts of Europe. Two ſpecies are common weeds in kitchen gardens: one of them [f] belongs to the fourth ſection, or thoſe which have trifid umbels: the ſubdiviſions of theſe are dichotomous: the *involucellæ* or *bractes* are ovate; and the leaves are quite entire, or without any notches about the edge; they are ovate in form, and attached to the ſtalk by ſhort petioles; each petal alſo has two little horns; the other [g] is of the ſixth ſection, having quinquefid umbels; each principal diviſion ſubdivides into three; the *involucellæ* are ſhaped as in the former; the leaves are wedge-ſhaped, and ſerrate about the edges; and the petals are round and entire. A third ſpecies [h], common in woods, is of the laſt ſection, with multifid umbels: it is a larger plant, and perennial; whereas the others are annual: the *involucellæ* are round and perfoliate; the leaves are very blunt at the end.

Spurges having little beauty, they are ſeldom cultivated in gardens. We muſt however except the *Euphorbia punicea*, a moſt ſplendid Jamaica plant, which flowers in the collection of the Marchioneſs of Rock-

[f] Euphorbia Peplus. Petty Spurge. Curtis, Lond. I. 35. Ger. 503. 19.

[g] Euphorbia helioſcopia. *Lin.* Sun Spurge. Curtis, Lond. I. 36. Ger. 498. 2.

[h] Euph. amygdaloides *Lin.* Wood Spurge. Mor. hiſt. ſ. 10. t. 1. f. 1. Ger. 500. 9.

ingham,

ingham, and is admirably figured in Dr. Smith's *Icones pictæ*. This belongs to the fifth ſection. One of the moſt common is a biennial ſpecies, of the ſame ſection, with the leaves oppoſite and quite entire, called *Broad-leaved Spurge* or *Cataputia*[i]. Its native place is Italy, and the ſouth of France: it grows three or four feet high; the flowers are of a greeniſh yellow, and the capſules being very elaſtic, the ſeeds are thrown to a conſiderable diſtance. A ſecond is perennial, and of the laſt ſection[k]; the *involucellæ* are heart-ſhaped; the petals are formed like a creſcent; and the capſules are ſmooth; ſome of the branches are barren, and others bear flowers and ſeed; on the firſt the leaves are narrow and ſetaceous; on the ſecond they are lance-ſhaped.

Sempervivum. There is a genus[l] of this claſs in which the number twelve prevails in all the parts. Having twelve ſtyles, it is of the order *Dodecagynia*. The calyx is divided into twelve parts; the corolla conſiſts of twelve petals; and the flower is ſucceeded by twelve capſules, containing many ſmall ſeeds. *Common Houſeleek*[m] is one of theſe, which, though ſo ſucculent a plant, flouriſhes on walls and

[i] Euphorbia Lathyris *Lin.* Mill. illuſtr.

[k] Euphorbia Cypariſſias *Lin.* Blackw. 163. f. 3.

[l] Sempervivum, nearly allied to the Sedums in the tenth claſs.

[m] Sempervivum tectorum *Lin.* Curtis, Lond. III. 29. Fl. dan. 601. Mill. illuſtr. Ger. 510. 1. Plate 17. of this work.

roofs.

roofs. The edges of the leaves are ſet with ſhort fine hairs; and they do not grow in a globular form, as ſome other ſpecies do, but ſpread open. From the centre of the heads of leaves ariſes a round, red, ſucculent flower-ſtalk, about a foot high, which at bottom has a few narrow leaves, and at top divides into two or three parts, each ſupporting a reflex range of flowers, with red corollas. Though the natural number in this genus be twelve, yet you will find it to vary exceedingly: nature being leſs conſtant in larger than in ſmaller numbers. With this ſhort ſketch, adieu, dear couſin, for the preſent.

LETTER

LETTER XXI.

THE CLASSES ICOSANDRIA AND POLYANDRIA.

June the 21ſt, 1775.

YOU have already, dear couſin, taken an imperfect view of the twelfth claſs, as far as it relates to fruit-trees [n]: you are not however to ſuppoſe, either that all theſe trees range in the claſs *Icoſandria*, or that no other but them are to be found there. No leſs than twenty-nine genera, and two hundred and ninety-four ſpecies, are included in this claſs, a conſiderable portion of which is trees or ſhrubs; many herbs however are found among them.

To diſtinguiſh this claſs and the next from the reſt, and from each other, remember always that it is not the number, but the ſituation of the ſtamens which furniſhes the claſſical character. In the next they ariſe, as generally in the other claſſes, from the receptacle; but in this they ſpring either directly, or with the parts of the corolla, from the calyx [o], which is of one leaf, and not flat but hollow: the corolla is moſt frequently of five petals.

[n] In Letter VII. [o] Plate 18. f. 1. c.

Of

Of the first order, *Cactus* is a very considerable genus, comprising the *Melon-thistles*, *Torch-thistles*, or *Cereuses*, and the *Opuntias* or *Indian Figs*. These all agree in a calyx, whole at the bottom, but yet consisting of several rows of leaves, and placed on the top of the germ: in a corolla which is double, or formed of several rows of petals: and in having a berry containing several seeds in one cell. Cactus.

The *Melon-thistles* are roundish bodies, without either leaf or stalk. The *Torch-thistles* have a long stem without leaves, which in many species is strong enough to support itself; but in some trails along the ground, or is supported by trees: these last are called *Creeping Cereuses*. *Opuntias* are composed of flat joints connected together.

These are all remarkable for a structure different from that of other plants; but some of the *Cereuses* are much esteemed for the beauty of the flowers, which are perhaps the more noticed, because they are the less expected from plants whose appearance is so unpromising. Those of the *Great-Flowering Creeping Cereus*[p] are near a foot in diameter, the inside of the calyx of a splendid yellow, and the numerous petals of a pure white: hardly any flower makes so magnificent an appearance during the short

[p] Cactus grandiflorus *Lin.* Mill. fig. pl. 90.

time

time of its duration, which is one night only; for it does not begin to open till ſeven or eight o'clock in the evening, and cloſes before ſun-riſe in the morning, unleſs it is gathered and kept in the ſhade, by which means I have prevented it from cloſing till about ten. This noble flower opens but once; but when, to the grandeur of its appearance, we add the fine perfume which it diffuſes, there is no plant that more deſerves your admiration. When it is not in blow, you will know it by the creeping ſtem, marked longitudinally with about five prominences.

Another ſpecies of *Creeping Cereus* [q] is more common, but ſcarcely leſs admirable for the beauty of its pink-coloured flowers, which the plant produces in greater quantity; they are alſo of longer duration, for they not only boldly ſhow their face to the ſun, but will even keep open three or four days. When it is not in flower, this ſpecies is diſtinguiſhed by its very ſlender branches, covered with ſpines, and marked with ten prominences. But you are well acquainted with this fine plant, which requiring little heat, forms one of the principal ornaments of your dreſſing-room, in the month of May.

There are many ſpecies of *Opuntia*, *Indian Fig*, or *Prickly Pear*, all natives of

[q] Cactus flagelliformis *Lin.* Ehret. pict. t. 2. Trew. Ehr. t. 30. Curtis Mag. 17.

America,

America, and kept rather for their ſingularity than their beauty, having no leaves, but a flat jointed ſtalk, ſet with knots of prickles, briſtles, or both. The *Cochineal Fig* [r], on which the inſect of that name feeds, is the only one that is unarmed: this has oblong joints; the common ſort [s] has roundiſh joints, with bruſhes of briſtles, but no prickles.

In this ſame order you will find the *Syringa* [t]. The natural number in the calyx, corolla, and capſule, is four; but ſometimes it is five. The taſte of the leaves like cucumbers, and the odour of its white flowers, like thoſe of the orange, ſufficiently diſtinguiſh this well known ſhrub from all others. The ſlight indentations about the edges of the leaf ſeparate it from another ſpecies, which has none. Philadelphus.

Here too will you find your favourite *Myrtle*, which has a calyx ſitting on the top of the germ, and generally cut into five ſegments; a corolla of five petals; and a berry for a fruit. Some ſpecies however have a quadrifid calyx, and then the corolla has four petals: others have an entire undivided calyx. The *Common Myrtle* [u], of which there are many varieties, has the Myrtus.

[r] Cactus cochinillifer *Lin.* Dill. elth. t. 297. f. 383.
[s] Cactus opuntia *Lin.* Mill. fig. t. 191.
[t] Philadelphus coronarius *Lin.* Duham arb. 83.
[u] Myrtus communis *Lin.* Mill. fig. 184.—Pl. 18. f. 1.

flowers coming out singly, and an *involucre* of two leaves upon the peduncle.

Cratægus. In the second order there is only the *Cratægus*, a genus comprehending several species of *Thorn*, and also two trees, the *Aria*, or *White Beam Tree* [v], and the *Maple-leaved Service* [w]. The generic characters are, a calyx cut into five segments, and sitting on the top of the germ; a corolla of five petals; and a berry containing two seeds. The first of the trees is readily known by the ovate shape of the leaves, with very prominent transverse veins, and unequal serratures about the edges; but particularly by the hoariness of their under surfaces: the second, by its leaves cut into many acute angles like those of the Maple; the divisions are five or seven; and the lowest lobes stand wider than the others. *Cockspur Hawthorn* [x] has the leaves ovate, and so deeply serrate, as to be almost lobate. *Virginian Azarole* [y] has oval leaves, wedge-shaped at the base, shining and deeply serrate. *Common Hawthorn*, or *White-thorn* [z], whose flower has obtained the name of

[v] Cratægus Aria *Lin.* Fl. dan. 302. Mill. illustr. Ger. 1327. 2. Hunt. Evel. silva. p. 173.

[w] Cratægus torminalis *Lin.* Ger. 1471. 2. Fl. dan. 798. Hunt. Evel. silva. p. 146.

[x] Cratægus coccinea *Lin.* Mill. fig. 179. Angl. hort. t. 13. f. 1.

[y] Crat. Crus-galli *Lin.* Mill. fig. 178. 2.

[z] Cr. Oxyacantha Jacqu. austr. 292. 1. Blackw. 149. 1. Ger. 1327. 1.

May,

May, from the month in which it appears, has obtuſe leaves, cut into three principal parts, and thoſe ſerrate. *True Azarole*[a] has leaves like the foregoing, but larger, paler, and with broad lobes: the flowers and fruit are alſo much larger. All theſe you will find in your plantations: as you will alſo two trees that are in the third order, under the genus *Sorbus*; viz. the *Mountain Aſh*[b] and the *Service*[c]; both which have pinnate or winged leaves, like the Aſh; ſmooth on both ſides in the firſt, but villous on the under ſurface in the ſecond; theſe alſo have the lobes broader, and not ſo much ſerrated. Their common characters are a quinquefid calyx, a pentapetalous corolla, and an inferior berry with three ſeeds. Sorbus.

The fourth Order (*Pentagynia*), beſides the *Apple*, *Pear*, and *Quince*, comprehended under one genus, *Pyrus*, has the *Medlar* with many other ſpecies of trees or ſhrubs in a ſecond[d]; and all the ſhrubs called *Spiræa*, in a third. Theſe genera agree in a quinquefid calyx, and a pentapetalous corolla; the germ is incloſed within the flower in the laſt; but is beneath it in the reſt:

[a] Cr. Azarolus *Lin.*

[b] Sorbus aucuparia *Lin.* Mill. illuſtr. Ger. 1473. Hunt. Evel. ſilva. p. 211.

[c] Sorbus domeſtica *Lin.* Edw. av. t. 211. Ger. 1471. 1.

[d] Meſpilus *Lin.*—germanica. Medlar. Ger. 1453. 1. Blackw. 154.

the fruit is the principal diſtinction; in *Pyrus* it is a *Pomum*—in *Meſpilus* a *Berry*—in *Spiræa* a ſet of *Capſules*.

Meſembryanthemum. This order boaſts a large and ſplendid genus of herbaceous ſucculent plants, called *Ficoides* or *Fig Marigolds* [e]. Fifty ſpecies all conſent in a quinquefid calyx on the top of the germ; a multifid corolla of narrow linear petals: and a fleſhy capſule divided into cells correſponding with the number of ſtyles, and containing many ſeeds. Though moſt of the ſpecies have five ſtyles, yet ſome have only four, and others have ten. This large genus is ſubdivided into three ſections, from the colour of the flowers, which being ſtriking and permanent, may here very well furniſh ſuch a diſtinction, though it is in moſt caſes a circumſtance not to be depended on. The corollas then, which are ſpecious, very large, and double, are in the firſt ſection *white*, in the ſecond *red*, and in the third *yellow*. The different forms of the ſucculent leaves afford, almoſt of themſelves, ſufficient ſpecific diſtinctions.

The moſt known ſpecies is that which is called *Diamond Ficoides*, or more commonly *Ice Plant* [f]. This has ovate, alternate, waving leaves, with white corollas; but it is chiefly regarded for the ſingularity of be-

[e] Meſembryanthemum *Lin.*

[f] Meſembryanthemum cryſtallinum *Lin.* Dill. elth. 180. f. 221. Bradl. ſucc. 5. t. 15. f. 48.

ing

ing covered with pellucid pimples, in the ſun appearing like cryſtalline bubbles. *Egyptian Kali* [g], eſteemed for making the beſt pot-aſh, is alſo of this genus; has alternate, roundiſh, obtuſe leaves, ciliate at the baſe, and white corollas.

Of the laſt order of this claſs the *Roſe* is a genus univerſally known; and, were it leſs ſo, would hold the firſt rank in the admiration of mankind. The diſtinctive characters are, a quinquefid calyx; a pentapetalous corolla; and a kind of pitcher-ſhaped, fleſhy berry, formed out of the calyx, terminated by the diviſions of it, and containing ſeveral oblong, rough ſeeds, growing to the calyx on every ſide. The ſpecies are diſtinguiſhed by the globoſe or ovate form of the fruit, by the ſituation of the ſpines on the different parts of the ſhrub, the infloreſcence, &c. The *Sweet-Briar* [h] has globoſe fruits beſet with crooked ſpines, and the leaves rubiginous or ruſty underneath. The *Dog-roſe* or *Wild-Briar* [i] has ovate fruit, but ſmooth, as are alſo the peduncles; the ſtalk however and the petioles are ſpinous, the petals are bluſh-coloured and bilobate, Roſa.

[g] Meſem. nodiflorum *Lin.* Mor. hiſt. ſ. 5. t. 33. f. 7. Several ſpecies of this beautiful genus are figured in Mr. Curtis's Magazine:—as M. dolabriforme in t. 32.—bicolorum 59.—pinnatifidum 67.—barbatum 70.—and many more in Dillenius's Hortus Elthamenſis.

[h] Roſa rubiginoſa *Lin.* Fl. dan. 870. Ger. 1269.

[i] Roſa canina *Lin.* Curt. Lond. 5. 34. Fl. dan. 555. Blackw. 8.

and there are two ciliate bractes, oppoſite each other, to every flower.

Fragaria. *Strawberry*, with all its various fruits, conſtituting only one ſpecies [k], is of this order. Here, though the corolla has only five petals, the calyx is cut into ten ſegments, alternately larger and ſmaller, and the ſeeds are diſperſed over the ſurface of a roundiſh, pulpy receptacle, vulgarly called a berry. Theſe are the generic characters. All the eatable Strawberries increaſe by runners; and by this circumſtance they are ſufficiently diſtinguiſhed from the barren ſort [l], which not only has a dry juiceleſs receptacle, but never throws out any of theſe runners.

THE CLASS POLYANDRIA.

The thirteenth claſs, *Polyandria*, has many ſtamens to the flowers [m] as well as the foregoing, but ſpringing from the receptacle along with the piſtil. Theſe two claſſes united would have formed too large a claſs for commodious examination; a difficulty to be avoided certainly in all caſes where we can; beſides, the plants contained in the one, are in general ſo different, both in their form and qualities, from thoſe of the other, that it would have been a pity to intermix beings ſo diſcord-

[k] Fragaria veſca. *Lin.* Mor. hiſt. ſ. 2. t. 19. f. 1. Ger. 997. Blackw. 77. 1.

[l] Fragaria ſterilis *Lin.* Curtis, Lond. III. 30. Ger. 998.

[m] From 20 to 1000.

ant,

ant, or to unite in the ſame claſs fruits which are ſo pleaſant to the palate, and wholeſome to the conſtitution, with herbs deſtructive to the human frame from their poiſonous qualities; as many of thoſe in the claſs *Polyandria* are known to be.

In the firſt order (*Monogynia*) you will find the *Poppy*, which is ſufficiently diſtinguiſhed by a calyx of two leaves [n]; a corolla of four petals; and a one-celled capſule, crowned with the ſtigma, under which it opens with many holes, to give exit to the numerous little ſeeds. Of this genus, four ſpecies have rough, and five have ſmooth capſules. The common *Corn Poppy* [o]; the ſpecies uſed in medicine, and which yields the Opium [p]; the Welch Poppy; and the Oriental ſort, now introduced as an ornament to the flower garden [q], are all of the latter diviſion. The firſt has the capſules almoſt globoſe; the ſtalk covered with hairs, and ſuſtaining ſeveral flowers of a fine high ſcarlet; and the leaves pinnatifid and cut. The ſecond has the calyx ſmooth, as well as the capſule, the leaves cut and embracing the ſtalk: that which is cultivated in the fields has white corollas, and oblate ſpheroidal Papaver.

[n] This falls off ſpontaneouſly when the flower expands.

[o] Papaver Rhæas *Lin.* Curtis, Lond. III. 32. Ger. 371. 1. Pl. 19. f. 2.

[p] Papaver ſomniferum *Lin.* Blackw. t. 483. Ger. 370.

[q] Papaver orientale *Lin.* Curt. Magaz. 57.

heads as big as an orange, with white ſeeds: the garden ſort has purpliſh corollas, very dark at the baſe, with ſmaller oblong heads and black ſeeds: this varies much in colour, and has ſometimes very large and very double flowers, then reſembling an immenſe Carnation. Some perſons are of opinion that the field and garden Poppy are different ſpecies; Linnæus makes them but one: I have given you the differences, but do not take upon me to decide. The capſules of the *Welch Poppy* [r] are oblong; the ſtalk ſmooth; the leaves winged and cut: the corollas large and yellow. The *Oriental Poppy* has rough leafy ſtalks, ſupporting one large, ſingle, red flower; the leaves are winged, and ſerrate about the edge. All the ſpecies of Poppy have a ſtrong diſagreeable ſmell.

The *Caper* [s] is of this firſt order; ſo is the *Tea-tree*, and the *Lime* [t]; the *Water-Lilies*, both *yellow* [u] and *white* [v], ſpreading their broad leaves on the ſurface of ſlow-moving ſtreams and ſtagnant pools, and raiſing their ample many-petalled corollas above it. Here alſo is the numerous and

Ciſtus. beautiful genus *Ciſtus*, known by a calyx of five leaves; two of which are leſs than

[r] Papaver cambricum *Lin.* Dill. elth. t. 223. f. 290.
[s] Capparis ſpinoſa *Lin.* Blackw. 417.
[t] Tilia Europæa *Lin.* Fl. dan. 553. Ger. 1483. Hunt. Ev. ſilva. p. 194.
[u] Nymphæa lutea *Lin.* Fl. dan. 603. Ger. 819. 2.
[v] Nymphæa alba *Lin.* Fl. dan. 602. Ger. 819. 1.

the

the other three; a corolla of five petals; and a capſule for a ſeed-veſſel. Of theſe there are forty-nine ſpecies, moſt of them ſhrubs, but ſome herbaceous; the corollas purple, white or yellow in the different ſorts.

Peony is of the ſecond order, which is a ſmall one: the characters of the genus are a calyx of five leaves, a corolla of five petals, and two or three germs, crowned immediately with ſtigmas, without the interpoſition of any ſtyles. Pœonia.

This, and ſome plants of the following orders, are ſtrictly united by one natural bond, under the name of *Multiſiliquæ* or *Many-podded*; having a fruit compoſed of ſeveral pericarps joined together. They agree likewiſe in having either no calyx, or at leaſt one very apt to fall off; a polypetalous corolla, and ſtamens exceeding the petals in number. Of theſe you are acquainted with the *Larkſpur* and *Aconite*, belonging to the third order; the *Columbines* to the fifth, and *Hellebore* to the laſt. None of them have any calyx; and they have all a corolla of five petals: the nectaries form the principal diſtinction of the genera [w]. This in *Larkſpur* is bifid, ſeſſile, and continued backwards into a horn or ſpur. *Aconite* has two recurved, pedunculate nectaries. *Columbine* has five of theſe

[w] See Pl. 34. f. 1, 2, 8.

horn-

horn-ſhaped nectaries, between the petals. *Hellebore* has many ſhort, tubulous nectaries, placed in a ring round the outſide of the ſtamens, each divided into two lips at top.

Delphinium. *Larkſpur* has alſo either one capſule or three, and the garden ſpecies [x] is diſtinguiſhed by its ſimple unbranched ſtem from the wild one [y], which has it ſubdivided: theſe both have the nectary of one leaf; in *Bee Larkſpur* [z] and the reſt it is of two.

Aconitum. *Aconite* has the upper petal arched; and three or five capſules. You have one ſpecies common in your flower-borders and plantations, with long ſpikes of large blue flowers, called *Monk's-hood* [a]; this is one of the ſpecies that have three capſules to a flower; and the leaves are multifid, with linear diviſions, broadeſt at top, and marked with a line running along them. *Wholeſome Wolfsbane* [b], as it is called, has five capſules, five ſtyles, and the flowers are ſulphur-coloured.

Aquilegia. *Columbine* has five diſtinct capſules: the common ſort [c] has bent nectaries: in its wild ſtate the flowers are blue, the petals ſhort, and the nectaries very prominent; in

[x] Delphinium Ajacis *Lin.* Ger. 1082.

[y] Delphinium Conſolida *Lin.* Fl. dan. 683. Ger. 1083. 5.

[z] Delphinium elatum *Lin.* Mill. fig. 250. f. 2.

[a] Aconitum Napellus *Lin.* Mill. illuſtr. Jacq. auſtr. 4. 381.

[b] Aconitum Anthora *Lin.* Mill. fig. pl. 12. Jacq. auſtr. 4. 382.

[c] Aquilegia vulgaris *Lin.* Fl. dan. 695. Mill. illuſtr. Ger. 1093, 1094.

the garden you obſerve not only a variety of colours, but that the petals are excluded, and the nectaries much multiplied. *Hellebore* has ſometimes more than five petals to the corolla: and always ſeveral capſules ſucceeding to each flower; theſe contain many round ſeeds, fixed to the ſuture of the capſule. The winter-flowering ſpecies, commonly called *winter Aconite* [d], is the only one that drops its petals; it bears one yellow flower ſitting on the leaf. *True Black Hellebore* or *Chriſtmas Roſe* [e] has one or two large white flowers upon a naked ſtalk, and fleſhy pedate leaves. *Stinking Black Hellebore* or *Bear's-foot* [f] ſuſtains many greeniſh flowers on one ſtalk, and pedate leaves on the ſtem, but none towards the root. This is not uncommonly wild, and you will find it flowering during winter under the trees in your plantations. Caution your poor neighbours againſt being too free in giving their children this plant againſt worms; for in too large a doſe it is certainly dangerous. Indeed all the herbs juſt now deſcribed are more or leſs poiſonous: *Aconite* is known to be highly ſo. Helleborus.

The laſt order of this claſs, *Polyandria*, contains alſo the *Tulip-tree*, which has a tri- Liriodendron.

[d] Helleborus hyemalis *Lin.* Curtis, bot. mag. 3.
[e] Helleborus niger *Lin.* Curtis, bot. mag. 8.
[f] Helleborus fœtidus *Lin.* Blackw. t. 57. Ger. 976. 4.

phyllous

phyllous calyx, ſix petals to the corolla, and many lance-ſhaped ſeeds lying one over another, and forming a ſort of *ſtrobile*. This tree is remarkable for the ſhape of its leaves, having the middle lobe of the three truncate, or cut tranſverſely at the end. The flowers are large and bell-ſhaped; the petals marked with green, yellow, and red ſpots [g]. Here alſo are the *Magnolias*, which have a calyx of three leaves like the laſt, but a corolla of nine petals; the fruit is a *ſtrobile* or ſcaly cone of bivalvular capſules, covering a club-ſhaped receptacle, each capſule containing a roundiſh ſeed, like a berry, hanging out by a thread. It is to be lamented that theſe fine trees, ſo beautiful both in leaf and flower, will not bear all the rigour of our climate.

Magnolia.

Anemone.

This order boaſts two numerous genera, much eſteemed among the floriſts—the *Anemone* and *Ranunculus*. The firſt has no calyx; a corolla of two or three rows, with three petals in each row: and many naked ſeeds, retaining each their ſtyle. You are now too far advanced in the ſcience, to need a caution againſt taking the fine flowers of your beds, upon which the gardener ſo much values himſelf, in order to examine the corolla of the *Anemone*; they are the children of art; not thoſe of nature, ſuch

[g] Liriodendron Tulipifera *Lin.* Trew, Ehr. t. 10. Cateſb. car. 1. t. 48.

as

as we are ſtudying. The early *Hepatica*[h] is of this genus; and is known by its three-lobed entire leaves. It is the only ſpecies which has any thing like a calyx; for it has a *perianth* of three leaves, which being remote from the flower, is rather an *involucre* than a calyx. The *Paſque-flower*[i], ſo called from its flowering about Eaſter, is alſo of this genus: it adorns ſome of our dry chalky hills with its beautiful bell-ſhaped, purple flowers; and though it has no calyx properly ſo called, yet the flower-ſtalk has a leafy multifid *involucre*; and the leaves are doubly winged, or *bipinnate*. Each plant bears but one nodding flower; and after that is paſt, the top of the plant is hoary with the tails, which adhere to the ſeeds. Another wild ſort is the *Wood Anemone*[k], bearing only one white or purpliſh flower on a plant; the leaves are compound, with cut lobes; and the ſeeds are pointed, but without tails. The garden Anemones, which are ſo ornamental to the flower-garden in the ſpring, are only of two ſpecies, notwithſtanding the great variety of their colours; red, white, purple, blue, with all the intermediate ſhades, and

[h] Anemone Hepatica *Lin.* Curtis, bot. mag. 10. Fl. dan. t. 610.

[i] Anemone Pulſatilla *Lin.* Relh. Fl. cantab. p. 208. Fl. dan. 153. Ger. 385. 1.

[k] Anemone nemoroſa *Lin.* Curtis, Lond. II. 38. Fl. dan. 549. Ger. 383. 2.

innumerable

innumerable variegations of them. Art, to increaſe their beauty, has rendered them very large and double; but we can ſtill diſtinguiſh the ſpecies by their leaves, which in one [l] are decompounded, dividing by threes; in the other [m] digitate: the ſtalk is leafy; and the ſeeds are tailed, in both ſpecies. The rival genus of the Anemone is the *Ranunculus*, which differs from it in having a calyx of five leaves, and a corolla of five petals: but the diſtinguiſhing mark of this genus is a honied gland juſt above the baſe of each petal, on the inſide [n]. Of forty-four ſpecies many are wild; and ſome extremely common in moſt parts of Europe, under the name of *Butter-flowers*, *Butter-cups*, and *King-cups*. Three ſorts particularly, which at one ſeaſon caſt a yellow veil over our meadows, are generally confounded and looked upon as one. However the *bulbous* [o] has the calyx bent back to the flower-ſtalk, whereas in the *creeping* [p] and *acrid* [q] it is open or ſpreading: in the firſt and ſecond the peduncle is furrowed; in the third it is round, without

Ranunculus.

[l] Anemone coronaria *Lin.* Mill. fig. pl. 31.

[m] Anemone hortenſis *Lin.* Curtis Magaz. 123.

[n] See Pl. 34. 4.

[o] Ranunculus bulboſus *Lin.* Curtis, Lond. I. 38. Ger. 953. 6.

[p] Ranunculus repens *Lin.* Curtis, Lond. IV. 38. Ger. 951. 1.

[q] Ranunculus acris *Lin.* Curtis, Lond. I. 39. Ger. 951. 2.

any

any channelling: beſides this, the leaves are very different upon inſpection; and the firſt has a bulbous root, the ſecond throws out abundance of runners which ſtrike root like thoſe of the ſtrawberry, and the third is a taller, genteeler, later-flowering plant. But not the meadows only are filled with *Ranunculi*; the woods [r], the corn-fields [s], the waters [t], have alſo their ſhare of them. One ſpecies, which flowers in moiſt meadows very early in the ſpring, is ſo diſtinct from its fellows, that ſome botaniſts have not ſcrupled to remove it from this genus, to form one by itſelf: for it has a calyx of three leaves only; but, to make amends, a corolla of more petals than five: it has heart-ſhaped, angular, petiolate leaves, one flower on a ſtalk, and tuberous or knobby roots [u]. But the *Perſian Ranunculus* [v] is the great rival of the Anemone, in the flower-garden, for the beauty and variety of the large, double corollas; which are ſo changed by art, that you muſt have recourſe, for

[r] Ranunculus auricomus *Lin.* Curtis, Lond. II. 41. Ger. 954. 7.

[s] Ranunculus arvenſis *Lin.* Fl. dan. 219. Ger. 951. 3.

[t] Ranunculus ſceleratus, hederaceus, aquatilis, &c. *Lin.*—ſceleratus Curtis, Lond. II. 42. Ger. 962. 4.—hederaceus, IV. 39. Fl. dan. 321,—aquatilis. Ger. 829. Fl. dan. 276.

[u] Ranunculus Ficaria *Lin.* Leſſer Celandine. Curtis, Lond. II. 39. Ger. 816.

[v] Ranunculus aſiaticus *Lin.* Mill. fig. 216.

the ſpecific diſtinction, to the leaves; theſe are ternate, and biternate, the lobes trifid and cut. The ſtalk is erect, round, hairy, and branching at bottom: the radical leaves are ſimple. With all this employment as a botaniſt, and amuſement as a floriſt, I leave you, dear couſin, for the preſent.

LETTER XXII.

THE CLASS DIDYNAMIA.

July the 1ſt, 1775.

HAVING now finiſhed more than half our courſe, we are arrived at a ſet of natural claſſes, with which you are ſo well acquainted, as to find no difficulty in aſſigning the proper place to any plant belonging to them.

The ſtructure of the flowers in the fourteenth claſs was explained at length in the fourth letter: but the proper and eſſential character of it is, the having four ſtamens, all in one row, and in pairs; the outer pair longer than the other, whence the name *Didynamia*; and one ſtyle: all included within an irregular monopetalous or ringent corolla.

This claſs has only two orders; which are not founded upon the form of the flower, as you might be led to ſuppoſe from what was ſaid in a former letter; nor upon the number of the ſtyles, as in the foregoing claſſes, becauſe none of the flowers have more than one; but upon the circumſtance of having four naked ſeeds, boſomed in the calyx; or elſe many fixed to a receptacle in the middle of a pericarp: the firſt of theſe is called *Gymnoſpermia*, the ſecond *Angioſpermia*.

This claſs contains one hundred and two genera, and ſix hundred and forty-three ſpecies; and each order forms a natural one—the firſt including the *Verticillate* plants, ſo called from the manner in which the flowers grow, in *verticilli* or *whorls*: they alſo agree in producing the leaves by pairs, and in having the ſtalks ſquare. The ſecond compriſing the *Perſonate* flowers; or ſuch as have moſtly a perſonate corolla, but always a pericarp, or veſſel incloſing the ſeeds.

THE ORDER GYMNOSPERMIA.

Glechoma.

The eſſential generic character of *Ground Ivy* [w] is at the ſame time beautiful and extremely diſtinctive, each pair of anthers forming an elegant little croſs, one above the other. The leaves are kidney-ſhaped, and notched about the edges. In this genus, in Hyſſop, Mint, Lavender, Bugle, Betony, Dead-Nettle, Cat-Mint, Savory, Horehound, &c. the calyxes are pretty regularly quinquefid. In Thyme, Baſil, Self-heal, Marjoram, Baum, &c. they are bilabiate. In *Mint* the corollas are hardly ringent; the filaments are ſtraight and diſtant. *Lavender* has the corollas, as it were,

[w] Glechoma hederacea *Lin.* Curtis, Lond. II. 44. Ger. 856. 1. Pl. 20. f. 1. of this work.

turned

turned *topſyturvy*; that which is the upper part in moſt others being the lower in this, and *vice verſa*; the calyxes alſo are ſupported by a *bracte*; and the ſtamens lie within the tube. *Teucrium* has no proper upper lip, but the corolla is ſlit quite through for the ſtamens to paſs. *Bugle* has Ajuga. the upper lip of the corolla remarkably ſhort, much ſhorter than the filaments; our common wild ſpecies [x] is known by its ſmoothneſs, and increaſing by runners. *Betony* has Betonica. the upper lip of the corolla flattiſh and riſing, with a cylindric tube; the ſegments of the calyx are prolonged into narrow thin points like awns; and the filaments extend not beyond the neck or opening of the tube. *Wood Betony* [y] is diſtinguiſhed by an interrupted ſpike, and by the middle ſegment of the lip being emarginate, or having one notch. *Cat-mint* has the middle divi- Nepeta. ſion of the lower lip crenate, or ſlightly notched; the edge of the chaps reflex; and the ſtamens cloſe. The flowers of the wild ſpecies [z] are in a ſpike, conſiſting of a ſet of whorls on ſhort peduncles; the leaves are heart-ſhaped, bluntly ſerrate and petiolate. If you have any doubt concerning this

[x] Ajuga reptans *Lin.* Curtis, Lond. II. 43. Ger. 631. 1.

[y] Betonica officinalis *Lin.* Curtis, Lond. III. 32. Ger. 714.

[z] Nepeta Cataria *Lin.* Fl. dan. 580. Mor. hiſt. ſ. 11. t. 6. f. 1. Ger. 682. 1.

plant preſent it to puſs, and ſhe will inform you by the careſſes which ſhe beſtows upon it, in common with Marum and Valerian; the firſt of which not growing wild, and the ſecond being ſo very different a plant, ſhe cannot lead you into an error. *Black Horehound* and *White Horehound* both have a calyx marked with ten ſtreaks; but the upper lip of the corolla, in the former, is arched and crenate: in the latter ſtraight, linear, and bifid. *Common Black Horehound*[a] is known by its whole, heart-ſhaped, ſerrate leaves, and ſharp-pointed calyxes: the corollas are red. *Common White Horehound*[b] has the diviſions of the calyx ending in ſetaceous hooked points: the corollas are white, and the whole plant has a white appearance from the nap that covers the ſtalks and leaves.

Ballota.

Marrubium.

Thymus.

Of the ſecond diviſion with bilabiate calyxes, *Thyme* has the opening of the tube cloſed with hairs. *Wild Thyme*[c] that ſmells ſo gratefully, and adorns dry ſheep-paſtures with its red flowers, is known by theſe flowers growing in a head; by the diviſions of the calyx being ciliate; the leaves ovate, flat, blunt at the end, dotted with little

[a] Ballota nigra *Lin.* Blackw. 136. Mor. hiſt. ſ. 11. t. 9. f. 14. Ger. 701. 1.

[b] Marrubium album *Lin.* Blackw. 479. Moris, t. 9. f. 1. Ger. 693. 1.

[c] Thymus Serpyllum *Lin.* Curtis, Lond. II. 47. Mor. hiſt. t. 17. f. 1.

glands,

glands, and ciliate at the baſe; and by its creeping ſtalks. *Garden Thyme*[d] is an erect plant, with its ovate leaves revolute, and the flowers in a ſet of whorls, all together making a ſpike. Of this there are ſeveral varieties, as there are alſo of the other. *Baſil* has an *involucre* of many narrow leaves immediately under the whorl of flowers. *Marjoram* is diſtinguiſhed by an *involucre* compoſed of ovate, coloured, imbricate *bractes*, forming all together a ſquare kind of ſpike or *ſtrobile*. *Wild Marjoram*[e] has the ſpikes rounded at the corners, conglomerate, and all together forming a panicle; the bractes longer than the calyxes. You will find this wild under hedges, and among buſhes. That which is in the kitchen garden, under the name of *Pot Marjoram*[f], differs not greatly from the next: the ſpikes are oblong, aggregate, and hairy; the leaves heart-ſhaped, and nappy; the ſtem woody, and the flowers white. *Sweet Marjoram*[g] has ovate leaves, blunt at the end, and roundiſh compact pubeſcent ſpikes. *Winter Sweet Marjoram*[h] has long, aggregate, pedunculate ſpikes, and the bractes the length of the calyxes. The corollas of this are Origanum.

[d] Thymus vulgaris *Lin.* Blackw. t. 211.

[e] Origanum vulgare *Lin.* Curt. Lond. 5. 39. Fl. dan. 638. Mor. hiſt. ſ. 11. t. 3. f. 12. Ger. 666. 4.

[f] O. Onites. Bocc. mus. 2. t. 38. Ger. 664. 2.

[g] Origanum Majorana *Lin.* Blackw. t. 319.

[h] Origanum heracleoticum *Lin.* Lob. ic. 492.

white; of the other red. *Dittany of Crete*[i] has the ſmall purple flowers collected in looſe, nodding heads, with imbricate bractes; the ſtalks are pubeſcent, purpliſh, and ſend out ſmall branches from their ſides by pairs; the leaves are round, thick, and ſo woolly as to be quite white: the whole plant has a piercing aromatic ſcent, and biting taſte. This is the celebrated plant with which Venus cured the wound of Æneas[k]. *Baum* has a dry, chaffy, angular calyx, flattiſh at top; the upper lip riſing: the caſque of the corolla is a little arched, and deeply notched or bifid; the lower lip is trifid, with the middle lobe heart-ſhaped.

Meliſſa. *Common Garden Baum*[l] has the flowers growing in ſmall looſe bunches from the wings of the ſtalk, in whorls, and the pedicles are ſimple or unbranched. There are two plants of this genus growing wild, that have the name of *Calamint*[m]. *Dracocephalum* is diſtinguiſhed principally by the great inflation, or wide opening of the chaps of the corolla, the upper lip alſo is arched, folded, and obtuſe. Of this genus is the very fine-ſmelling plant vulgarly called *Baum of Gilead*[n], which has compound

Dracocephalum.

[i] Origanum Dictamnus *Lin.* Blackw. t. 462.

[k] Virgil Æneid. XII.

[l] Meliſſa officinalis *Lin.* Blackw. t. 27.

[m] Meliſſa Calamintha & Nepeta *Lin.* Blackw. t. 166, & 167.

[n] Dracocephalum canarienſe *Lin.* Mor. hiſt. ſ. 11. t. 11. fig. laſt.

leaves,

leaves, confifting of three or five oblong, pointed, ferrate leaflets; and flowers coming out in thick, fhort fpikes: the corollas are pale blue. *Self-heal* is known immediately by its forked filaments, with the anthers inferted below the top: the ftigma alfo is emarginate or bifid. *Wild Self-heal*[o], fo common in paftures, has all the leaves of an oblong ovate form, ferrate about the edge, and petiolate. *Scutellaria* is abundantly diftinct from all the other genera of this order by its fructification; for the calyx is entire at the mouth, and after the flower is paft, clofes with a kind of lid; fo that the whole bears a refemblance to a helmet, whence the names of *Caffida*, *Skull-cap*, and *Hooded Willow-herb:* and the feeds being hereby inclofed in a kind of capfule, this genus forms the connecting link between this order and the next. The fpecies common on the banks of rivers, by ditch fides, and other watery places[p], has lance-fhaped leaves, hollowed at the bafe, notched about the edge, and wrinkled on the furface; the flowers are blue, and proceed from the axils, or angles formed by the leaves or fubdivifions of the main ftem.

Prunella.

Scutellaria.

[o] Prunella vulgaris *Lin.* Curtis, Lond. IV. 42. Ger. 632. 1.

[p] Scutellaria galericulata *Lin.* Curtis, Lond. III. 36. Ger. 477. 10.

THE ORDER ANGIOSPERMIA.

The corollas in all the genera of the firſt order, with very few exceptions, are open-mouthed, *Labiate*, or *Ringent*, properly ſo called. In the ſecond order, which you are now going to ſurvey, many of them are *Perſonate*, or Labiate, with the lips cloſed; ſome however have open bell-ſhaped, wheel-ſhaped, or irregular corollas. To have ſeeds incloſed in a *Pericarp* is common to all, and hence the name of the order *Angioſpermia*. In moſt of the genera the calyxes are quinquefid; in ſome however they are bifid, in one trifid, in many quadrifid, and in two multifid.

Orobanche. Of thoſe with bifid calyxes, you have the *Orobanche* or *Broom rape*; which has an open corolla, divided at top into four ſegments, and nearly regular; there is a gland at the baſe of the germ; and the capſule is unilocular and bivalvular. The common ſpecies [q] has a pubeſcent ſtalk, abſolutely undivided; the ſingular *feuillemort* hue of this plant is alone ſufficient to betray it to you at firſt ſight.

Rhinanthus. Among ſuch as have quadrifid calyxes, are *Rhinanthus*, *Yellow Rattle*, or *Cock's-comb*, and *Eyebright*: theſe have *Perſonate*

[q] Orobanche major *Lin.* Curtis, Lond. IV. 44. Ger. 1311. 2.

corollas:

corollas: the firſt has the calyx ſwelling; and an obtuſe, compreſſed bilocular capſule. The wild ſort[r], common in moiſt meadows, is known by the ſhortneſs and compreſſed form of the upper lip of the corolla; the colour is yellow: the calyx is very large, being an early flowering plant; this part grows dry before the time of mowing, and makes a craſhing or ratling ſound under the ſcythe. *Euphraſy*, or *Eyebright*, once celebrated as fit "to purge the viſual ray," has the calyx cylindric; the anthers ſpinous at the baſe of one of their lobes: and the capſules of an oblong ovate form, and bilocular. The officinal ſpecies[s] has ovato-linear leaves, ſharply indented about the edges. It is an humble, neat plant, growing in dry paſtures and heaths; and the corolla, on a near view, is very elegant. Euphraſia.

In the largeſt ſection, with quinquefid calyxes, you will find the *Antirrhinum* genus comprifing forty-ſeven ſpecies. The corolla is perſonate, prolonged at the baſe into a bag or ſpur; and the ſeed veſſel is a bilocular capſule. Of two ſpecies formerly mentioned to you, *Toadflax*[t] has linear leaves inclining to lanceolate, growing many toge- Antirrhinum.

[r] Rhinanthus Criſta galli *Lin.* Curtis, Lond. V. 43. Mor. hiſt. ſ. 11. t. 23. f. 1. Ger. 1071. 1.

[s] Euphraſia officinalis *Lin.* Curtis, Lond. V. 42. Mort. hiſt. t. 24. f. 1. Ger. 663.

[t] Antirrhinum Linaria *Lin.* Curtis, Lond. I. 47. Ger. 550. 1.

 ther

ther upon an erect ſtalk ; the flowers grow cloſe in ſeſſile ſpikes, terminating the ſtem ; the under lip of the corolla is hairy within, the chaps are orange-coloured, but the reſt is of a pale yellow, and it ends in a long ſpur. It is now in flower, or will ſoon be ſo. Accident has produced a ſtrange variation in this plant, by changing the corolla from perſonate with four didynamous ſtamens, to regularly pentapetalous with five, the reſt of the plant remaining the ſame [u]. Varieties partaking of the nature of two ſpecies are not uncommon [v], but as they are generally found among annual plants, and never produce ſeed, they are loſt almoſt as ſoon as they come to perfection. Whereas this being perennial, and creeping much at the roots, has been preſerved as an example of monſters in vegetable nature. *Snapdragon* [w] has the leaves of the calyx rounded at top, the flowers growing in a ſpike, and the corollas ſpurleſs ; the colours of theſe are red with white or yellow mouths, or entirely white, or elſe white with yellow mouths : the leaves are lance-ſhaped and petiolate. Several ſpecies of *Antirrhinum* are wild on walls and in corn fields ; and ſeveral others are not uncommon in gardens :

[u] This is deſcribed at length under the name of *Peloria* in the firſt volume of *Amœn. Acad.*

[v] Theſe are called *Hybridous* plants, or *Mules.*

[w] Antirrhinum majus *Lin.* Mill. fig. t. 42. Ger. 549. 1, 2, 3.

as

as *Three-leaved Toadflax*[x], an annual plant, having ovate, ſmooth, gray leaves, generally ternate, as the name implies, but ſometimes only in pairs: the flowers grow in ſhort ſpikes at the top of the ſtalks, and are ſhaped like thoſe of common Toad-flax, only the tubes are not ſo long; they are yellow, with ſaffron-coloured chaps. Two or three perennial ſpecies, with handſome ſpikes of blue flowers, and ſome of them ſmelling ſweet[y], are uſually in large borders, among flowering-ſhrubs, and other perennials.

Scrophularia or *Figwort* is another of these; the corolla is of the topſyturvy kind, almoſt globular in its form; the two upper diviſions are the largeſt and erect; the two ſide-ones ſpread open, and the fifth below is reflex. In many ſpecies, under the topmoſt diviſion, in the chaps of the corolla there is a little flap reſembling a lip: the flower is ſucceeded by a bilocular capſule. Two ſpecies are ſufficiently common; one in woods and hedge-rows[z], with the angles of the ſtem blunted, and heart-ſhaped leaves, much prolonged at the tip, and marked with three riſing nerves: the other by river Scrophularia.

[x] Antirrhinum triphyllum *Lin.* Bocc. ſic. t. 22.

[y] Antirrhinum purpureum, repens & monſpeſſulanum, &c. *Lin.* 1. Riv. mon. 82.—2 Dill. elth. 198. t. 163. f. 197.—3. Dill. elth. 199.

[z] Scrophularia nodoſa *Lin.* Blackw. t. 87. Mor. hiſt. ſ. 5. t. 8. f. 3. Ger. 716. 1.

ſides,

ſides, and in other watery places[a], with a membrane running along the ſtalk at the angles, and heart-ſhaped leaves blunted at the end. Theſe plants have a duſky ſhade ſpread over their green, and their flowers are of a dull red.

Digitalis. *Foxglove*, one of the moſt ſhowy of our wild plants, has an open corolla, divided into four ſegments at top, and ſwelling out below, ſhaped like the fingers of a glove; the capſule ovate and two celled. *Wild* or *purple Foxglove*[b] is diſtinguiſhed by having the leaves of the calyx ovate and acute, with the ſegments of the corolla obtuſe, and the upper lip entire: the inſide of the corolla is beautifully ſprinkled with ſpots reſembling eyes; and the leaves are large and wrinkled: red is the colour of the flower in its wild ſtate; when cultivated in gardens it varies to white and yellow.

Bignonia. *Bignonia* has a *cyathiform* calyx, narrow at bottom, and ſpreading wide at top; a bell-ſhaped corolla, ſwelling out below, and divided into five ſegments at top; and a two-celled *ſilique* for a ſeed-veſſel, containing winged ſeeds lying cloſe over each other. The *Trumpet-flower*[c] of Virginia and Canada, with its trailing branches, putting

[a] Scrophularia aquatica *Lin.* Curt. Lond. V. 44. Fl. dan. 507. Blackw. t. 86. Ger. 715.

[b] Digitalis purpurea *Lin.* Curtis, Lond. I. 48. Fl. dan. t. 74. Ger. 790. 1.

[c] Bignonia radicans *Lin.* Mill. fig. pl. 65. Pl. 20. f. 2.

out

out roots from the joints, to acquire ſupport and nouriſhment from trees, has pinnate leaves, the leaflets of which are cut: the large trumpet-ſhaped flowers are orange coloured. The *Catalpa*[d] is a large tree with leaves remarkably ſimple, and heart-ſhaped: the flowers are produced in great branching panicles; they are of a dirty white, with a few purple ſpots, and faint ſtripes of yellow; but, what is moſt remarkable, they have only two perfect ſtamens, with ſmall rudiments of three others; the calyx alſo is not barely quinquefid, but divided quite to the bottom.

Acanthus, the leaves of which are ſaid to have given the firſt hint of the elegant Corinthian capital, is alſo of this order, but of that ſection which has bifid calyxes: it has an irregular corolla, without any upper lip; the lower one has three lobes; the anthers are villous, and the capſule is two-celled. Acanthus.

I cannot help remarking to you, ſince it has ſtruck me, that the greater part of the genera in the principal ſection of this order, is dedicated to the memory of eminent botaniſts. Here ſtands the great Linnæus himſelf; the celebrated Arabian Avicenna; thoſe fathers of the ſcience Geſner and Columna: in Italy, Creſcentio, Tozzi, Vandelli, Durante, Cirillo; the illuſtrious Frenchmen,

[d] Bignonia Catalpa *Lin.* Duham. arb. 1. t. 41. Cateſb. car. 1. t. 49.

Bignon,

Bignon, Barrelier, Ruellius, Cornutus, Dodart; Celſius, Toren, Brovall, Swedes; Brunfelſius, Buchner, Bontius, Volkamer, Loeſel, Beſler, Hebenſtreit, Lindern, Gmelin, and other Germans; Oviedo the Spaniard; and of England old venerable Gerard, Millington, and in more modern times, Lord Petre and two contemporary profeſſors of Oxford and Cambridge. The illuſtrious, the indefatigable Baron Haller, occupies a ſection alone, as he well merits, being himſelf an hoſt. This plan, of conſecrating newly diſcovered plants to perpetuate the memory of perſons who have been eminent in the ſcience, appears to me well imagined. Ladies have had this honour[e], as well as the men; and I have no doubt, dear couſin, but that you will one day merit a nich in this temple.

[e] See Strelitzia Reginæ in Hort. Kew. 1. 285. Curt. magaz. 119, 120. John Miller's plates, t. 5, 6. Portlandia grandiflora in Dr. Smith's Icones pictæ. Monſonia ſpecioſa. Curt. magaz. 73.

LETTER XXIII.

THE CLASS TETRADYNAMIA.

August the 4th, 1775.

BEFORE any idea of ſyſtem or arrangement had gone abroad, every ſcientific eye perceiving a ſimilitude between the Cabbage and Turnip, the Stock and Radiſh in the fructification, there was an univerſal agreement among authors to place theſe plants, and others like them, in the ſame ſection or diviſion of their books, and to treat of them all together. You have already ſeen[f] the nature of this ſimilitude, and are not at any loſs in claſſing the *Cruciform* tribe: you have only to learn that the fifteenth claſs (Tetradynamia) in the ſyſtem of Linnæus, contains the ſame plants as you have been accuſtomed to call *Cruciform*; and to recollect that it has the long Greek name from four of the ſtamens being more powerful or longer than the remaining two; the circumſtance on which Linnæus founds the character of the claſs; and which diſtinguiſhes it from the ſixth, wherein the ſix ſtamens are of equal length, or at leaſt not of that regular, proportional inequality that we diſcover in the claſs now before you.

[f] In Letter II.

It

It will ſuffice to examine a few of the genera and ſpecies, which are not extremely numerous [g], and therefore my preſent letter will not extend to that frightful length that ſome of the former have done.

THE ORDER SILICULOSA.

The *Siliculoſe* or ſhort-podded order leads the way, and is ſubdivided into two ſections; the firſt containing thoſe which have the ſilicle entire, and the ſecond ſuch as have the ſilicle notched at top. From the firſt ſubdi-

Lunaria. viſion I ſhall ſelect *Honeſty* for your obſervation, becauſe it is common in gardens, and has larger parts than moſt of theſe flowers. The ſilicle is oval, entire, quite flat, and ſtands on a pedicle; the valves are equal to the partition, parallel and flat: the leaflets of the calyx are bagged. The brilliant whiteneſs of theſe ſilicles has occaſioned this plant to be called *White Sattin*: and from the ſhape of them it is named *Lunaria* and *Moonwort*. Linnæus mentions but two ſpecies; the *annual* [h] differing from the *biennial* [i] in having larger flowers of a lighter purple, and the pods longer and narrower: they have both heart-ſhaped leaves, indented on their edges, are a little hairy, and end in

[g] The genera are 32, and the ſpecies 287.

[h] Lunaria annua *Lin.* Mill. illuſtr. Beſl. eyſt. 7. f. 1.

[i] Lunaria rediviva *Lin.* Beſl. eyſt. 7. f. 2.

acute

acute points: the lower ones are on long petioles, but the upper ones ſit cloſe to the ſtalk.

Of the ſecond ſubdiviſion is the *Candy-tuft* or *Iberis*, known by an irregular corolla with the two outer petals larger than the two others. *Red Candy-tuft*[k] is an annual herbaceous plant with red flowers growing in a kind of umbel; your gardener ſows it in patches about the borders of your flower garden; it has lance-ſhaped leaves drawn to a point: the lower ſerrate, the upper ones quite entire: the flowers of this are ſometimes white, and then it is confounded with the bitter ſpecies[l], which however has the leaves not ſo ſharp-pointed, and with only few indentations: the flowers alſo grow in a raceme, and the plant is more branched. Iberis.

In this ſubdiviſion alſo ranges *Scurvy-graſs* and *Horſe-radiſh*, agreeing in a heart-ſhaped, turgid, rugged ſilicle, the valves of which are gibbous and obtuſe. *Officinal* or *Garden Scurvy-Graſs*[m] has a branching ſtalk; the lower leaves roundiſh and hollowed next the petiole; the ſtem-leaves oblong and ſubſinuous: the white flowers are produced in cluſters at the ends of the branches. *Engliſh* Cochlearia.

[k] Iberis umbellata *Lin.* Riv. tetr. 225. Curt. mag. 106.

[l] Iberis amara *Lin.* Riv. tetr. 112. Ger. 263. 5.

[m] Cochlearia officinalis *Lin.* Fl. dan. 135. Ger. 401. 1.

Sea Scurvy-Graſs[n] has longer leaves, and all of them ſinuate. *Horſe-radiſh*[o], which few beſides botaniſts obſerve in flower, has the radical leaves lance-ſhaped, and notched about the edges, the ſtem-leaves gaſhed.

THE ORDER SILIQUOSA.

The ſecond order, containing the *Cruciform* flowers, ſucceeded by a *ſilique* or long pod, is alſo ſubdivided into two ſections; in the firſt of which the leaflets converge at top, in the ſecond they gape. *Radiſh*, *Eryſimum*, *Stock*, *Wall-flower*, *Rocket*, *Arabis*, *Cabbage*, *Turnep*, &c. range in the firſt ſection: *Woad*, *Sea-Colewort*, *Cardamine*, *Muſtard*, *Charlock*, *Water-Creſs*, &c. in the ſecond. Raphanus. *Radiſh* has a cylindric, jointed, toroſe or ſwelling ſilique; and one pair of glands between the ſhorter ſtamens and the piſtil, with a ſecond pair between the longer ſtamens and the calyx. Eryſimum. *Eryſimum* has a columnar ſilique with four equal ſides. Of this there are ſeveral wild ſpecies: as firſt, the *common*[p], growing by road ſides, well diſtinguiſhed by its runcinate leaves, and ſiliques preſſed cloſe to

n Cochlearia anglica *Lin.* Fl. dan. 329. Ger. 401. 2.

o Cochlearia Armoracia *Lin.* Mor. hiſt. ſ. 3. t. 7. f. 2. Ger. 241. 1.

p Eryſimum officinale *Lin.* Curtis, Lond. V. 50. Fl. dan. 560. Ger. 254. 1.

to the ſtalk: ſecondly, *Winter Creſs*[q] with lyrate leaves, the outmoſt lobe roundiſh; and ſpikes of yellow flowers, growing by ditch-ſides: and thirdly, the *garlick-ſmelling*, call-ed thence *Sauce-alone*, and from the uſual place of its growth, *Jack-by-the-hedge*[r], has heart-ſhaped leaves: the flowers are white, but the ſmell betrays it at once.

Stock and *Wall-flower* have two leaflets of the calyx gibbous at the baſe; the germ has a glandular toothlet on each ſide; and the ſeeds are flat. The two ſpecies are thus diſtinguiſhed. *Wall-flower*[s] has acute, ſmooth leaves, with angular branches. *Stock*[t] has obtuſe hoary leaves, with flatted ſiliques truncate at top: both have ſhrubby ſtems, and lance-ſhaped entire leaves. The *Annual* or *Ten-week Stock*[u] differs in having an herbaceous ſtalk, the leaves ſomewhat toothed, the petals notched, and the ſiliques cylindric and acute at the end. *Rocket*[v] has the petals obliquely bent; a gland on each ſide within the ſhorter ſtamens; the ſtigma forked, with the parts converging at top; and the ſilique ſtiff and upright. Cheiranthus.

[q] Eryſimum Barbarea *Lin.* Mor. hiſt. t. 5. f. 11, 12. Ger. 243.

[r] Eryſimum Alliaria *Lin.* Curtis, Lond. II. 48. Ger. 794.

[s] Cheiranthus Cheiri *Lin.* Mor. ſ. 3. t. 8. f. 15. Ger. 456.

[t] Cheiranthus incanus *Lin.* Mill. illuſtr. Ger. 458.

[u] Cheiranthus annuus *Lin.*

[v] Heſperis *Lin.*

Arabis. *Arabis* has four glands, within the leaflets of the calyx, like reflex ſcales. Some of the ſpecies are wild[w], and the *Alpine* ſort[x] is now common in many gardens: the leaves of this embrace the ſtalk, and are toothed about the edges; it bears white

Braſſica. flowers in looſe *corymbs*. *Cabbage*[y], *Turnep*[z], *Coleſeed*[a], &c. All agree in having the glands diſpoſed as in the radiſh; the leaflets of the calyx are erect: the claws of the corollas hardly ſo long as the calyx; the ſilique is roundiſh, a little flatted on each ſide, with the valves ſhorter than the partition; and filled with ſeveral globoſe ſeeds.

Iſatis. Of the ſecond ſection, *Woad* has a lance-ſhaped, bivalve, one-celled ſilique, containing one ſeed only, and deciduous; the valves are boat-ſhaped. The ſpecies cultivated for dying[b], has the radical leaves notched and petiolate; the ſtem-leaves ſagittate or ſhaped like the head of an arrow, and embracing the ſtalk; and oblong ſilicles. It is a large plant, with corymbs of ſmall

Crambe. yellow flowers. *Sea-Colewort* has a globoſe ſilique, or rather dry berry, which is deciduous, and contains one ſeed; but its moſt

[w] Arabis thaliana, Curtis, Lond. II. 49. ſtricta, Turrita *Lin.* Jacq. auſtr. t. 11. but the laſt has glands as in *Braſſica*.

[x] Arabis alpina *Lin.* Fl. dan. 62.

[y] Braſſica oleracea *Lin.*

[z] Braſſica Rapa *Lin.*

[a] Br. Napus *Lin.*

[b] Iſatis tinctoria *Lin.* Blackw. 246. Mor. hiſt. ſ. 3. t. 15. f. 10, 11. Ger. 491.

4 remarkable

remarkable character is, that the four long filaments are forked at the end, and the anthers are borne on the outer forks. Our species[c] has the stalk and leaves smooth.

Cardamine, Cuckow-flower or *Lady's Smock*, (forgive the vulgar name) has the calyx gaping a little: two glands, one on each side, between the shorter stamens and the calyx; and an elastic silique, the valves rolling back with force when the seeds are mature, and thus throwing them off to some distance. There are many species wild, but that which is common in moist meadows, and on the banks of brooks[d], has pinnate leaves, the folioles on the radical leaves roundish, on the stem-leaves lance-shaped. The allusions to the whiteness of the corollas will not always hold, since in some countries they are purple. Cardamine.

Mustard has the claws of the corollas straight, and the glands as in the Cabbage genus, to which it is very nearly allied; differing from it only in the circumstance first mentioned, and in having the leaflets of the calyx spreading: the silique indeed is different; being torose and rough, with the partition usually very long; but this is reserved for the specific distinction. The wild species, a weed so common among corn, Sinapis.

[c] Crambe maritima *Lin.* Fl. dan. 316. Ger. 315. 15.

[d] Cardamine pratensis *Lin.* Curtis, Lond. III. 40. Ger. 259. 1, 2.

and generally called *Charlock*[e], has many angled, torose, smooth siliques, longer than the two-edged beak. *Black* or *common Mustard*[f] has smooth siliques pressed to the raceme, or common bunch of the fructification. *White Mustard*[g] has the siliques hispid, terminated by a very long, oblique, sword-shaped beak. If you suffer some of the plants which your gardener sows for small sallad to grow up and flower, you will find it to be the last named species. Common Mustard is a much larger plant, growing four or five feet high; the lower leaves large and rough, like those of the Turnep. Charlock does not grow more than two feet in height; the leaves, which are also rough, are sometimes jagged, and sometimes entire.

Sisymbrium. *Water-Cress* is of a numerous genus, there being twenty-nine species of *Sisymbrium*. The corolla is spreading as well as the calyx in this genus; and the silique gapes with straightish valves. The specific characters of *Water Cress*[h] are, short, declining siliques, and pinnate leaves, with the lobes a little heart-shaped. The flowers are

[e] Sinapis arvensis *Lin.* Curtis, Lond. V. 47. Fl. dan. 753. Mor. hist. f. 3. t. 3. f. 7. Ger. 233. 2.

[f] Sinapis nigra *Lin.* Blackw t. 446.

[g] Sinapis alba *Lin.* Curtis, Lond. V. 46. Blackw. 29. Ger. 244. 4.

[h] Sisymbrium Nasturtium *Lin.* Curtis, Lond. II. 61. Fl. dan. 690. Ger. 257. 1. and pl. 21.

white,

white, and grow in a corymb[i]. There is another ſpecies, called *Flixweed*[k], not uncommon on dunghills, where rubbiſh is thrown out, by way-ſides, and in uncultivated places: this has decompound pinnate leaves, and very ſmall corollas, the petals being leſs than the calyx: the ſilique is very long and ſlender, filled with ſmall, roundiſh ſeeds: the leaves are as finely cut as Roman Wormwood; and the ſmall yellow flowers are produced on looſe corymbs, at the top of the ſtalks.

The ſeaſon, dear couſin, is now in its wane, and a journey I muſt make on affairs of buſineſs, obliges me to leave the completion of my plan to another ſummer. If leiſure and health are then granted me, I ſhall with pleaſure reſume the employment which you honour with your attention. In the mean time you and your fair daughter have enough to amuſe you for the autumn, and even till winter confines you to the arrangement of your ſummer's labours within.

i See more in Letter XVII.
k Siſymbrium Sophia *Lin.* Fl. dan. 528. Ger. 1068.

LETTER XXIV.

THE CLASS MONADELPHIA.

June the 1ſt, 1776.

SOME neceſſary occupations, dear couſin, have prevented me from reſuming my pleaſing taſk ſo ſoon as I had wiſhed. But the ſpring has not been unprofitably employed by you, in the examination of ſuch plants as were paſt flowering, before you received my former letters. You have done well by marking in your pocket-book the names of all thoſe which have either wholly eſcaped your ſearch, or have preſented themſelves to you in a ſtate unfit for complete examination. You are not ſo unreaſonable as to expect that all Nature ſhould be open to your view at once. On the contrary, I am charmed with your patience and aſſiduity in awaiting the proper ſeaſons of flowering and fruiting; marking the times which authors have ſet down; and repeating your examinations in order to view plants in their different ſtates, when they ſometimes put on appearances ſo different, that to a leſs informed eye they might ſeem to be diſtinct ſpecies.

We are now arrived at a claſs, of which you have had no previous information in the introductory letters, deſigned to give you

you a general knowledge of the moſt natural. The claſs *Monadelphia* however is a natural, as well as a moſt beautiful one. The union of the filaments at bottom into one body, or brotherhood as it were, is the leading character, and the occaſion of the name. You will recollect that hitherto the ſtamens have been ever free and diſtinct from each other, how many ſoever you may have found in a ſingle flower; you will alſo recollect having been informed, that in the ſixteenth and ſucceeding claſſes, they are united, either at top or bottom, into one body or more. In this, as I obſerved before, the filaments all join below, next the receptacle, ſome higher than others; all of them, together with the anthers, being ſtill entirely ſeparate at top.

If then you have met with a plant which has five, ten, or eſpecially many ſtamens, and you have not been able to aſſign it a place in the fifth, tenth, or thirteenth claſſes, examine it a little more attentively, and conſider whether it has not a peculiar port or ſtructure, announcing it to be a natural tribe. It may perhaps have a permanent calyx; but if it is alſo double you may be almoſt certain that it ranges here. The corolla of your flower may perhaps have five heart-ſhaped petals, the ſide of one embracing or at leaſt touching that which is next to it, in a direction contrary to the ſun's apparent motion. The filaments perhaps,

haps, connected at bottom only, whether ſlightly, or for a conſiderable portion of their lengths, are gradually ſhorter as they recede from the middle; and the anthers are incumbent, or lie along over the top of them. You find the receptacle of the fructification prominent in the centre of the flower; the top of this receptacle ſurrounded by erect germs forming a jointed ring: all the ſtyles united below into one body with the receptacle; but diſtinguiſhed at top into as many filaments as there are germs: theſe germs becoming a capſule conſiſting of as many cells as there are piſtils in the flower: and frequently conſiſting of as many connected *Arils*. In each of theſe cells lurks a kidney-ſhaped ſeed.

If you have not already divined this riddle, take the flower of a wild Mallow, an Althæa, Lavatera, or other plant reſembling theſe; examine it by the characters juſt laid down, and you will have a perfect idea of the claſs *Monadelphia*. From the circumſtance of the receptacle ſtanding up in the middle of the flower, like a column, theſe have alſo the name of *columniferous* plants.

The orders are five, taken from the number of the ſtamens, which you remember determined the claſs in the firſt thirteen claſſes; but being now no longer uſed for that purpoſe, may ſerve very well for the other.

The

The fruit was formerly taken for discriminating the genera. This being found insufficient, succeeding nomenclators had recourse to the leaves; but Linnæus has, for this purpose, wisely adopted the calyx, which is always present, and is remarkable for its structure in this class. The illustrious Swede has ever shown great sagacity in seizing that part of the plant which is most constant, and furnishes the greatest choice of permanent variations, whereon to found the essential characters of his genera and species [1].

THE ORDERS PENTANDRIA AND DECANDRIA.

Not having taken the pistil for the distinction of the orders, that part remains to assist us in characterising the genera. Accordingly in the first order of this class, in which the flowers have five stamens, two genera have one, and two have five styles; the number of cells in the capsules serves to complete the generic character. Thus *Hermannia* has five styles, and a five-celled capsule; to which we may add that the five petals of the corolla are rolled spirally in a direction contrary to the sun's apparent motion; and that their claws have

Hermannia.

[1] Genera 35, and species 256, in this class.

a little

a little membrane on each ſide uniting to form a cowled tube. Though there are many ſpecies of this genus, yet perhaps none of them may offer themſelves to your view. We will proceed therefore to a favourite genus, that ranges in the ſecond order, or that which has ten ſtamens: I mean *Geranium*, which, out of its eighty-two ſpecies, will furniſh you ample matter for examination, eſpecially as I know you cultivate ſo many of them. Before you determine the circumſtances in which they differ, let us ſee in what they all agree; this is in having one ſtyle terminated by five ſtigmas; and a fruit compoſed of five grains, and beaked; whence its names of *Geranium* and *Craneſbill*. We may add that the calyx is ſingle and five-leaved, as well as the corolla; that the filaments are alternately longer and ſhorter, but all ſhorter than the corolla; and very ſlightly connected in thoſe which have a regular corolla; that the ſtyle is longer than the ſtamens, and permanent; and that each of the five ſeeds is terminated by a tail or awn, aſſiſting to form the beak, and which when the ſeed is ripe becomes ſpiral, and thus detaches the ſeed from the plant.

Geranium.

The African ſpecies, of which we have ſo many from the Cape of Good Hope, have the five parts of the calyx united at bottom; the petals unequal; and ſeven only of the filaments furniſhed with anthers; the

the flowers grow many together in a kind of umbel; the feeds are naked, with a feathered awn, and the leaves grow alternate upon the ftalk, which is fhrubby.

In this firft fection you find, among many others, the *Fulgid*[m], with a flefhy ftem, putting out but few branches; the leaves three-parted and gafhed, the middle fegment much larger than the others; frequently falling off, fo as to give the ftalks an appearance of being dead during the fummer; the flowers are produced on fhort footftalks, in a fort of double umbel, each fuftaining but two or three flowers, remarkable for their deep fhining fcarlet colour.

The well known *Scarlet*[n], which would be at leaft as much efteemed as the Fulgid, were it not more common. The leaves are almoft orbicular, except that they are hollowed next the petiole; they are notched about the edge, but not gafhed or lobed; their furface is downy; and they ftain the fingers if handled roughly, whence the trivial name of *inquinans* or ftaining. This is a much loftier plant than the laft, growing as high as eight or ten feet; and fends out abundance of erect branches: the flowers in the umbels are numerous, and are produced on very long peduncles.

[m] Geranium fulgidum *Lin.* Dill. elth. t. 130. f. 137.
[n] Geranium inquinans *Lin.* Mill. illuftr. Dill. elth. t. 125. f. 151, 151. Mart. cent. 3.

The

The *Papilionaceous* [o], ſo called, becauſe the corollas have ſomething the appearance of butterfly or pea-bloſſom flowers, the two upper petals, which are large, turning up like the banner or ſtandard in thoſe flowers; theſe are finely variegated, but the three under petals being reflex and ſmall are ſcarcely obſerved, but on a near inſpection; the flowers are many in each umbel: the leaves are large, angular, rough, and ſtand on long petioles.

The *Hollow-leaved* [p] has roundiſh leaves contracted on the ſides ſo as to ſtand hollow; the edges are ſharply indented; the flowers are large, and produced in large looſe umbels; the corollas are purple: it is a plant of large ſtature, and very hairy.

There is another ſort, or variety, very like this; but it has leaves of a thicker ſubſtance, and divided into ſeveral acute angles: the branches are not ſo irregular, and the bunches of flowers are not ſo large.

The *Horſe-ſhoe* [q] is perhaps the ſpecies moſt commonly known of all the Africans; the dark or purpliſh mark, in ſhape of a horſe-ſhoe upon the leaves, ſhows this Geranium to the eye at firſt ſight; but it is

[o] Geranium papilionaceum *Lin.* Dill. elth. t. 128. f. 155. Mart. cent. 15.

[p] Geranium cucullatum *Lin.*—cowled. Dill. elth. t. 129. f. 156. Mart. cent. 28.

[q] Geranium zonale *Lin.* Comm. præl. 51. t. 1.—See the flower in pl. 22. f. 3.

not abſolutely permanent; for we have varieties without it; we muſt have recourſe therefore to the form of the leaves, as a more certain diſtinction: they are orbicular, hollowed next the petiole, divided on the circumference into ſeveral obtuſe ſegments, each of which is ſlightly indented. This ſort is very branching: the flowers are produced in large, cloſe umbels, on long peduncles, and vary from a light purple to a high ſcarlet.

The *Vine-leaved*[r] has ovate, aſcending pubeſcent leaves, having the ſmell of Baum, when rubbed; the flowers grow in a cloſe head, on long peduncles, riſing much higher than the branches; they are ſmall, and pale blue.

The *Roſe-ſcented*[s] has alſo lobed leaves, waved and villous; like the laſt, the flowers grow in cloſe heads; they are of a purpliſh blue: the branches are very irregular and weak: and the whole is weaker and grows taller than the former: the leaves when rubbed ſmell like dried roſes.

The plants of the ſecond ſection have many things in common with thoſe of the firſt; but differ in being herbaceous, and having the leaves oppoſite. Of theſe the *Odorous*[t] is remarkable for its powerful ſcent,

[r] Geranium vitifolium *Lin.* Dill. elth. t. 126. f. 153.

[s] Geranium capitatum *Lin.* Riv. pent. 326.

[t] Geranium odoratiſſimum *Lin.* Dill. elth. t. 131. f. 138.

ſomething

ſomething like Aniſeed: this has a very ſhort fleſhy ſtem, with long branches, and heart-ſhaped leaves extremely ſhort: the flowers are produced from the ſide of long proſtrate ſtalks, upon ſlender peduncles, three, four, or five together; they are white, and very ſmall.

The *Night-ſcented*[u] has ſeſſile calyxes, and bifid one-leafed ſcapes: the leaves are hairy, and almoſt as finely divided as the carrot; the ſtalks are about a foot high, and have two or three ſmaller leaves that are ſeſſile; hence ariſe two or three naked peduncles, terminated by an umbel of yellowiſh flowers, marked with dark purple ſpots, ſmelling very ſweet after ſun-ſet. Linnæus has taken his trivial name from the dulneſs of the colour in the flower.

The third ſection contains ſuch Geraniums as have only five of the ſtamens anther-bearing; five-leaved calyxes, and fruits hanging down. The corollas of theſe are leſs irregular; and the ſeeds are naked, terminated by a hairy awn.

Of this ſection we have ſome European ſpecies, as *Hemlock Craneſbill*[v], common in ſandy ſoils: this has a branching ſtalk, pinnate leaves, with the ſegments gaſhed and obtuſe, and many flowers on a peduncle.

[u] Geranium triſte *Lin.* Com. can. t. 110. Breyn. cent. t. 58.

[v] Geranium cicutarium *Lin.* Curtis, Lond. I. 51. Ger. 945. 3.

Very

Very like this is *Muſk Craneſbill* [w], but it is a larger plant, much leſs common, and eaſily known by its muſky odour: the diviſions of the leaves are pinnatifid. Some ſpecies [x] of this ſection are remarkable for the largeneſs of their beaks, and furniſh a good idea of the name of the genus.

In the three remaining ſections, all the ten filaments are topped with anthers; the calyxes are five-leaved; the corollas regular; the ſeeds covered with an *aril*, and terminated by a ſmooth awn. In the fourth ſection, the flowers are conjugate; that is, there are two always on every peduncle: the plants are perennial.

Some of the largeſt and handſomeſt of the European ſorts range in this ſection; as *Spotted Craneſbill* [y], with the peduncles and leaves alternate, the calyxes a little awned, the petals waved, and the ſtem erect. The leaves are divided into five or ſix lobes, laciniate on their edges; thoſe near the root ſit on long petioles, but on the upper part of the ſtalk they are ſeſſile. The flowers are of a dark purple. There is a variety of this with light purple corollas.

Meadow Craneſbill [z] has the leaves divided

[w] Geranium moſchatum *Lin.* Riv. pent. 110. Ger. 941.

[x] Geranium arduinum, gruinum, ciconium *Lin.*

[y] Geranium phæum *Lin.* Ger. 942. 3. Park. 704. 3.

[z] Geranium pratenſe. Curtis, Lond. IV. 49. Ger. 942. 1.

into ſix or ſeven lobes, cut into ſeveral acute ſegments; they are wrinkled, and rather peltate; the petals are entire, and of a fine blue.

The Geraniums of the fifth ſection differ from thoſe of the fourth only in being annual. Moſt of the common European ſorts are of this diviſion: as *Herb Robert*[a], known by its hairy, pointed, ten-angled calyxes. The leaves are doubly pinnate, with the end-lobes confluent; they are generally hairy, the ſtalks red, and the whole plant has a ſtrong hircine ſmell. *Shining Craneſbill*[b] has the calyxes pyramidal, angled, elevated and wrinkled; the leaves rounded and five-lobed; the whole plant is ſmooth and ſhining; the ſtalks are red.

The *common Dove's-foot* or *ſoft Craneſbill*[c] has the peduncles and floral leaves alternate; the petals bifid or rather obcordate; the calyxes awnleſs, but ending in a ſhort point; and the ſtem rather erect. The ſtipules are alſo bifid: the leaves are very ſoft, kidney-ſhaped, divided half-way into five or ſeven parts, and each of theſe lobes trifid and blunt. This is very common, eſpecially in ſandy ſoils. Another[d],

[a] Geranium Robertianum *Lin.* Curtis, Lond. I. 52. Ger. 939. & 945. 5.

[b] Geranium lucidum *Lin.* Fl. dan. 218. Mor. t. 15. f. 6. Park. 707. 9.

[c] Geranium molle *Lin.* Curtis, Lond. II. 50. Fl. dan. 679.

[d] Geranium rotundifolium *Lin.* Blackw. 58. Vaill. par. t. 15. f. 1. Ger. 938. Park. 706. 2.

very like it in many reſpects, but more partially diſtributed, has entire petals, ſcarcely longer than the calyx; and the ſtem more proſtrate. *Long-ſtalked Craneſbill*[e] has peduncles longer than the leaves, which are divided into five multifid lobes acute at the end; the calyxes are awned, and the arils are ſmooth. The peduncle is very long, and the lobes of the leaves are doubly trifid. *Jagged Craneſbill*[f] has the leaves divided into five parts, and each of thoſe into three acute ſegments; the petals are of the length of the calyx, and notched, and the arils are villous: this has the leaves more and finer cut than any of the others.

Of the laſt ſection, with one-flowered peduncles, we have a handſome ſort wild, but not common, with orbicular leaves, divided into five or ſeven parts, and each of thoſe into three: the flowers ſtand on long hairy peduncles, the corollas are large, and of a deep purple[g]. Many more ſpecies are known to the curious[h]; but I have only ſelected ſuch as the fields, the garden, and your little conſervatory, are moſt likely to afford.

[e] Geranium Columbinum *Lin.* Vaill. par. t. 15. f. 4. Petiv. 64. 8.

[f] Geranium diſſectum *Lin.* Vaill. par. t. 15. f. 2. Petiv. 64. 6.

[g] Geranium ſanguineum *Lin.* Bloody Craneſbill. Ger. 945. 2. Petiv. 64. 9.

[h] See ſome figured in Curtis's Magazine, n. 18, 20, 55, 56, 95, 103, 136.

I have mentioned that Linnæus has ſubdivided this unwieldy genus from the number of effective ſtamens. A celebrated modern author has, from this circumſtance, made three diſtinct genera out of this one. 1. *Erodium*, containing the *Myrrhina* of Linnæus, or the Geraniums with five perfect ſtamens only. 2. *Pelargonium*, comprehending the *Africana* of Linnæus, or ſuch as have ſeven perfect ſtamens. 3. *Geranium*, taking in the remaining ſpecies, which anſwer exactly to the character of the order in having all the ten ſtamens with anthers, and which Linnæus had called *Batrachia*. Rivinus long ſince ſeparated this natural genus into two, from the regularity or irregularity of the corolla. I ſhall not diſpute whether all this be right or not. It is my deſign to explain the ſyſtem of the illuſtrious Swede as he left it.

Brownea. In this claſs we find a ſingular plant, which has naturally eleven ſtamens; a number which you did not find among the claſſes. Having the Monadelphic character, it here forms the order *Endecandria*, and ſtands alone. Being a plant little known, I inſiſt no longer on it [i].

The laſt order *Polyandria* is much the moſt conſiderable in number of genera and ſpecies. You have here *Silk-Cotton* [k], the

[i] Brownea coccinea *Lin.* [k] Bombax *Lin.*

True

True Cotton[l], ſo much uſed in our manufactures, the numerous genus of *Sida* or *Indian Mallow*, *Althæa* or *Marſh-Mallow*, *Alcea* or *Hollyhock*, *Mallow*, *Lavatera*, *Hibiſcus*, &c. The two firſt, with Sida and Hibiſcus, have one piſtil only; the reſt have many. Sida and Bombax have a ſingle calyx, but all the others have it double. The exterior calyx in Cotton and Lavatera is trifid; in Mallow conſiſts of three leaflets; in Alcea is ſexfid; in Hibiſcus octofid; in Althæa novemfid. Lavatera, Mallow, Alcea and Althæa, agree in having many ſeeds in a ring round a column, each covered with its proper aril. The ſeed-veſſel of Hibiſcus is a capſule compoſed of united cells including many ſeeds.

The officinal[m] ſpecies of Marſh-Mallow Althæa. is known by its ſimple downy leaves, hoary to the ſight, and very ſoft to the touch; they are angular, but not divided to the bottom, and therefore ſimple. The flowers are like thoſe of the Mallow, but ſmaller and paler.

Of Mallow there are many ſpecies: that Malva. which is ſo very common[n], has an erect herbaceous ſtem; five or ſeven-lobed acute leaves with both petioles and peduncles

[l] Goſſypium *Lin.*

[m] Althæa officinalis *Lin.* Fl. dan. 530. Mor. hiſt. ſ. 5. t. 19. f. 12. Ger. 933. 1. Park. 304. 1.—Pl. 22. f. 1.

[n] Malva ſylveſtris *Lin.* Curtis, Lond. II. 51. Ger. 930. 1. Pl. 22. f. 2.

 hairy.

hairy. *Dwarf Mallow* [o] has a proſtrate ſtem; orbiculate leaves hollowed next the petiole, obſcurely five-lobed; the fruit-bearing peduncles declining. This is every way a ſmaller plant. *Vervain Mallow* [p] has an erect ſtem, rough with ſpreading hairs in bunches, many-parted roughiſh leaves, the lobes of which are obtuſe and indented; the flowers large, and light purple. Another wild ſpecies called *Muſk Mallow* [q], is very like this, but has the radical leaves kidney-form and gaſhed; the ſtem-leaves five-parted, and the diviſions finely cut into narrow ſegments: the flowers have a muſky ſmell, and the ſtem has ſingle erect hairs ſitting on a prominent point. *Cape Mallow* [r] has an arboreſcent ſtem ten or twelve feet high, and the leaves five-lobed and hollowed at the baſe. The whole plant is hairy, and theſe hairs exude a viſcid aromatic juice. The flowers are deep red, and ſmaller than thoſe of the common Mallow. The trivial name informs us of its country, and conſequently that it ſtands in need of protection from you.

Alcea. The gigantic, the gaudy *Hollyhock* is of the genus *Alcea:* there are many varieties

[o] Malva rotundifolia *Lin.* Curtis, Lond. III. 43. Fl. dan. 721. Ger. 930. 2. Park. 299. 1.

[p] Malva Alcea *Lin.* Blackw. 309.

[q] Malva moſchata *Lin.* Curtis, Lond. IV. 50. Mor. hiſt. ſ. 5. t. 18. f. 4.

[r] Malva capenſis *Lin.* Dill. elth. t. 169. f. 206.

with

with double flowers, and different colours, as white, red of all hues from pale carnation to almoſt black, and yellows of different ſhades; but there are only two ſpecies [s], the firſt having roundiſh leaves, cut at the extremity only into angles; the ſecond palmate, cut deeply into ſix or ſeven ſegments, like the fig-leaf. Of the firſt there is a dwarf variety with variegated flowers, much eſteemed, and called *Chineſe Hollyhock*.

The ſhrub vulgarly named *Althæa Frutex* is an Hibiſcus; a very numerous genus, comprehending no leſs than thirty-ſix ſpecies, moſt of them inhabitants of either India, and not generally known here. The *Althæa Frutex* [t] however is a native of Syria, and bears the rigour of our climate, though it is very late ere it produces its flowers. The ſpecific characters are, an arboreous or woody ſtem, and wedge-ſhaped leaves, divided at top into three lobes, and ſtanding on ſhort petioles. The flowers are bell-ſhaped, and of various colours—pale or bright purple with dark bottoms, white with purple bottoms, variegated with dark bottoms, and yellow with the ſame: theſe flowers being large, gay, and numerous, make a handſome appearance, and give the completeſt idea of the claſſical character. Hibiſcus.

China Roſe alſo, notwithſtanding its name,

[s] Alcea roſea Mill. illuſtr.—& ficifolia *Lin.*
[t] Hibiſcus ſyriacus *Lin.* Curt. Magaz. 83.

is no Roſe, but an *Hibiſcus*[u], with a woody ſtem, and ovate, ſharp-pointed leaves, ſerrate about the edges; the colour, ſize, and appearance of the flowers, when they are double, gave occaſion to the name of Roſe: they frequently appear on Chineſe paintings and paper, and are certainly very ornamental. The *Muſk plant*[v] of the Weſt Indies is another ſpecies of *Hibiſcus*; its kidney-ſhaped ſeeds have a very ſtrong ſmell of muſk. The bark of ſome ſpecies[w] is formed of fibres ſtrong enough for cordage. One of them is cultivated in the Weſt Indies for its pods, which they put into their ſoups[x]. But all this we have nothing to do with as botaniſts.

[u] Hibiſcus Roſa Sinenſis *Lin.* Rheed. mal. 2. t. 17.
[v] Hibiſcus Abelmoſchus *Lin.* Mer. Surin. t. 42.
[w] Hibiſcus vitifolius & Sabdariffa *Lin.*
[x] Hibiſcus eſculentus *Lin.* Sloan. jam. 1. t. 133. f. 3.

LETTER

LETTER XXV.

THE CLASSES DIADELPHIA AND POLYADELPHIA.

June the 4th, 1776.

AFTER a ſhort excurſion, we are returned, dear couſin, among your old acquaintance; and you have only to apply to the term *Diadelphia*, which is the name of the ſeventeenth claſs in Linnæus's ſyſtem, all the knowledge you firſt acquired from the letter on Papilionaceous flowers [y], and which you have ſince increaſed ſo much by your obſervation and experience. You have admired the ſingularly admirable and beautiful ſtructure of theſe flowers, in which all the plants of this claſs agree: you will now not be diſpleaſed to accompany me in an enquiry into their generic and ſpecific differences. The number of genera in this claſs is 57, of ſpecies 695. The orders are four, taken from the number of ſtamens, which in the firſt order is five, in the ſecond ſix, in the third eight, and in the fourth ten. In the order *Pentandria* however there is only one genus; in the order *Hexandria* two; and in the order *Oc-*

[y] Letter III.

tandria three; ſo that you perceive the laſt *(Decandria)* abſorbs the far greater part of the claſs; and what you have learnt of Papilionaceous flowers belongs indeed principally to this order. Of the three firſt orders there are only two genera, which you will have an opportunity of obſerving; and we will begin if you pleaſe with them.

Fumaria. *Fumitory* has two filaments, each of them terminated by three anthers; it has the claſſical character therefore, and muſt be of the order *Hexandria*. This genus has, beſides this, a two-leaved calyx, a ringent rather than a papilionaceous corolla, the upper lip however anſwering to the banner, the lower lip to the keel, and the bifid chaps to the wings: the baſe of each lip is prominent, but the upper one the moſt; and one filament is incloſed in each. *Common Fumitory*[z] which you will readily meet with as a weed in your kitchen garden, is known by a weak, diffuſe, branching ſtem, multifid leaves dividing into three, and the lobes trifid: the flowers growing in a raceme, and each being ſucceeded by a round or rather obcordate one-ſeeded pericarp.

Polygala. *Milkwort* has eight filaments, each terminated with an anther, and all united at bottom: it appertains therefore to the order *Octandria* of this claſs. The characters of the genus are, a five-leaved calyx, with

[z] Fumaria officinalis *Lin.* Curtis, Lond. II. 52. Ger. 1088. 1. Park. 287. 1.

two of the leaflets like the wings of the papilionaceous flower, and coloured: the banner of the corolla is cylindric; the legume is obcordate, or inverſe-hearted, and two-celled. Many of the ſpecies have a beard, creſt, or pencil-formed appendage to the keel; thoſe which have none are called beardleſs: and hence a commodious ſubdiviſion of this large genus; the laſt are ſubdivided into ſhrubby and herbaceous; the herbaceous again into ſimple and branched. Of thirty-eight ſpecies we have only one wild, and that is common on dry paſtures and heaths [a]: it is of the creſted diviſion, and bears the flowers in a raceme; the ſtem is herbaceous, ſimple, and procumbent, and the leaves are linear. This is a lowly plant, with pretty flowers, blue, red or white. There is a beautiful ſpecies [b] in the green-houſe, from the Cape, with a ſhrubby ſtem; oblong, ſmooth leaves, blunt at the end; and handſome flowers, large, white on the outſide, but bright purple within; the keel creſted, and ſhaped like a half moon. *Senega* [c] root, ſo famous among the American Indians as an antidote to the bite of the rattle-ſnake, is from a ſpecies of this genus.

The plants of the order we are now to

[a] Polygala vulgaris *Lin.* Fl. dan. 516. Ger. 564. 5. Park. 1332. 2.

[b] Polygala myrtifolia *Lin.* Mill. illuſtr.

[c] Polygala Senega *Lin.* Mill. Dict.

examine

examine are obvious, not only by their papilionaceoûs flowers, but by their compound leaves, which in the greater part are pinnate, winged, or feathered, but in others trifoliate [d]. In ſome genera the pinnate leaves have the leaflets in pairs only [e], but it is more common to have them terminate in an odd one [f]. Many of this pulſe tribe have ſtems too weak to ſuſtain themſelves, they fly therefore to ſome ſtronger plant or other prop for ſupport, and they are furniſhed with the neceſſary means of helping themſelves, either by twining their ſtalks about and embracing their friend [g], or elſe by throwing out ſlender threads, like the vine, called *claſpers* or *tendrils*, by which they lay faſt hold [h].

Moſt of theſe plants having fruits that are eſculent either to us, to quadrupeds or to birds, produce flowers in great abundance, and cloſe bunches; in ſome of the genera they grow in a kind of umbel [i], much like

[d] As in Trifolium or Trefoil, which has its name from this circumſtance, Lotus, Medicago, Erythrina, Geniſta or Broom, Cytiſus, Ononis, Trigonella, Phaſeolus or Kidney Bean, Dolichos and Clitoria.

[e] Orobus, Piſum or Pea, Lathyrus or Everlaſting Pea, Vicia or Vetch, Ervum and Arachis.

[f] Biſerrula, Aſtragalus, Phaca, Hedyſarum, Glycyrriza or Liquorice, Indigofera or Indigo, Galega, Colutea, Amorpha and Piſcidia.

[g] Phaſeolus, Dolichos, Clitoria, Glycine.

[h] Piſum, Lathyrus, Vicia, Ervum.

[i] Lotus, Coronilla, Ornithopus, Hippocrepis, Scorpiurus.

thoſe

thoſe of the ſecond order of the fifth claſs. I mention theſe circumſtances, not as claſſical characters, but as leading features that may give you a ſhrewd ſuſpicion, rather than a certain aſſurance. When you find a plant endued with ſome of theſe ſubordinate characters, you, I am certain, will not determine it at once upon them: no, they will only lead you to a more ſtrict examination. Neither pinnate or trifoliate leaves, weak twining or climbing ſtems, nor even papilionaceous flowers, will ſatisfy your diſcerning eye, till you have ſeen the union of the filaments at bottom. If you can procure any ſpecies of *Sophora* [k], you will be convinced of this; for without ſuch caution you would infallibly have been miſled; this genus agreeing with the pulſe tribe in every reſpect, except in having the ten filaments diſtinct.

The proper character of this claſs, you know, is to have the filaments in two diſtinct bodies; and the character of the order *Decandria* is to have nine filaments united at bottom into a membrane ſurrounding the germ, and the tenth ſingle, filling up the opening which is left for the germ to diſengage itſelf, when it has arrived at a ſtate proper to paſs into a pod or legume. I muſt advertiſe you however that this is not ſtrictly

[k] A genus of the claſs *Decandria* and the order *Monogynia*. Anagyris, Cercis, &c. have alſo the ſame appearance.

true

true of all the genera; there are no fewer than eighteen out of fifty, which have all the ten filaments connected, so that the germ cannot grow into a legume without tearing asunder the membrane formed of the filaments. You must not therefore be deterred from setting down a plant as of the Pulse tribe, and of the class *Diadelphia*, when you find the ten filaments united into *one*, inclosed within a papilionaceous flower, and furnished with the other marks of the class. Of those which answer regularly to the classical character, some have a pubescent stigma[l], and the rest are distinguished by their legumes, as we shall now see; for we are going to examine their distinctive marks more narrowly.

Spartium. You will observe in this class some trees, and many shrubs, with papilionaceous flowers, as *Common*[m] and *Spanish*[n] Broom; both of a genus in which the ten filaments are all united, and form a membrane adhering close to the germ: the stigma grows along the upper side of the top of the style, and is villous; the calyx is continued downwards, and is marked beneath with five little notches at the tip. *Spanish Broom*, with some other species, has simple leaves, in the rest they are ternate, trefoil, or three-leaved.

[l] Colutea, Phaseolus, Dolichos, Orobus, Pisum, Lathyrus, Vicia.

[m] Spartium scoparium *Lin.* Curt. Lond. V. 52. Fl. dan. 313. Blackw. 244. Ger. 1311. 1. Park. 229. 1.

[n] Spartium junceum *Lin.* Curt. Magaz. 85.

leaved. In *Common Broom* however there is a mixture of both. In the firſt alſo the leaves are lance-ſhaped, and the ruſh-like branches are oppoſite, round, and produce the flowers from the top, in a looſe ſpike. In the ſecond the branches are angular, and the flowers come out ſingly for a conſiderable length towards the top. They are large, and of a bright yellow in both ſpecies. There is alſo a Spaniſh Broom with a white flower [o]; which has leaves like the other, but the branches ſtriated, and the flowers in ſhort ſpikes or cluſters on the ſides of them; they are ſucceeded by large oval pods containing one ſeed, whence the trivial name. *Portugal Brooms* with trifoliate leaves and yellow flowers, differing little from ours: and a ſort with prickly branches, thence called *Prickly Cytiſus* [p].

We have ſome wild ſhrubs of an humbler growth, ſomewhat reſembling theſe, but of another genus called *Geniſta*; the characters of which are a two-lipped calyx, the upper lip two-toothed, the lower three-toothed; the banner of the corolla oblong and turning downwards from the piſtils and ſtamens; the piſtil depreſſing the keel, and the ſtigma involute. *Dyer's weed*, called alſo *Wood-waxen* and *Baſe Broom* [q], which Geniſta.

[o] Spartium monoſpermum *Lin.*

[p] Spartium ſpinoſum *Lin.*

[q] Geniſta tinctoria *Lin.* Fl. dan. 526. Ger. 1316. 1. Park. 229. 7.

grows

grows in paſtures and headlands, has ſmooth lance-ſhaped leaves, and erect, round, ſtreaked branches. *Needle Furze* or *Petty Whin* [r], which you will find wild on heaths, has ſmall lance-ſhaped leaves, ſlender branches armed with long, ſimple ſpines; the flower branches are ſhort, have no ſpines, and have five or ſix flowers in a cluſter at the end of them: the colour of the corolla in both ſpecies is yellow; and you would at firſt ſuppoſe that the former was a *Spartium*, and the latter a Furze, or

Ulex. of the genus *Ulex*; which however differs from both in having a two-leaved calyx, with the legume ſo ſhort as ſcarcely to emerge from it. We have only one ſpecies, than which nothing, as you know, is more common on all our heaths; it has the three different names of *Furſe*, *Gorſe* and *Whins* [s], in different parts of the kingdom.

Ononis. *Reſtharrows* are a lowly kind of ſhrubs, or rather underſhrubs, with purple flowers, growing on commons, barren paſtures, and headlands of corn-fields; they have the name from the ſtrength and matting of the roots, which circumſtance has induced the Dutch to ſow them on their ſea-banks. The cylinder of filaments is quite entire at bottom, without any fiſſure, in this genus;

[r] Geniſta anglica *Lin.* Fl. dan. 619. Ger. 1320. 4. Park. 1004. 4.

[s] Ulex europæus *Lin.* Fl. dan. 608. Ger. 1319. 1. Park. 1004. 1.

the

the calyx is parted into five linear divisions; the banner of the corolla is striated; and the legume, a section of which is a rhomb, is turgid and sessile. We have two sorts, one[t] with prickly smooth branches, and the flowers in a raceme, but coming out singly: the other[u] with villous leaves and branches, but without spines; the flowers in a raceme, but generally two together; both have ternate leaves, except that towards the top they are simple.

In *Anthyllis* the calyx is turgid, and includes the legume, which is small and roundish, containing one, or at most two seeds. The only species we have wild is herbaceous, is called *Ladies-Finger* or *Kidney-Vetch*[v], and is not uncommon in chalky pastures; it has unequally pinnate leaves, and a double head of yellow flowers, but this latter character is not constant. The leaves are pubescent, and consist of three or four pair of leaflets; except two under the umbel, which are digitate. There are several flowering-shrubs of this genus; as that which is generally called *Jupiter's beard* or *Silver bush*[w], from the splendid whiteness of the leaves, which is owing to a fine nap Anthyllis.

[t] Ononis spinosa *Hudsoni*. Common, smooth, or prickly Restharrow. Blackw. t. 301. Ger. 1322. 1.

[u] Ononis inermis *Hudsoni*. Hairy Restharrow. Ger. 1322. 3.

[v] Anthyllis Vulneraria *Lin*. Rivin. t. 18. Ger. 1240. 1.

[w] Anthyllis Barba Jovis *Lin*. Mill. fig. t. 41. f. 2.

or down that covers them; they are equally pinnate: the flowers are produced at the extremity of the branches, in ſmall heads, and are yellow.

Lupinus. *Lupins*, which are ſo well known in the flower-garden, agree in a two-lipped calyx, in having five of the anthers round, and five oblong, and in the ſhell of the legume being coriaceous or leathery. The common *white* [x] ſort, which is cultivated as a pulſe in moſt of the ſouthern parts of Europe, has the flowers growing alternate, without appendages; the upper lip of the white corolla is entire, the lower three-toothed: the ſeeds are orbiculate and flatted. There are three ſorts with blue flowers: the *Perennial* [y], which is the only one that is not annual, with alternate, unappendaged flowers; the upper lip of the corolla notched, the lower one entire. This is an American plant: the digitate leaves are compoſed of ten or eleven leaflets, whereas thoſe of the former have no more than ſeven or eight: the flowers grow in long looſe ſpikes, and are pale blue. The *great blue* [z], with alternate appendaged flowers; the upper-lip two-parted, the lower three-toothed. This has a ſtrong ſtem, covered with a ſoft browniſh down; the leaves have nine, ten, or eleven hairy, ſpatulate leaf-

[x] Lupinus albus *Lin.* Riv. tetr. Blackw. 282.
[y] Lupinus perennis *Lin.* Mill. fig. 170. 1.
[z] Lupinus hirſutus *Lin.*

lets:

lets: the flowers are in whorls, forming a ſort of ſpike; they are large, and of a beautiful blue: the pods are very large, and have three roundiſh compreſſed ſeeds, very rough and of a purpliſh brown. *Narrow-leaved* or *tall blue Lupin* [a], has the flowers alternate and appendaged or pedunculate; the upper lip of the corolla two-parted, the lower three-toothed: the lobes of the leaves are linear. The *Varied* [b] is not very different in appearance from this: the flowers grow in half whorls, and are appendaged; the upper lip is bifid, and the lower ſlightly three-toothed: the corollas are light blue or purple. It is ſhorter than the laſt; the leaves have fewer leaflets, and ſtand on ſhorter petioles. The *Hairy* [c] has the flowers in whorls and appendaged, with the upper lip two-parted, like the Great Blue Lupin; which it much reſembles in ſtature and appearance; but the corollas are fleſh-coloured with the middle of the banner red, the lower lip is entire; the plant is hairy all over, and the leaves are lance-ſhaped, and a little obtuſe at the end. The *Yellow* [d] is eſteemed for the ſweetneſs of its flowers: they grow in whorls and on peduncles; the upper lip of the corolla is two-parted, the lower three-toothed. Thus

[a] Lupinus anguſtifolius *Lin.* Riv. tetr.
[b] Lupinus varius *Lin.*
[c] Lupinus piloſus *Lin.*
[d] Lupinus luteus *Lin.* Riv. tetr.

have you a hiſtory of the whole genus of Lupin; for theſe are all the ſpecies hitherto known: and as you may eaſily have them growing together, you may compare them at leiſure, and aſcertain all their agreements and differences: could we do this in every genus, how clearly might we diſtinguiſh the ſpecies! but remember that culture may produce fictitious characters, which miſlead unwary botaniſts.

In all the genera hitherto examined, the filaments have made one body at bottom; in the reſt, which I ſhall now offer to your conſideration, nine only are united, and the tenth is free, according to the proper character of the claſs. We will begin with ſome genera, diſtinguiſhed (as I mentioned

Phaſeolus.

before) by a pubeſcent ſtigma. *Phaſeolus* or *Kidney Bean*, in having the keel with the ſtamens and ſtyle ſpirally twiſted, poſſeſſes one obvious character, that diſcriminates it ſufficiently from all its congeners. Some of the ſpecies have an outer calyx, conſiſting of two roundiſh leaflets, which

Lathyrus.

may more properly be called *bractes*. *Lathyrus* or *Everlaſting Pea* has a flat ſtyle, villous above, growing broader upwards: in this it differs from the *Pea*, which has a triangular ſtyle keeled above: both genera have the two upper diviſions of the calyx ſhorter than the other three, and, in other reſpects, are very nearly allied. Some ſpecies of *Lathyrus* have one flower only on a peduncle:

peduncle: of theſe we have two wild ones; one with yellow flowers, ſupporting itſelf among the corn by leafleſs tendrils, and having broad ſtipules ſhaped like the head of an arrow: the other with crimſon flowers, long narrow leaves difficult to be diſtinguiſhed from the graſs among which it grows, and ſmall, ſubulate or awled ſtipules. The firſt is called *Yellow Vetchling*[e]; the ſecond, *Crimſon Graſs Vetch*[f]. *Sweet Scented Pea*[g], with ſome few others, has two flowers on every peduncle; each tendril has a pair of oblong ovate leaves, and the legumes are rough. The banner of the corolla is dark purple, the keel and wings light blue; but there are varieties; one all white, and another with a pink banner, wings of a pale bluſh, and a white keel; this is called *Painted Lady Pea*. *Tangier Pea*[h], another of the *biflorous* ſection, has the two leaves alternate, lance-ſhaped and ſmooth; the ſtipules ſhaped like a creſcent. The flowers grow on ſhort peduncles; have a purple banner, with wings and keel of a bright red, and are ſucceeded by long

[e] Lathyrus Aphaca *Lin.* Mill. fig. pl. 43. Curtis, Lond. V. 51. Ger. 1250. Park. 1067.

[f] Lathyrus Niſſolia *Lin.* Ger. 1249. 2. Park. 1079. 4.

[g] Lathyrus odoratus *Lin.* Curtis magaz. 60.

[h] Lathyrus tingitanus *Lin.* Jacq. hort. t. 46. Curt. magaz. 100.

jointed pods. *Everlasting Pea*[i] is of the last division, having many flowers produced on one peduncle: this has also conjugate leaves, that is, growing in pairs, furnished with a tendril or clasper; the form of the leaves is elliptic or oval; and the stems, which climb very high, have membranaceous wings on each side between the joints; the flowers are red. There is a variety of this in the gardens, with broader leaves, larger and deeper coloured flowers. There is another sort not very different from this[k], having sword-shaped leaves; and a third[l], growing in woods, bogs, and wet meadows, which has many-leaved tendrils, and lance-shaped stipules: the leaflets are six; and there are from three to six flowers on each peduncle; the corolla is blue, with the greatest part of the wings and keel white. One species of this section[m], with yellow flowers, two-leaved tendrils, which are extremely simple, and lance-shaped leaves, is very common in pastures, hedges, and woods.

Vicia. *Vetch* or *tare* is sufficiently distinguished by having a stigma transversely bearded on the

[i] Lathyrus latifolius *Lin.* Mill. fig. pl. 160. Mill. illustr. Fl. dan. 785. Pl. 23.

[k] Lathyrus sylvestris *Lin.* Fl. dan. 325. Mor. hist. s. 2. t. 2. f. 4. Ger. 1229. 1.

[l] Lathyrus palustris *Lin.* Fl. dan. 399.

[m] Lathyrus pratensis *Lin.* Curtis, Lond. III. 44. Ger. 1231. 6. Park. 1061. 1.

the under ſide. The ſpecies, which are eighteen in number, may be ranged under two diviſions, the firſt comprehending ſuch as have flowers in bunches on peduncles; the ſecond, thoſe which are axillary, or have the flowers ſitting almoſt cloſe to the ſtem, and coming out from the angle which the leaves form with it. Of the firſt diviſion we have the *Tufted*[n] and *Wood Vetch*[o] wild: both having flowers in bunches many together, but in the firſt imbricate; in this alſo the leaflets or component leaves are lance-ſhaped and pubeſcent, and the ſtipules entire; in the ſecond, the leaflets are oval, and the ſtipules ſlightly toothed. The cultivated, and ſeveral wild ſorts, are of the ſecond diviſion. The firſt[p] has erect, ſeſſile legumes, moſtly two together; the leaves are retuſe, and the ſtipules ſpotted. Of the others, *Spring Vetch*[q], which is very nearly related to the former, has however the legumes generally ſingle; the lower leaflets retuſe, the upper ones narrow, and almoſt linear: the leaflets are from four to ten; and the ſtipules are ſpotted, as in the former. *Buſh Vetch*[r]

[n] Vicia Cracca *Lin.* Curtis, Lond. V. 54. Fl. dan. 804. Mor. hiſt. ſ. 2. t. 4. f. 1.

[o] Vicia ſylvatica *Lin.* Fl. dan. 277.

[p] Vicia ſativa *Lin.* Fl. dan. 522. Mor. t. 4. f. 12. Ger. 1227. 1, 4.

[q] Vicia lathyroides *Hudſ.* Fl. dan. 58.

[r] Vicia dumetorum *Lin.* Riv. tetr. 50.

has about four erect legumes growing together on short pedicles: the leaflets are ovate, and quite entire; they decrease in size towards the end of the leaf: it ramps in hedges. The *Bean*[s] is placed by Linnæus in the Vetch genus; and very justly, since it agrees with them in the characters of the fructification, and differs only in having a stouter stalk that supports itself, and therefore is not furnished with tendrils. Its native place of growth is supposed to be not far from the Caspian Sea, on the borders of Persia. All the different sorts of Bean are in reality but varieties from the same original stock: you understand me to speak of Beans properly so called, in exclusion of Kidney Beans and others, which are not merely specifically different, but also of another genus.

Colutea. Of the same section, with pubescent stigmas, is a genus of well known shrubs called *Colutea:* distinguished by their quinquefid calyx; and inflated legume, opening from the base by the upper suture; the English name of *Bladder-Sena* is taken from the latter character. *Common Bladder-Sena*[t] has an arboreous stem, and inversely-hearted leaves. It grows twelve or fourteen feet high; its winged leaves have four or five pair of grayish leaflets; the flowers

[s] Vicia Faba *Lin.*
[t] Colutea arborescens *Lin.* Curt. Magaz. 81.

come

come out from the axils, two or three together, on ſlender peduncles; they are yellow with a dark-coloured mark on the banner. This grows wild in the ſouthern countries of Europe. There is another, which comes from the Eaſt, and has flowers like this, only of a brighter yellow; differing in being a much lower ſhrub, and in having nine pair of ſmall, oval, entire leaflets to each leaf. A third, about the ſame height with the ſecond, but with branches ſtill more ſlender, comes from the ſame country: the leaves of this have five or ſix pair of ſmall heart-ſhaped leaflets; the flowers are ſmaller, and of a dark red, marked with yellow. It is a doubt whether theſe be ſpecifically different from the firſt [u]: there is however one from Æthiopia, with ſcarlet flowers, which is very diſtinct [v]: for it is a low, weak ſhrub, with leaves compoſed of ten or twelve pair of oblong-ovate, hoary leaflets: the flowers are long, owing to the length of the keel, for the banner is ſhorter than that, and the wings are minute. You will eaſily ſuppoſe, from its country, that it cannot ſtand the cold of a ſevere winter with us; it does not ſhrink however from a mild one, in a dry ſoil and warm ſituation. There is alſo an herbaceous ſpecies [w], with ſmooth

[u] Figured in Comm. rar. t. 11. and Mill. fig. 100.
[v] Colutea fruteſcens *Lin.* Mill. fig. pl. 99.
[w] Colutea herbacea *Lin.* Comm. hort. 2. t. 44.

linear

linear leaflets; but this is an annual plant of little beauty, and therefore rarely cultivated.

Cytiſus. There are ſeveral other ſhrubs of the Pea-bloom tribe: as the different ſpecies of *Cytiſus*, of which *Laburnum* [x] is one. This is known by yellow flowers hanging in large ſimple racemes, and three oblong-ovate leaflets to each leaf. There is a variety with narrower leaves, and longer bunches of flowers, more common in ſhrubberies than the firſt, which is a larger tree, and comes to excellent timber; but this making a better appearance when in flower, is preferred in ornamental plantations. *Seſſile-leaved Cytiſus* [y], vulgarly called *Cytiſus ſecundus Cluſii*, has the flowers in ſhort, erect racemes, at the ends of the branches; each flower has a little triple bracte at the baſe of the calyx; the leaves on the flowering branches are ſeſſile, but the others are petiolate. The flowers are of a bright yellow, and the pods are ſhort, broad, and black. *Evergreen Cytiſus* [z] has the flowers coming out ſingly from the ſide of the ſtalk, with very hairy, trifid, obtuſe, oblong, ſwelling calyxes; the ſtalks extremely hairy; the leaves alſo hairy, eſpecially underneath. The flowers are pale yellow; and the pods long, narrow, and rough.

[x] Cytiſus Laburnum *Lin.* Jacq. auſtr. 4. t. 306.
[y] Cytiſus ſeſſilifolius *Lin.* Duham. arb. 1.
[z] Cytiſus hirſutus *Lin.* Jacq. obſ. 4. 96.

All

All theſe, and the reſt of the ſpecies, agree in a two-lipped calyx, the upper lip bifid, the lower three-toothed; and a legume attenuated at the baſe; and pedicled, with ſeveral ſeeds in it. The leaves are ternate.

Robinia has a quadrifid calyx; an expanding, reflex, roundiſh banner; and a gibbous, elongate legume, containing ſeveral ſeeds. The tree which you admire for its long racemes of ſweet-ſmelling white flowers, hanging down like thoſe of Laburnum, is of this genus: I mean the *Baſtard Acacia* [a], called in North America, its native country, *Locuſt-tree*. The leaves are pinnate, conſiſting of eight or ten pair of oval leaflets terminated by an odd one; all entire, and ſitting cloſe to the mid-rib: the ſtipules are armed with ſtrong, crooked thorns; and the flowers come out ſingly, or only one on a pedicle in the racemes. The *Caragana* [b], a Siberian ſhrub, has leaves abruptly pinnate, that is, winged, not terminated by an odd leaflet; they have four or five pairs of oval leaflets: this has no ſpines, and the yellow flowers come out ſingly from the axils. There are ſeveral other trees and ſhrubs of this genus; but theſe are the moſt known. Robinia.

[a] Robinia Pſeudacacia *Lin.* Seba mus. 1. t. 15. f. 1. Duham. arb. 2. t. 42.

[b] Robinia Caragana *Lin.* Duham. arb. 3.

Coronilla

Coronilla. *Coronilla* is another genus of ſhrubs, comprehending however ſome herbaceous plants. They all agree in a two lipped calyx; the upper lip having two, the lower three little teeth; the ſuperior teeth conjoined; in a banner ſcarcely longer than the wings; and in a very long, ſtraight legume, contracted between the ſeeds, and, inſtead of opening by the ſutures, falling off in joints.—*Scorpion Sena* [c] is a ſpecies of this genus very common among ſhrubs: it is immediately known, by having the claws of its yellow corollas three times as long as the calyx; two or three flowers come out together upon long peduncles from the ſides of the branches, which are ſlender, and angular: the leaves are pinnate, and compoſed of three pair of leaflets terminated by an odd one: the legumes are long, ſlender, taper, and pendulous; the ſeeds cylindric. There are ſeveral beautiful ſhrubs of this genus, but too tender to bear the open air in our climate.

Indigofera. The plants from which indigo is made [d] are of this claſs; and many of the kindred genera reſemble them in quality as well as outward form and character. Scorpion Sena in particular, it is ſaid, will yield a dye nearly equal to indigo, if the leaves are fermented in a vat in the ſame manner as is

[c] Coronilla Emerus *Lin.* Mill. fig. 132.
[d] Indigofera *Lin.* Mill. fig. 34.

practiſed

practiſed with thoſe plants; and you remember complaining perhaps, that the yellow flowers of the Lotus would turn blue in drying, unleſs you took care to keep them ſeparate from other plants, and to change them often.

Liquorice is alſo of the ſame claſs: it has a two-lipped calyx, with the upper lip divided into three parts, and the lower abſolutely ſimple and undivided; the legume is ovate and compreſſed, with very few kidney-ſhaped ſeeds. The ſpecies which is cultivated for the ſake of its roots [e] has ſmooth legumes, no ſtipules, and pinnate leaves conſiſting of four or five pairs of leaflets, terminated by an odd one, which is petiolate. It is a lofty plant for an herbaceous one, the ſtalks being from four to five feet high; the flowers come out in erect ſpikes from the axils, and are pale blue. Glycyrrhiza.

Hedyſarum is a moſt numerous genus, containing no fewer than ſixty-ſeven ſpecies, all however conſpiring in having the keel tranſverſely obtuſe, and the legume jointed, with one ſeed in each joint. The genus is ſubdivided into four ſections, from the leaves; which in the firſt are ſimple; in the ſecond, conjugate; in the third, ternate; and in the fourth, pinnate. I ſhall preſent you only two ſpecies, and they of Hedyſarum.

[e] Glycyrrhiza glabra *Lin.*

the

the laſt ſection. One tranſplanted from Italy into the gardens; and the other from a wild ſtate to a cultivated one. The firſt is the *French Honeyſuckle* [f], which is diſtinguiſhed from the reſt by a diffuſed ſtalk, and by its jointed, prickly, naked, ſtraight legumes; its pinnate leaves point it out to be of the fourth ſection: they have five or ſix pair of leaflets, terminated by an odd one; and from their baſe comes out a long peduncle, ſuſtaining ſpikes of beautiful red flowers. The other is the *Saintfoin* [g]; the characters of which are an elongated ſtem; the wings of the corolla equalling the calyx, and one-ſeeded prickly legumes: this has alſo, of courſe, pinnate leaves. It adorns the chalky hills with its beautiful ſpikes of red flowers; and contributes largely among many others of this claſs to feeding of cattle. For this the *Trefoils* are moſt juſtly celebrated; there are forty-ſix ſpecies of them, all having the flowers growing in a head; and the legume very ſhort, ſcarcely emerging from the calyx, not opening, but falling off entire, and containing but one, or at moſt two ſeeds. Though this be a genus eaſily diſtinguiſhed by its habit, yet the characters are by no means conſtant, and perhaps there is not one com-

Trifolium.

[f] Hedyſarum coronarium *Lin.*

[g] Hedyſarum Onobrychis *Lin.* Rivin. tetr. t. 2. Ger. 1243. 1. Park. 1082. 1.

mon to all the ſpecies. *White Trefoil*, commonly called *Dutch Clover* [h], has a creeping, perennial ſtem; the heads umbelled; and the legumes covered and four-ſeeded. *Purple Trefoil*, *Honeyſuckle Trefoil*, or *Red Clover* [i], has the flowers growing in globular ſubvillous ſpikes, girt with oppoſite membranous ſtipules; and the corollas all of one petal. There are many wild ſpecies of this genus; but the Yellow Trefoil, cultivated under this name, or that of *Noneſuch*, is of another genus, as we ſhall ſee preſently.

Lotus has a tubular calyx; the wings of the corolla clapping cloſe together upwards longitudinally; and an upright cylindric legume. The wild ſpecies is called *common Bird's-foot* [k], and is diſtinguiſhed by its decumbent ſtems, many flowers growing together in depreſſed heads; and exactly cylindric, ſpreading legumes. The corollas are of a bright yellow. Lotus

Lucerne [l] is of the genus *Medicago*, the character of which is that the keel of the corolla bends down from the banner, and that the legume is flatted, and ſpiral or wreathed like the ſhell of a ſnail. The Medicago.

[h] Trifolium repens *Lin.* Curtis, Lond. III. 46. Ger. 1185. 1.

[i] Trifolium pratenſe *Lin.* Blackw. t. 20.

[k] Lotus corniculatus *Lin.* Curtis, Lond. II. 56. Ger. 1190. 5.

[l] Medicago ſativa *Lin.* Mor. hiſt. ſ. 2. t. 16. f. 2. Ger. 1189. 2. Park. 1114. 1.

ſpecific

ſpecific character is this—the ſtem is erect and ſmooth, the flowers grow in a raceme; and the legumes are contorted: the colour of the corollas is blue. The ſpecies cultivated under the name of *Trefoil* or *Nonſuch* [m] has the ſtems procumbent; the flowers in oval ſpikes; and the legumes kidney-form, with one ſeed only in each; the corollas are ſmall and yellow. In a cultivated ſtate the ſtems draw each other up, and loſe, in a great meaſure, their natural procumbency, as does alſo *Bird's-foot Trefoil*, when it has other plants about it, as in graſs-fields, &c. There is a ſpecies of *Medicago* called *polymorphous* or *many-form* [n], from the variety of appearances it puts on, or from the change of figure in the pod. We have one variety very common wild [o], called *Heart-Clover* from the form of the leaves, which are alſo generally ſpotted: each head conſiſts of four or five little yellow flowers; the legumes are globoſe, ſpiral, and covered with very diverging ſpines: and in the garden you have the vegetable *Snails* [p], with large, ſpiral, globoſe legumes, naked, or not covered with ſpines; and the *Hedge-*

[m] Medicago lupulina *Lin.* Curtis, Lond. II. 57. Ger. 1186. 5. Park. 1105. 6.

[n] Medicago polymorpha *Lin.*

[o] Medicago polymorpha arabica *Lin.* Curtis, Lond. III. 47. Ger. 1190. 4. Park 1115. 6.

[p] Med. polym. ſcutellata *Lin.* Mor. hiſt. ſ. 2. t. 15. f. 4.

hogs,

hogs [q], whofe legumes are clofely armed with long fpines pointing every way. Thefe all have the ftem diffufe; the ftipules toothed, and the legumes fpiral. This clafs has alfo its vegetable *Caterpillars*, but they are of another genus [r].

I fear you will think I have already made this letter too long. However, as it may be fome time before you hear from me again; as the next clafs is a very fmall one, and completes the fet of plants with united filaments, I will trefpafs on your patience whilft I go through it.

THE CLASS POLYADELPHIA.

The Clafs *Polyadelphia*, then, comprehends all fuch flowers as have the filaments united at bottom into more than two parcels. The filaments are in bunches, or pencilled, as one might call it, fince they are collected into bodies refembling a camel's hair pencil. If you were not to attend to this character, you might eafily fuppofe thefe plants to belong to the clafs *Polyandria*, for they have no ftriking appearance, like the pulfe tribe and fome others, announcing them immediately to range under this clafs.

There are four orders, taken from the

[q] Med. polym. intertexta. Mor. f. 7, 8, 9.
[r] Scorpiurus. Riv. tetr. 210.

ſtamens; *Chocolate*[s] is in the firſt, *Pentandria*, a genus called *Monſonia* in the ſecond: *Citron*, comprehending *Oranges* and *Lemons*, in the third; and eight genera in the fourth. The whole number of ſpecies is only ſixty-five.

Citrus. The beautiful, odoriferous, well known, and deſervedly eſteemed genus of *Citrus* has theſe characters—a ſmall calyx five-toothed at top; a corolla of five oblong petals; about twenty ſtamens, placed cylindrically round the germ, with the filaments connected rather ſlightly, ſometimes into more, ſometimes into fewer parcels; one piſtil, and, for a fruit, a berry generally nine-celled, with a bladdery pulp, in which the ſeeds are lodged.

You will have pleaſure in examining at leiſure the three elegant ſpecies of this genus, and in regaling your ſenſes, whilſt your mind imbibes inſtruction. When they are in fruit, you diſtinguiſh them immediately; but when they are not, you will find that the *Citron*[t] has the petioles linear or all of a ſize, like moſt other petioles; whereas the *Orange*, *Lemon*, and *Shaddock*, have the petioles winged in ſhape of a heart; ſo that the main leaf ſeems to grow out of a ſmaller one. Linnæus makes the

[s] Theobroma Cacao *Lin.* Sloan. jam. 2. t. 160. Merian. ſurin. t. 26. and 63. Cateſb. car. 3. t. 6.

[t] Citrus Medica *Lin.* Virg. georg. edit. Mart. p. 135.

Orange and *Lemon* [u] to be of one ſpecies, and to be diſtinguiſhed by pointed leaves from the *Shaddock* [v], which has them obtuſe, and emarginate or notched at the end: not to mention the great ſize of the fruit, the flowers of this grow more in racemes, which are alſo a little nappy or woolly. I dare preſume that you are by this time ſo great an adept in Botany as readily to admit, in ſpite of the information of your taſte to the contrary, that the Seville and China Oranges may be varieties of the ſame ſpecies, owing all their difference to climate. Neither perhaps do you find much difficulty in perſuading yourſelf, that the large and generous Lemon may not be ſpecifically different from the little, round, ſour Lime; notwithſtanding ſome little difference in the leaves, and the ſpines on the branches of the latter But I much doubt whether you will be able to perſuade your fair daughter to admit that the auſtere, long, pale Lemon, is not a ſpecies totally diſtinct from the round, deep-coloured Orange, the flavour of whoſe juice ſhe enjoys with ſo much delight. I will conſent that ſhe ſhould enjoy her incredulity, at leaſt if ſhe can diſtinguiſh theſe trees when they are deſtitute of fruit. The poſition of the ſtamens informs you that this genus is of the order *Icoſandria.*

[u] Citrus Aurantium *Lin.* Mill. illuſtr.

[v] Citrus decumana *Lin.* Rumph. amb. 2. t. 24. f. 2.

Hypericum. The genus *Hypericum*, in the laſt order (*Polyandria*) of this claſs, has many more ſpecies than all the other genera put together. Several of them are wild, and ſeveral others are commonly cultivated among ſhrubs: they are not however all ſhrubs, for many ſpecies are herbaceous. All plants do not exhibit the claſſical mark, in this or any other claſs, with equal evidence; in this genus the numerous ſtamens will eaſily ſeparate from the receptacle in pencils or parcels, and thus evidently ſhow what is their proper place in the ſyſtem. Being thus certified that your plant does not belong to the claſs *Polyandria*, but to this, you will eaſily diſtinguiſh it from its congeners, by its five-parted calyx including the germ; by its corolla of five petals; by the abundance of ſtamens, uſually forming five ſquadrons; and by the ſeed-veſſel being a capſule, divided into as many cells as there are ſtyles to the flower; theſe are either one, two, three, or five in number; and hence a ſubordinate diviſion of the genus into four ſections: there is however only one ſpecies with one ſtyle, and there are only two ſpecies with two; the far greater number have three: and among theſe are all the European ones.

Common St. John's wort [w] has two characters ſo remarkable that it cannot well be

[w] Hypericum perforatum *Lin.* Curtis, Lond. 1. 57. Mill. illuſtr. Ger. 539. 1. Park. 573. 1.

miſtaken,

miſtaken, as ſoon as they are underſtood: for it has an *ancipital* or two-edged ſtem, that is, roundiſh, or a little flatted, and running out longitudinally into two little edges or membranes oppoſite to each other: and its obtuſe leaves are punctured all over their ſurface, ſo as to appear, when held up againſt the light, as if they had been pricked with a pin. Another wild ſort not near ſo common, growing in moiſt hedges and woods, and called *Saint Peter's wort*[x] has ſquare ſtalks; it is about the ſame ſize with the other, but does not branch ſo much: the leaves are ſhorter and broader, and have none of the pellucid dots which are ſo remarkable in the former. *Trailing Saint John's Wort*[y] is a pretty little plant, found on dry paſtures and heaths: it has two-edged, proſtrate, filiform ſtems; ſmooth leaves; and axillary, ſolitary flowers. *Upright Saint John's wort*[z] is an elegant ſpecies, growing in woods and heaths; with columnar ſtems: ſtem-claſping, ſmooth, heart-ſhaped leaves; and ſerrated calyxes with the teeth glandular.

The two moſt common ſorts, cultivated among other ſhrubs, are the *ſtinking ſhrubby*[a]

[x] Hypericum quadrangulum *Lin.* Curtis Lond. IV. 52. Fl. dan. 640. Ger. 542. Park. 575.

[y] Hypericum humifuſum *Lin.* Curtis, Lond. III. 50. Fl. dan. 141. Ger. 541. 4.

[z] Hypericum pulchrum *Lin.* Curtis, Lond. I. 56. Fl. dan. 75. Petiv. 60. 6.

[a] Hypericum hircinum *Lin.*

and *Canary*[b] *St. John's worts.* They have both a rank ſmell, reſembling that of a goat, which however, in ſome circumſtances, and at certain diſtances, ſeems to be ſweet, at leaſt to ſome perſons; both alſo have three piſtils: but the firſt is a much lower plant, and has the ſtamens longer than the corolla; whereas in the ſecond they are ſhorter. *Garden Tutſan* is evidently of this genus: it is one of thoſe which have five piſtils; the ſtems are low, ſimple, herbaceous, and quadrangular; the leaves ſmooth, and quite entire: the roots creep extremely, and the flowers are very large. *Wild Tutſan*, or *Tutſan Saint John's wort*[d], called alſo *Park-leaves*, has a ſhrubby two-edged ſtem; three piſtils, and a berried fruit, or ſoft, coloured pericarp: the flowers of this are ſmall, and the ſtamens extend beyond the corollas. It grows wild in woods, and ſometimes in moiſt hedges. Of the more rare and tender ſorts, the *Majorca Saint John's wort*[e] is very diſtinguiſhable by the warts all over the ſlender red branches; the leaves alſo are *repand* or waved on their edges, have ſmall protuberances on their under ſurface, and at the

[b] Hypericum canarienſe *Lin.* Comm. hort. 2. t. 68.

[c] Hypericum Aſcyron *Lin.* Gmel. ſibir. 4. t. 69. Pl. 24.

[d] Hypericum Androſæmum. *Lin.* Curtis, Lond. III. 48. Ger. 543. 1.

[e] Hypericum balearicum *Lin.* Mill. fig. pl. 54. Curt. Mag. 137.

baſe

baſe embrace the ſtalk: the flowers are large, with the ſtamens a little ſhorter than the corolla, and five piſtils. Laſtly, *Chineſe Hypericum* [f], which ſtands alone, as having one piſtil only, has a ſhrubby ſtem, coloured calyxes, ſtamens longer than the corolla, and is one of the moſt beautiful of this genus, ſo gay with its yellow corollas, and abundant crop of ſtamens.

With this large harveſt, I leave you, dear couſin, till I ſhall have found leiſure to prepare the extenſive and moſt difficult tribe of compound flowers for your inſpection.

[f] Hypericum monogynum *Lin.* Mill. fig. pl. 151. f. 2.

LETTER XXVI.

THE CLASS SYNGENESIA.

August the 24th, 1776.

THOUGH this letter, dear cousin, will arrive late in the season, yet it will be in time for you to examine the far greater part of the class *Syngenesia*, or tribe of compound flowers, which blow chiefly in the autumn. You are well aware that the essential character of this class is the union of the anthers. You are perfect mistress of the structure of a compound flower, and of the different florets that compose it[g]. And lastly, the several orders into which the class is divided are familiar to you, and the foundation of them well understood[h]. Very little therefore remains to premise, before we proceed to the examination of the genera and species.

This is by much the most numerous of the natural classes[i]; and therefore it should, in all probability, be more difficult to find sufficient generic and specific distinctions here than in any other: such however

[g] See letter VI.

[h] See letter X.

[i] The number of genera being 116, and of species 1247.

has

has been the ſagacity and induſtry of Linnæus, that I hope you will not find any great difficulty, even in the two firſt orders, which contain above two thirds of all the genera.

THE ORDER POLYGAMIA ÆQUALIS.

To facilitate the inveſtigation, in the firſt order, *Polygamia Æqualis*, it is ſubdivided into three battalions, eaſily diſtinguiſhed by the moſt obvious characters. The firſt contains the flowers compoſed wholly of ligulate florets, which are the Semifloſculous flowers of Tournefort: the ſecond contains the *capitate* or headed flowers: and the third the *diſcoid* flowers. So that there are no radiate flowers in this order: the flowers of the firſt ſection are wholly made up of ſuch florets as compoſe the ray of theſe: in the two other ſections there are none of theſe ligulate corollas or ſemiflorets, but the compound flower is wholly made up of tubulous corollas, or florets properly ſo called: in the ſecond ſection theſe are long, and the calyx bulges out at bottom, as in the thiſtles; in the third, the flowers reſemble a Daiſy or other radiate flower, with the ray pulled off.

The calyx, the receptacle, and the crown of the ſeed will in general be found ſufficient

to

to furniſh the generic diſtinctions in this order[k].

Tragopogon. Thus *Tragopogon* or *Goat's-beard* is known by its ſimple calyx, naked receptacle, and feathered ſtipitate down: and theſe three circumſtances are ſufficient to diſtinguiſh this genus from all others; provided you have firſt aſſured yourſelf, by the rules already laid down, that your flower is of the compound tribe, that each floſcule has the anthers united into a cylinder, which the piſtil, terminated by two revolute ſtigmas, perforates; and that the corollas are all ligulate: for thus it is that you come at the claſs, order, and ſection. I cannot ſuppoſe that you have any difficulty in diſtinguiſhing a natural compound flower from a double one, the creature of art and culture, though the ſimilarity may miſlead thoſe who are not

[k] The calyx is ſingle, or ſimple in *Seriola*, *Geropogon*, *Andryala*, *Tragopogon:* calycled, or furniſhed with a ſecond ſet of leaflets at the baſe, in *Cichoreum*, *Picris*, *Crepis*, *Chondrilla*, *Prenanthes*, *Lapſana*, *Hyoſeris*; in the reſt imbricate. The receptacle is villous in *Scolymus*, *Cithoreum*, *Catananche*, *Seriola*, *Hypochæris*, *Geropogon*; in the reſt it is naked, that is, has neither hairs nor chaffs between the floſcules. *Scolymus* and *Lapſana* have no *pappus* or down: in *Seriola*, *Andryala*, *Crepis*, *Prenanthes*, *Lactuca*, *Hieracium*, *Sonchus*, the down is ſimple; in *Hypochæris*, *Geropogon*, *Tragopogon*, *Picris*, *Leontodon*, *Scorzonera*, *Chondrilla*, it is feathered; in *Cichoreum* the crown of the ſeed is five-toothed, in *Catananche* five-awned, in *Hyoſeris* crowned with a calycle. In ſome genera this down ſits cloſe to the ſeed, in others it is *ſtiped* or *ſtipitate:* that is, has a ſtem interpoſed between it and the ſeed.

accuſtomed

accuſtomed to obſervation; becauſe I am certain that if you have the leaſt doubt, you will pull out a floſcule, in order to ſee whether it has a ſeed, ſtamens, and piſtil, or is only a mere flat petal. But to return to our plant.—*Yellow* or *Common Goat's-beard*[l], which grows wild among the graſs in meadows, is diſtinguiſhed by entire upright leaves, and by the ſegments of the calyx at leaſt equalling in length the outer floſcules. Towards noon you will not eaſily find this plant, becauſe the flowers are then always cloſed: after the flower is paſt, *Goat's-beard* is very apparent, on account of the large globe formed by the down of the ſeeds, till the wind has at length torn them from the receptacle, and wafted them ſeparately to diſtant places.

Salſafy[m], which your gardener will furniſh you with from the kitchen garden, has the ſegments of the calyx much longer than the floſcules, and the peduncles ſwell out remarkably under the flower; which is large, and of a fine blue.

Another plant of this tribe which you may alſo have from the kitchen garden, is the *Scorzonera*, of a genus nearly allied to the laſt; agreeing with it in having a naked receptacle and a feathered ſtipitate down, Scorzonera.

[l] Tragopogon pratenſe *Lin.* Mor. hiſt. ſ. 7. t. 9. f. 1. Ger. 735. 2.

[m] Tragopogon porrifolium *Lin.* Mor. t. 9. f. 5. Ger. 735. Fl. dan. 797. Pl. 25. f. 1.

but

but differing from it by an imbricate calyx, with the ſcales membranaceous about the edge. The cultivated ſpecies [n] has a branching ſtem, and entire, ſtem-claſping leaves, ſlightly ſawed on their edges; the flowers are of a bright yellow.

Sonchus, & Lactuca.

Sowthiſtle and *Lettuce* agree in a naked receptacle, an imbricate calyx, and a ſimple down to the ſeed. But in the firſt the calyx is gibbous, or ſwelling at the baſe; in the ſecond it is cylindric, with membranous edges: the firſt has a ſeſſile down; in the ſecond it is ſtipitate, and the ſeeds are poliſhed. You will always find it uſeful, where you can, thus to bring together and compare plants of nearly allied genera, in order to conſider well their ſimilitudes and differences, and to give you a readineſs in making thoſe minute but important diſtinctions, ſo neceſſary to diſcrimination in natural tribes, wherein all ſeems alike to the untutored eye, as the ſheep of the flock to the ordinary paſſenger; whereas the ſhepherd knows each by its proper marks, and calls them all by their names.

Of the *Sowthiſtle* [o], that vulgar weed of the kitchen garden, there are many varieties; the rough and the ſmooth; with lacerate leaves and ſimple ones, &c. which I

[n] Scorzonera hiſpanica *Lin.* Blackw. 406.

[o] Sonchus oleraceus *Lin.* Curtis, Lond. II. 58. Ger. 292.

mention

mention only that you may not be led to ſearch for them as diſtinct ſpecies; in reality theſe differences are owing merely to accident and ſituation.

Hieracium or *Hawkweed* is a numerous genus of this order and ſection; the calyx is ovate and imbricate, the receptacle naked, and the down ſimple and ſeſſile. There are many ſpecies wild in this country; one[p], which is a large plant, on walls and banks and in woods, with a branching ſtem, the radical leaves oval and toothed, and a ſmaller leaf on the ſtalk: and another very common indeed in dry paſtures, called *Mouſe-ear Hawk-weed*[q], from the long hairs upon the leaves, which are ovate, and abſolutely entire; this ſort throws out runners, and the flowers come out ſingly on naked ſtalks. There are other ſpecies, vulgarly called *Hawkweeds*, which range under other genera, as the *Crepis*, which differs from *Hieracium*, in having the calyx only calycled, with deciduous ſcales. Hieracium.

I ſhall conclude the firſt ſection with *Succory* or *Endive*; which has the calyx calycled, a few chaffs between the floſcules on the receptacle, and the crown of the ſeed moſtly five-toothed and obſcurely hairy. *Wild Succory* Cichoreum.

[p] Hieracium murorum *Lin.* Mor. hiſt. ſ. 7. t. 5. f. 54. Ger. 304.

[q] Hieracium Piloſella *Lin.* Curtis, Lond. IV. 54. Ger. 638. 2. Park. 690. 1, 2.

cory [r] has runcinate leaves, and generally two ſeſſile flowers coming out together: *Endive* [s] has ſolitary, peduncled flowers, and entire leaves, only notched about the edge. Both have flowers of a fine blue; but the firſt is perennial, and the ſecond only biennial. Curled Endive, though differing ſo remarkably from its parent in the leaves, is but a variety of the laſt.

Carduus. The greater part of the ſecond ſection, in this firſt order of the nineteenth claſs, is occupied by the Thiſtles, a moſt untractable genus, not at all adapted to the delicate fingers of our lovely *Flora*. The calyx is all imbricate with thorny ſcales [t]; and how will ſhe tear this aſunder, to diſcover that the receptacle has hairs between the ſeeds; yet theſe two circumſtances form the character of the genus; and ſhe muſt obſerve that there are ſome plants commonly called *Thiſtles*, which are not of the genus *Carduus*. For inſtance, the *Common Way-Thiſtle* [u] not having ſpines to the ſcales of the calyx, which alſo is cylindric in ſhape, whereas in the *Cardui* it bulges out at bottom, and the receptacle being

[r] Cichoreum Intybus *Lin.* Curtis, Lond. IV. 56. Ger. 284. 1. Park. 776. 2.

[s] Cichoreum Endivia *Lin.*

[t] See Pl. 25. f. 2.

[u] Serratula arvenſis *Lin.* Curt. Lond. n. 63. under the name of Carduus. Fl. dan. 644. Mor. hiſt. ſ. 7. t. 32. f. 14. Ger. 1173. 4.

naked,

naked, is not a *Carduus* in Linnæus's idea, but a *Serratula*. So likewiſe *Cotton-Thiſtle*[v] having a honey-combed receptacle, is ſeparated on account of that circumſtance. Indeed the genus would have been too vaſt and unmanageable, without an attention to theſe marks, which might ſometimes appear otherwiſe too minute. You have perhaps even heard it ſaid that the *Artichoke*[w] is nothing but a Thiſtle. It differs indeed very little; having a hairy receptacle, only the hairs being ſtiffer, it may be called briſtly; and the ſtructure of the down being the ſame, they differ principally in the calyx, for the ſcales in the Artichoke are *ſcarioſe* or ragged, fleſhy, and terminated by a channelled appendicle, emarginate and pointed—a character which you may examine at your leiſure at table. If you would ſpeculate on the blue flowers; which being ſo large, will give a good idea of florets; at the ſame time that it is alſo an excellent inſtance of the order *Polygamia-Æqualis*, and the *Capitate* or *Headed* ſection of it; you muſt prevail on your gardener to let ſome heads ſtand long after the time that they ſhould be cut for the table. Cynara.

The *Burdock*, whoſe heads ſometimes faſten themſelves to your clothes as you paſs, is in the ſame diviſion with the Thiſtles: Arctium.

[v] Onopordon Acanthium *Lin.* Curt. Lond. V. 57. Mor. t. 30. f. 1. Ger. 1149. 1.
[w] Cynara Scolymus *Lin.* Blackw. 458.

the

the globoſe form of the calyx, together with the hooked tops of the ſcales which compoſe it, are the eſſential characters of the genus. The common wild ſpecies [x] has very large woolly heart-ſhaped leaves, petiolate, and unarmed.

Eupatorium.

Of the third ſection, with *Diſcoid*, or, as ſome call them, *naked diſcous* flowers, few are at hand. The banks of rivers and ditches will furniſh a ſpecies of *Eupatorium* [y], a large plant with digitate leaves: uſually there are three leaflets to each leaf, which are hairy, and ſharply ſerrate, the middle one the largeſt; ſometimes the ſide leaflets are wholly wanting, and the leaf becomes ſimple: the ſtalks are lofty, rough, and quadrangular; and bear large bunches of ſmall purple flowers on their tops, with about five florets in each calyx. The characters of the genus are an oblong, imbricate calyx, a naked receptacle; a feathered down, and a very long ſtyle, divided half way the length.

Bidens.

The ſame ſituations will produce you the *Bidens*; which has alſo an imbricate calyx: but the receptacle is chaffy; the corolla is ſometimes furniſhed with one floret alternately radiant; and the ſeeds are crowned

[x] Arctium Lappa *Lin.* Curtis, Lond. IV. 55. Ger. 809.

[y] Eupatorium cannabinum *Lin.* Fl. dan. 745. Mor. hiſt. ſ. 7. t. 13. f. 1. Ger. 711. 2. Common Hemp-Agrimony. See Pl. 25. f. 3.

with

with two erect, rugged awns, which being hooked make the seeds adhere to any thing that comes near them. We have two wild species, the *trifid*[z], so called from its trifid leaves; with erect seeds, and leafy calyxes: and the *nodding*[a], with lance-shaped, stem-clasping leaves, nodding flowers, and erect seeds. The corollas of both are yellow; but those of the last, which is the least common, are most specious.

THE ORDER POLYGAMIA SUPERFLUA.

The second order of the class Syngenesia, entitled *Polygamia superflua*, being scarcely less numerous than the first, is subdivided into two sections, the first containing the discoid, and the second, the radiate flowers: there is only one genus in this order with semiflosculous flowers.

Of the first section, with discoid flowers, you have the *Tansy*; which you find to have an imbricate, hemispheric calyx; the corollas of the ray, or on the outside, trifid; the others quinquefid; the seeds naked, being only slightly edged; and the receptacle naked. Sometimes in this genus there are no imperfect flowers. Our *common* Tanacetum.

[z] Bidens tripartita *Lin.* Water Hemp-Agrimony. Curtis, Lond. IV. 57. Ger. 711. 1.

[a] Bidens cernua *Lin.* Nodding Water Hemp-Agrimony. Curtis, Lond. III. 55. Fl. dan. 841.

Tansy[b], which not only the kitchen-garden, but dry, upland pastures will furnish you with, has bipinnate, or twice-feathered leaves, which are gashed, and serrate about the edges.

Artemisia. *Southernwood*, the *Wormwoods* and *Mugwort*, all range under the genus *Artemisia*; which has a calyx imbricate, with rounded, converging scales; naked seeds; and a receptacle either naked or with few hairs: the flowers have no ray whatever, but are strictly discoid. *Southernwood*[c] is shrubby, erect, and has setaceous leaves, very much branched: there is a *field* or *wild Southernwood*[d], with procumbent, twiggy stems, and multifid, linear leaves. *Common* and *Roman Wormwoods* and *Mugwort* have erect herbaceous stems, and compound leaves. The *Common*[e] species has the leaves multifid, the flowers subglobular and pendulous, and the receptacle hairy. *Roman Wormwood*[f] has the leaves many-parted, and downy underneath, the heads of flowers roundish and nodding, as in the other; but the receptacle naked. *Mugwort*[g] has pin-

[b] Tanacetum vulgare *Lin.* Fl. dan. 871. Mor. hist. 3. 6. t. 1. f. 1. Ger. 650. 1.

[c] Artemisia Abrotanum *Lin.* Blackw. 555.

[d] Artemisia campestris *Lin.* Ger. 1106. 5. Park. 94. 7.

[e] Artemisia Absinthium *Lin.* Blackw. t. 17. Ger. 1096. 1.

[f] Artemisia pontica *Lin.* Jacq. austr. 1. t. 99.

[g] Artemisia vulgaris *Lin.* Blackw. t. 431. Ger. 1103. 1.

natifid, flat, gaſhed leaves, downy underneath: the flowers are borne in ſimple, recurved racemes, and have a ray of five flowers. *Common Sea Wormwood*[h] has procumbent ſtems; many-parted downy leaves, nodding racemes, and three flowers in the ray.

Gnaphalium, comprehending many wild *Cudweeds* and the *Immortal* flowers, or *yellow* and *white Everlaſtings*, has an imbricate calyx, with the ſcales rounded, ſcarioſe, and coloured; a naked receptacle, and feathered down. There are ſeveral ſpecies both of yellow and white Everlaſtings; the moſt known of the firſt, is common in Portugal, where they adorn their churches with the flowers, which are alſo ſent annually to England: it is ſuppoſed to have been brought originally from India[i]: the leaves are linear-lanced, and ſeſſile: the flowers are borne in a compound corymb, on elongated peduncles; and the ſtem is ſubherbaceous. One of the latter[k] is very common in the gardens, and is originally of North America; this has leaves like the former, ſharp-pointed, and alternate; the ſtems herbaceous, and branched above, the flowers in corymbs, with level tops. This Gnaphalium.

[h] Artemiſia maritima. Ger. 1099. 1. Petiv. 2c. 2.
[i] Gnaphalium orientale *Lin.* Comm. hort. 2. t. 55. Mor. hiſt. ſ. 7. t. 10. f. laſt.
[k] Gnaphalium margaritaceum *Lin.*

 has

has a very creeping root; and the ſtalks and leaves are woolly: the ſilvery calyxes, as well as the golden ones, of the former, if gathered before they are too open, will continue in beauty many years.

Xeranthemum. *Xeranthemum*, or *Eternal flower*, has an imbricate calyx, with the inner ſcales membranaceous, ſhining, and forming a ſet of coloured rays to crown the flower; the receptacle is moſtly naked; and the down is either briſtly or feathered. *Annual Xeranthemum*[l] is an exception to the general character, in having a chaffy receptacle; it is alſo the only one which has a down of five briſtles: it is herbaceous, has lance-ſhaped ſpreading leaves; the outſide florets have a ſimple ſtigma, with a naked ſeed; thoſe in the middle have a ſub-bifid ſtigma. The colour of the corolla is either purple or white. There is a ſort from the Cape with yellow flowers[m].

Tuſſilago. The ſecond diviſion of this order, with *Radiate* flowers, is much the largeſt. *Tuſſilago* or *Colt's-foot* has a cylindric calyx, with equal ſcales, from fifteen to twenty in number, as long as the diſk of the flower, and a little membranous; a naked receptacle, and a ſimple or hairy down. *Common wild Colt's-foot*[n] has angular leaves, rather

[l] Xeranthemum annuum *Lin.* Mill. illuſtr. Jacq. auſtr. 4. 388.

[m] Xeranthemum ſpecioſiſſimum. Seba 2. t. 43. f. 6.

[n] Tuſſilago Farfara *Lin.* Curtis, Lond. II. 60. Ger. 811. Park. 1220.

heart-

heart-ſhaped, with ſlight indentations about the edges, underneath white; and one yellow flower on a ſcape, which is imbricate or covered with ſcales, *Butter-bur* [o] has vaſt leaves ſhaped much like thoſe of the *Colt's-foot*; many (from ten to twenty) purpliſh flowers, collected into an ovate thyrſe, on the top of a purpliſh ſcape ſet with ſcales of the ſame colour: there are ſometimes from two to ſix imperfect, white, ligulate florets, with ſcarcely any corolla, among the others. You will not be able to examine all the ſpecific characters of theſe two plants at once; for the naked ſtem which bears the flowers puſhes up alone very early in the ſpring; and the leaves do not ſucceed till the flowers are paſt.

Senecio, or *Groundſel*, is a very numerous genus [p], having a cylindric calycled calyx, with the ſcales *ſpacelate* or ſeeming mortified at top; a naked receptacle, and a ſimple down. Moſt of the ſpecies have radiate flowers, eight of them however have not, and among theſe is the *Common Groundſel* [q], ſo vulgar a weed in kitchen-gardens. *Stinking Groundſel* [r], a plant not very unlike this, has however radiate corollas, with the ſe- Senecio.

[o] Tuſſilago Petaſites *Lin.* Curtis, Lond. II. 59. Ger. 814.

[p] Fifty-nine ſpecies.

[q] Senecio vulgaris *Lin.* Curtis, Lond. I. 61. Ger. 278. 1.

[r] Senecio viſcoſus *Lin.* Dill. elth. t. 258. f. 336.

miflorets of the ray revolute; the ſcales of the calyx are looſe; and the leaves are pinnatifid and viſcid. This grows in hedgerows and on heaths, and is a much taller plant than the laſt.

Common Ragwort[s] has alſo radiate corollas, with the ray however not revolute but expanding: the ſtem of this is erect; the leaves pinnatifid, approaching to lyrate, with the diviſions a little jagged. This is very common by road-ſides and in paſtures. The gardens have a *purple African Groundſel*[t] from the Cape; an annual plant with a yellow diſk, and purple rays: it agrees with Ragwort in having radiate corollas with the ray expanding; the leaves are pinnatifid, equal, and very ſpreading, with a thickened recurved margin; and the ſcales of the calyx are thinly ciliated. A ſingular plant of this genus came up one year in my garden, which I took at firſt to be a new ſpecies; but, on more accurate examination, it proved to be a hybridous plant or mule, produced from this and the common Groundſel; it had the radiate flowers of the one, ſmall indeed and ſlightly tinged with purple, and the herb of the other: being annual, and producing no ſeed, this variety paſſed away with the ſeaſon.

[s] Senecio Jacobæa *Lin.* Mor. hiſt. ſ. 7. t. 18. f. 1. Ger. 280. 1. Park. 668. 1.

[t] Senecio elegans *Lin.* Comm. hort. 2. t. 30. Seba mus. 1. t. 22. f. 1.

The

The two genera of *Aster* and *Golden-rod* furnish abundance of flowers that enliven the autumnal season, and continue till the severity of frost puts an end to them. They both agree in an imbricate calyx, a simple down, and a naked receptacle: but the inferior scales in the calyx of the *Aster* are spreading, and have a ragged appearance; whereas in the *Golden-rod* they are close: all the species also of the *Aster* have more than ten semi-florets in the ray, but the *Golden-rods* have only about five or six remote ones. Some of the *Asters* are shrubby, but most of them are tall herbaceous plants, dying down to the ground at the approach of winter, and rising again from the same root the ensuing spring: many are confounded under the vulgar title of *Michaelmas Daisies*. The *Amellus*, or *purple Italian Starwort*[u], is one of the lowest species, but has large purple flowers, growing in a corymb on naked peduncles, with the scales of the calyx obtuse; the leaves are lance-shaped, obtuse, rugged, entire about the edges, and marked underneath with three nerves. The greater part of the perennial American Asters have scaly peduncles; some have entire, and others have serrate leaves; hence a convenient subdivision of the genus: there are however some few species with serrate leaves Aster.

[u] Aster Amellus *Lin.* Jacq. austr. 435. Virg. georg. edit. Mart. p. 368.

and naked ſmooth peduncles. *Large flowering* or *Cateſby's Starwort*[v], is one of the handſomeſt; the flowers being large and of a deep purple; the calyx is ragged; the peduncles are ſcaly, and ſuſtain only one flower; the leaves are quite entire, tongue-ſhaped, and claſp the ſtem. *Chineſe Aſter*[w] is an annual plant, with ovate, angular leaves, toothed about the edge, and petiolate; the flowers terminate the branches, and have ſpreading leafy calyxes. The variety of colour, and ſize of the corolla, have made this ſpecies very generally cultivated: their being frequently double, will not induce you to miſtake a double radiate for a natural ligulate flower; which, to an unobſerving eye, it perfectly reſembles. The ſalt-marſhes on the ſea-coaſt of Europe furniſh one ſpecies, called *Sea-Starwort*[x]: this has lance-ſhaped, entire, fleſhy, ſmooth leaves; the branches are unequal; and the flowers in a corymb.

Solidago. Of the *Golden-rods* we have only one European ſpecies[y], unleſs we diſtinguiſh the *Welſh Golden-rod*[z], which ſeems but an

[v] Aſter grandiflorus *Lin.* Mart. cent. 19. Mill. fig. 292.

[w] Aſter chinenſis *Lin.* Dill. elth. t. 34. f. 38.

[x] Aſter Tripolium *Lin.* Fl. dan. 615. Mor. hiſt. ſ. 7. t. 22. f. 36. Ger. 413. 1. Park. 674.

[y] Solidago Virgaurea *Lin.* Fl. dan. 663. Mor. t. 23. f. 4. Ger. 430. 2.

[z] Solidago cambrica *Hudſ.* Dill. elth. t. 306. f. 303. Petiv. herb. Brit. t. 16. f. 11.

humble

humble variety. The ftem is a little flexuofe or winding; and the flowers grow in erect, crowded, panicled racemes. The Welfh variety has the leaves a little hoary underneath, and roundifh cluftered fpikes at the top of the ftalk, with larger flowers appearing earlier than the common fort: in lofty fituations and dry foils, a ftem will fometimes produce one flower only. North America has furnifhed abundance of fpecies, whofe golden racemes of flowers mix happily with the purple corymbs of the Afters; and thus they jointly enliven plantations of fhrubs in the latter feafon.

Inula, of which *Elecampane*[a] is the leading fpecies, has the following characters— a naked receptacle; a fimple down; and the anthers ending at the bafe in two briftles: this ftructure of the anthers is *unique*— the cylinder is compofed of five fmaller linear anthers, each ending in two briftles, of the length of the filaments. The true *Elecampane*[a] is diftinguifhed by its large, ftem-clafping, ovate, wrinkled leaves, downy underneath; and by the ovate form of the fcales of the calyx. The ftalks are three feet high, and divide towards the top into feveral fmaller branches, each of which is terminated by one large yellow flower. The Inula.

[a] Inula Helenium *Lin.* Fl. dan. 728. Mor. hift. f. 7. t. 24. f. laft. Ger. 793.

Flea-

Flea-banes middle [b] and *leſs* [c] are of this genus; the firſt is common in moiſt meadows, and has ſtem-claſping, oblong leaves, hollowed next the petiole; a villous ſtem terminated by yellow flowers in panicles; and the ſcales of the calyx briſtly. The ſecond [c] has alſo ſtem-claſping leaves, but waved; proſtrate ſtems; and ſubglobular flowers, eaſily known by the ſhortneſs of the ray. The place of this is by road-ſides, and where water ſtands in winter.

Doronſ-cum. *Doronicum*, or *Leopard's-bane*, a wild plant of the Alps, and now common among the perennials of the garden, has the ſcales of the calyx in two rows, equal, and longer than the diſk, the ſeeds of the ray naked or deſtitute of down; thoſe of the diſk crowned with a ſimple down; the receptacle naked. The common ſpecies, above alluded to [d], has heart-ſhaped leaves, ſlightly indented about the edge, and obtuſe at the end; thoſe at the root petiolate, thoſe above ſtem-claſping. The ſtalks are channelled and hairy, near three feet high: theſe put out a few ſide branches, each of which is terminated by a large yellow flower. A ſecond ſpecies [e] has ovate, acute leaves,

[b] Inula dyſenterica *Lin.* Curtis, Lond. III. 56. Ger. 482. 3.

[c] Inula pulicaria *Lin.* Curtis, Lond. III. 57. Ger. 482. 4.

[d] Doronicum pardalianches *Lin.* Mill. fig. 128. Jacq. auſtr. 4. t. 350. and Pl. 26. of this work.

[e] Doronicum plantagineum *Lin.*

ſlightly

ſlightly indented, and alternate branches. A third[f] has a naked, ſimple ſtem ending in one flower: and theſe make up the whole genus.

Tagetes has a one leafed, five-toothed, tubular calyx; five permanent florets to the ray; the ſeeds are crowned with five erect awns; and the receptacle is naked. *French*[g] and *African*[h] Marigolds, two of the gaudy annuals of the flower-garden, are of this genus. The firſt is diſtinguiſhed by a ſubdivided ſpreading ſtem; the ſecond, by an erect, ſimple ſtem, with naked, one-flowered peduncles. Of both theſe, as you well know, there are many varieties in colour, from pale brimſtone to deep orange; and the more double they become, ſo much the more does your gardener value himſelf on his ſkill or good fortune. Tagetes.

Chryſanthemum, ſo named from its golden-coloured flowers, is known by its hemiſpheric, imbricate calyx, formed of cloſe ſcales, the inner ones gradually larger, and the inmoſt membranous or chaffy; there is no down to the ſeeds, but they are only edged or margined; the receptacle is naked. Some of the ſpecies are improperly termed *Chryſanthema*, having white rays to the flowers: of theſe we have an inſtance in Chryſanthemum.

[f] Doronicum Bellidiaſtrum *Lin.* Jacq. auſtr. 4. t. 400.
[g] Tagetes patula *Lin.*
[h] Tagetes erecta *Lin.*

the

the *Ox-eye Daiſy*[i], a plant common among ſtanding graſs in meadows, and having oblong, ſtem-claſping leaves, ſawed above, and toothed below. *Corn Marigold*[k], which is a weed among the corn in ſandy lands, has yellow rays, and ſtem-claſping leaves, jagged above, and toothed below; they are ſmooth, and of a glaucous hue. Leſt you ſhould think the colour of more importance than it really is, I will put you in mind, that the ſpecies ſo commonly cultivated in flower-gardens under the name of *Chryſanthemum creticum*[l], has both yellow and white rays: theſe flowers are eſteemed in proportion as they deviate from nature; but the plant may always be known, by the pinnate, gaſhed leaves, growing broader towards the end.

Matricaria. The three genera of *Matricaria*, *Cotula*, and *Anthemis*, are nearly allied. The firſt has a hemiſpheric, imbricate calyx, with the marginal ſcales ſolid, and rather acute; the ſeeds have no down; and the receptacle is naked. The ſecond has a convex calyx; the florets of the diſk quadrifid; thoſe of the ray have only a germ with its ſtyle and ſtigmas, without any corolla: there is no down, but the ſeed is margined: and the receptacle

[i] Chryſanthemum Leucanthemum *Lin.* Curt. Lond. V. 62. Blackw. t. 42. Mor. hiſt. ſ. 6. t. 8. f. 1. Ger. 634. Park. 528. 1.

[k] Chryſanthemum ſegetum *Lin.* Curt. Lond. n. 63. Mor. t. 4. f. 1. Ger. 743. 1. Park. 1370. 1.

[l] Chryſanthemum coronarium *Lin.* Mor. t. 4. f. 2, 3.

is naked, or nearly ſo. The third has a hemiſpheric calyx, with the ſcales nearly equal; more than five ſemiflorets in the ray; no down; and a chaffy receptacle. There are plants vulgarly known by the name of *Mayweed* or *Camomile*, in each genus. *Common Fever-few*[m] alſo is a ſpecies of *Matricaria*: the leaves are compound and flat, the diviſions are ovate, and gaſhed, and the peduncles are branched: it grows upon banks, has a ſtrong, unpleaſant ſcent, the leaves are of a yellowiſh green, and the rays of the flower are white: admitted into gardens, it has generally double flowers. *Common* or *true Camomile*[n] is an *Anthemis*; and has compound pinnate leaves; the diviſions linear, acute, and a little villous. It ſometimes covers a conſiderable extent of ground on dry ſandy commons, trailing along, and putting out roots from the ſtalks; its agreeable odour betrays it as we tread upon it: that which is found in gardens, has uſually loſt all character by cultivation. Anthemis.

Achillea or *Milfoil* has an oblong-ovate imbricate calyx; from five to ten ſemiflorets in the ray; no down; and a chaffy receptacle. *Common wild Milfoil* or *Yarrow*[o] has Achillea.

[m] Matricaria Parthenium *Lin.* Fl. dan. 674. Ger. 652 1.

[n] Anthemis nobilis *Lin.* Blackw. 298. 1. Ger. 755. 4.

[o] Achillea Millefolium *Lin.* Curt. Lond. n. 63. Fl. dan. 737. Mor. hiſt. ſ. 6. t. 11. f. 6, 14. Ger. 1072. 2. A. Ptarmica, Curt. Lond. V. 60.

bipinnate

bipinnate naked leaves, the divisions of which are linear and indented; the stems are furrowed above. It is a vulgar plant in pastures, and particularly by way sides; for it seems to delight in being trod upon, and in such places spreads itself abundantly. The usual colour of the flower is white, but it sometimes varies to a fine purple. Other foreign species are yellow.

The four remaining orders of this class being much less numerous than the two which we have already examined, there is not the same occasion for subdivisions; and accordingly Linnæus has not made any.

THE ORDER POLYGAMIA FRUSTRANEA.

The third order of *Frustraneous Polygamy* comprehends no more than seven genera, from which I shall select two—*Helianthus* and *Centaurea*. The first has an imbricate calyx, rather squarrose, or having a ragged appearance from the spreading of the tips of the scales; a two-leaved or two-awned crown to the seeds; and a flat chaffy receptacle. Every species of this genus is a native of America alone, and on the discovery of the new world, some of them were vaunted as miracles of nature, though they are now become so common as almost to be disregarded.

Helianthus.

garded. The *annual Sun-flower*[p] however it muſt be acknowledged is a flower of wonderful magnificence, and owes the diminution of regard to the facility of its propagation: the ſpecific characters are heart-ſhaped leaves, marked with three principal nerves; peduncles thickening immediately under the calyx; and the flowers nodding. No flower is more proper than this, from its great ſize, to give you an idea of a compound flower, and its component floſcules, or florets and ſemiflorets; only you will remember not to expect ſeeds from thoſe of the ray, that being the character of the order. This plant had its name from the form of the flower, not from any power it poſſeſſes of turning towards the ſun: there is uſually but one flower on a ſtalk, but I had four in my garden on a ſingle ſtem, looking to the four cardinal points. *Perennial Sun-flower*[q] is yet more common than the laſt, becauſe it ſpreads much at the root, and requires no care in the cultivation: the inferior leaves of this are heart-ſhaped and three-nerved, but the upper ones ovate. The flowers, though much ſmaller than thoſe of the laſt, are yet the largeſt and moſt ſightly of the perennial ſorts, and the ſame plant produces abundance of them. You will be on your guard againſt double flowers. The perennial ſorts ſeldom

[p] Helianthus annuus *Lin.* Mill. illuſtr.

[q] Helianthus multiflorus *Lin.* Pluk. phyt. 159. f. 2.

produce

produce feeds in our climate: whereas the annual, which can be propagated no otherwife, has them in plenty. *Jerufalem Artichoke*[r] is alfo a fpecies of *Helianthus*; the leaves are *ovato-cordate*, or egg-fhaped, only hollowed at the bafe; they are alfo marked with three principal nerves: this frequently does not even flower, but it is cultivated not for the fake of thefe, but the tuberous or knobbed roots, refembling in form the potatoe, but in tafte an artichoke bottom. There is a fpecies which has the common or trivial name of *giganteus* or *giant*: *Jerufalem Artichoke* juftly merits the fame title, for I have meafured ftems of it twelve feet high.

Centaurea. *Centaurea* is a moft numerous genus of the fame third order, containing no lefs than fixty-fix fpecies. The corollas of the ray are funnel-form, or tubular, longer than thofe of the difk, and irregular; the down is fimple; and the receptacle has briftles between the florets. This otherwife unwieldy genus is commodioufly fubdivided into fix fections, by the variations of the calyx, which you obferve make no part of the generic character. I. Plants commonly called *Jaceas*, with fmooth, unarmed calyxes. II. *Cyanufes*, with the fcales of the calyx ferrate and ciliate. III. *Rhaponticums*, with dry, fcariofe fcales, like chaff, or as if parched. IV. *Stoebes*, with the fpines of the calyx

[r] Helianthus tuberofus *Lin.* Jacq. hort. 2. t. 161.

palmate.

palmate. V. *Calcitrapas*, with the ſpines of the calyx compound or ſubdivided. VI. With the ſpines ſimple or wholly undivided. To the firſt ſection belongs the *Sweet Sultan*[s], which has a roundiſh calyx with ovate ſcales; and lyrate leaves, indented about the edge. It is an annual plant, with purple flowers, of a ſweetneſs ſo powerful as to be offenſive to many perſons; they come out ſingly on long naked peduncles, and frequently vary to fleſh colour and white. There is a yellow Sweet Sultan, which differs not only in the colour of the flowers, and in having a milder odour, but alſo in having the edges of the leaves ſerrate: it is doubtful however whether it be a diſtinct ſpecies from the former. The *Great* or *Officinal Centaury*[t] is alſo of this ſection: the ſcales of the calyx are ovate; the leaves are pinnate; the diviſions ſerrate and decurrent. The plant is large and tall, and the flowers are purple.

Of the ſecond ſubdiviſion we have three plants commonly wild, and one little leſs common in gardens. *Common* or *Black Knap-weed*[u], perhaps more properly *Knob-Weed*, which the country people in ſome places call *Hard-heads*, is found in almoſt all paſtures, and is one inſtance, among many

[s] Centaurea moſchata *Lin.* Mor. hiſt. ſ. 7. t. 25. f. 5.

[t] Centaurea Centaureum *Lin.* Blackw. 93.

[u] Centaurea nigra *Lin.* Ger. 727. 1. Park. 468. 1.

 others,

others, of the vile weeds which are ſuffered to occupy graſs fields with impunity; the ſcales are ovate, with erect, capillary cilias: the leaves are lyrate and angulate; and the flowers are floſculous. *Great Knapweed*[v] has pinnatifid leaves, with the lobes lanceolate. This grows in corn fields and on balks. The flowers of both are red; but thoſe of the latter are much the largeſt and moſt ſpecious. *Blue-Bottle*[w], the third wild plant of this ſection, which every body knows for an univerſal weed among corn, and whoſe beautiful blue colour would have attracted regard, had it been rare, has linear leaves, which on the ſtem are quite entire; towards the ground they are broader, indented about the edges, and ſometimes pinnate. *Mountain Blue-bottle*[x], which has migrated from the Swiſs mountains into our gardens, is very nearly allied to this, but its flowers are much larger: the leaves alſo are lance-ſhaped and decurrent, and the ſtem is quite ſimple, whereas the wild ſort is branched. *Carduus Benedictus*, or *Bleſſed Thiſtle*[y], is an inſtance of the fourth ſection: it has doubly ſpined, woolly calyxes, furniſhed with an *involucre*; the leaves are ſemi-decurrent, in-

[v] Centaurea Scabioſa *Lin.*

[w] Centaurea Cyanus *Lin.* Mor. t. 25. f. 4. Ger. 732. 2. Park. 482. 2.

[x] Centaurea montana *Lin.* Mill. fig. 114. Curt. mag. 77. Pl. 27. f. 1.

[y] Centaurea benedicta *Lin.*

dented,

dented, and prickly: this is a ſmall annual plant with yellow flowers. We have a wild ſpecies of this ſection—the *Star-thiſtle*[z], growing by road-ſides, and in dry paſtures, but not every where: it has ſeſſile flowers, with the calyxes rather doubly ſpined: the leaves pinnatifid, linear, and toothed; the ſtem hairy, and much branched: the ſpines of the calyx are white, and the flowers red. Of the other ſections none are likely to meet your eye; indeed the roughneſs and vulgarity of their habit, in which they much reſemble Thiſtles, have occaſioned the numerous ſpecies to be little cultivated.

THE ORDER POLYGAMIA NECESSARIA.

The *Marigold* of the kitchen garden will furniſh a familiar inſtance of the fourth order—*Polygamia Neceſſaria.* The genus is known by a calyx of many equal leaves; by the ſeeds having no down, and thoſe of the diſk being membranous; and by the receptacle being naked. The *common* or *officinal*[a] ſpecies is diſtinguiſhed in having all the ſeeds boat-ſhaped, bent inwards and muricate. Calendula.

[z] Centaurea Calcitrapa *Lin.* Ger. 1166. 1.

[a] Calendula officinalis *Lin.* Mill. illuſtr. Pl. 27. f. 2.

THE ORDER POLYGAMIA SEGREGATA.

Echinops. In the *Segregate* order, besides the calyx or perianth common to the whole flower, there is a secondary one, including several floscules, or sometimes one only; this forms one character of the genera. *Echinops* has only one flower to each partial calyx: besides this, the floscules are tubular, and complete; the seeds have an obscure down; and the receptacle is bristly. *Common Globe-thistle* [b] is so called from the flowers growing in globular heads: the leaves are sinuous and pubescent, the jags ending in spines; the flowers are blue, and sometimes white.

THE ORDER MONOGAMIA.

We have now done with the natural tribe of compound flowers, but there remains yet one order of the class *Syngenesia*, in which the flowers are totally different, except in the common character of the union of the five anthers; they are simple, like the flowers of other classes, or have only one corolla inclosed within the calyx, without any common perianth. The *Violet* will furnish you with a number of notorious examples of this order. All the species,

Viola.

[b] Echinops sphærocephalus *Lin.* Mill. illustr. & Pl. 28.

which are twenty-eight, agree in a five-leaved calyx; a five-petalled irregular corolla, produced into a horn or ſpur behind; and in a three-valved, one-celled capſule, above the receptacle, or incloſed within the calyx, the *Sweet Violet* [c], that ſcents the banks, hedges, and borders of woods, in the ſpring, with its fragrant purple flowers, is one of thoſe which have no ſtalks, except the ſcape which ſupports the flower, and the runners by which they are propagated; the leaves are heart-ſhaped. The corollas are ſometimes white, and the gardens boaſt a large double variety. This is one of the few wild plants, whoſe allowed merit has ſecured it a place in every cultivated ſpot. The later ſpecies without ſcent, commonly called *Dog Violet* [d], is one of the cauleſcent or ſtalky kind, the more adult ſtems aſcending; the leaves are heart-ſhaped, but drawn to a point at the end: the corolla is paler than that of the Sweet Violet, and having leaves proceeding from a ſtalk, cannot be miſtaken for that in which they grow immediately from the root, even if the odour were not attended to. *Heart's-eaſe* or *Panſies* [e], the univerſal favourite

[c] Viola odorata *Lin.* Curtis, Lond. I. 63. Ger. 850. 1. Pl. 29.

[d] Viola canina *Lin.* Curtis, Lond. II. 61. Ger. 851. 6.

[e] Viola tricolor *Lin.* Curtis, Lond. I. 65. Fl. dan. 623.

favourite of the more ſimple, unrefined ages, is one of thoſe which have pinnatifid ſtipules, and an urceolate or pitcher-ſhaped ſtigma; it has alſo a three-cornered, diffuſe ſtem; and oblong gaſhed leaves. Such are the characters of a plant, which every child becomes acquainted with as ſoon as he can walk into a garden: but it is not therefore wholly uſeleſs to mention it, becauſe it may at leaſt ſerve to explain ſeveral terms to you, and to aſſiſt you in the examination of plants with which you are not ſo well acquainted.

When we compare the diminutive and almoſt colourleſs Panſy, which we find wild among the corn, with the ample rich-coloured corolla, that boaſts the tiſſue of velvet, ſuch as we ſee in ſome curious gardens; we cannot but allow that human art has made a conſiderable improvement; and we ſurvey it with the more pleaſure becauſe it is not at the expenſe of the natural characters of the flower; and you may enjoy it both as a botaniſt and a floriſt.

Impatiens. That beautiful flower called *Balſam* is of this order. Linnæus names the genus *Impatiens*, becauſe the capſule when ripe is

623. Ger. 854. 1. This has numberleſs provincial names, bearing ſome alluſion to love.

> " Yet markt I where the bolt of Cupid fell.
> " It fell upon a little weſtern flower,
> " Before milk white, now purple with Love's wound,
> " And maidens call it *Love in Idleneſs*."
>
> Midſum. Night's Dream, II. 2.

impatient

impatient of the touch, eaſily burſting, and thus throwing out its ſeeds. It has an irregular corolla of five petals like the violet, when it has not been improved into beautiful duplicity by culture; but the calyx is two-leaved; the nectary or horn is cucullate or cowl-ſhaped; and the capſule is five-valved. *True Balſam*, or, more properly, *Balſamine*[f], has the leaves lance-ſhaped, thoſe on the upper part of the plant alternate; the flowers come out three or four together, from the joints of the ſtalk, only one on each ſlender peduncle; and the nectary is ſhorter than the flower: the varieties of colour—white, red, purple and variegated, are well known. That which comes from the Eaſt-Indies has larger, finer flowers than what comes from the Weſt, moſt beautifully variegated with ſcarlet and white, or purple and white. We have a wild ſpecies called *Yellow Balſam*, and alſo by the familiar names of *Quick in hand*, or *Touch me not*[g]: one long ſlender peduncle comes out from the axils, which ſubdivides into ſeveral others, each ſuſtaining a yellow flower; the leaves are ovate; and the ſtem ſwells at the knots. This is a local plant, being obſerved only or chiefly in Weſtmoreland and Yorkſhire, in moiſt

[f] Impatiens Balſamina *Lin.* Mill. fig. pl. 59.

[g] Impatiens noli tangere *Lin.* Fl. dan. 582. Ger. 446. Park. 296. 5.

ſhady places, or by the ſides of lakes and rivers.

You have now abundant amuſement for your autumnal walks; and as the ſeaſon for examination will be over before I ſhall have leiſure to prepare you freſh matter for future amuſement, I take leave of you till the enſuing ſpring; when, if health and leiſure permit, we ſhall travel through the few remaining claſſes.

LETTER

LETTER XXVII.

THE CLASS GYNANDRIA.

May the 1ſt, 1777.

I RENEW our purſuit as early as poſſible, my dear couſin, in order that I may be able to accompliſh my purpoſe of completing our original ſcheme during the courſe of the preſent ſeaſon.

The twentieth claſs, which falls now under our conſideration, is entitled *Gynandria*, from a circumſtance peculiar to it, which is that of having the ſtamens ſituated upon the ſtyle itſelf. You have remarked, that in every claſs hitherto examined, theſe two parts are entirely independent, ſo that we can at any time remove the one from a flower, and leave the other; but in the claſs *Gynandria* this is not permitted us; the ſtamens uſually growing out of the piſtil itſelf; but in ſome caſes upon a receptacle, produced or lengthened in form of a ſtyle, which bears both piſtil and ſtamens. This claſs has nine orders, founded on the number of ſtamens in the flowers of each; the genera are 33, and the ſpecies 275.

The firſt order, called *Diandria*, from there being two ſtamens only to the flowers in it, is perfectly natural; that is, contains a tribe a tribe of plants agreed upon by all the world to be in ſtrict alliance; or ſuch, as when an eye properly informed has ſeen one of them, it immediately refers any of the others to the ſame tribe, clan, or family, as ſoon as they occur. Indeed the alliance between the greater part of theſe plants is ſo ſtrict, that ſome nomenclators have been induced to refer them to one genus, or one family properly ſo called: for the genera differ hardly in any thing elſe from each other but in the ſhape of the nectary. Some former nomenclators had eſtabliſhed the genera upon the roots, which are certainly the part leaſt proper for this purpoſe, becauſe you cannot examine the character, without deſtroying the plant. But they were induced to it, from the ſingular form of the roots in this tribe: which in ſome ſpecies are a pair of ſolid bulbs; in others a ſet of oblong fleſhy bodies tapering to the extremities, and ſpreading out like the fingers, whence they have the name of *palmate* or *handed*.

Having ſaid ſo much of this tribe, it is almoſt time, you think, to be acquainted with the ſingular perſonages that compoſe it. The far greater number of them then have the common appellation of *Orchis*, a name I am perſuaded you are not wholly unacquainted with.

Orchis. Take one of theſe flowers, of any ſort you can meet with; or, if no ſpecies is yet in in blow, you will not have long to wait for ſome of them. You will find an oblong, writhed germ, below the flower, which has no proper calyx, but only ſpathes or ſheaths: the corolla is made up of five petals, the two innermoſt of which uſually join to form an arch or helmet over the top of the flower; the lower lip of the corolla forms the nectary, taking the place of the piſtil and a ſixth petal: the ſtyle adheres to the inner edge of the nectary, ſo that, together with its ſtigma, it is ſcarcely diſtinguiſhable: the filaments are very ſhort, and each of them is terminated by an anther, that has no covering, but has the texture of the pulp of oranges or lemons; each is lodged in a cell opening downwards, and adhering to the inner margin of the nectary; ſo that without this information you might have been at a loſs where to find the ſtamens, unleſs they happened to have burſt from their cells: the germ in time becomes a capſule, of three valves, opening at the angles under the carinated ribs; within is only one cell, and a great number of ſmall, irregular ſeeds, ſhaped like ſaw-duſt, are affixed to a linear receptacle on each valve. I have been more particular on the character of this tribe, becauſe the flowers have rather a ſtrange and unuſual appearance, owing to the ſingular poſition of the parts of fructification. There is a connexion between this and the liliaceous tribe;

tribe; both having but one lobe to the ſeed, ſucculent roots, entire leaves, and a naked corolla: they differ however in the number of ſtamens, the form of the corolla and nectary, the ſituation of the germ, the number of cells in the capſule, the ſhape and arrangement of the ſeeds: this tribe alſo bears its flowers on a ſpadix, and has bractes interpoſed between them.

The principal genera of this tribe are thus diſtinguiſhed:

Nectary horn-ſhaped. *Orchis.*
——— bag-ſhaped. *Satyrium.*
——— ſlightly keeled. *Ophrys.*
——— ovate, gibbous underneath. *Serapias.*
——— pedicelled. *Limodorum.*
——— inflated. *Cypripedium.*
——— turbinate or top-ſhaped. *Epidendrum.*
——— connate with the ringent corolla. *Arethuſa.*

The *Orchis* is the largeſt genus, there being no leſs than fifty ſpecies, of which eleven are found wild in England. The greater number have double bulbs; in the reſt the roots are either palmate or faſciculate.

Of thoſe with double bulbs, woods and buſhy paſtures produce the *Butterfly Orchis*,

chis[h], which has the lip of the nectary lance-ſhaped[i] and quite entire: the horn very long; and the petals ſpreading out wide. The flowers of this ſmell ſweet, particularly in an evening, and very early in the morning. There are only two, or at moſt three large leaves: the ſtem is a foot, or eighteen inches high: the ſpike is long, but the flowers are thinly ſpread in it; the bractes are large, and of the length of the germ: the flowers are of a greeniſh white; the ſpur is twice as long as the germ, very ſlender, and tranſparent enough for you to diſcern the nectar through it. There is a ſmaller variety, but differing no otherwiſe than in ſize.

Pyramidal Orchis[k], found in paſtures where the ſoil is chalky, is another of thoſe which have double bulbs: the lip of the nectary is two-horned, trifid, the ſegments nearly equal, the middle one being rather the narroweſt; all of them are quite entire; the horn, or ſpur, is cylindric, ſlender, and longer than the germ; and the petals are nearly lance-ſhaped. This is an elegant ſpecies, having ſix or more radical

[h] Orchis bifolia *Lin.* Fl. dan. 235. Vaill. par. t. 30. f. 7. Mor. hiſt. ſ. 12. t. 12. f. 18. Ger. 211. 2. Park. 1351. 7.

[i] Haller ſays linear.

[k] Orchis pyramidalis *Lin.* Raii ſyn. t. 18. Jacq. auſtr. t. 266. Vaill. t. 31. f. 38. Hall. helv. t. 35. 1. Ger. 210. 4. Park. 1349. 4.

leaves;

leaves; the ſtem a foot, or eighteen inches high; the ſpike of flowers ſhort, of a broad conical form, and very thick ſet at firſt; the bractes at leaſt equal in length to the germs, lance-ſhaped, and ending in a point; the corolla bright purple.

Two of the moſt common ſorts with double bulbs, are called *Male* and *Female Orchis* fooliſhly, becauſe there is no diſtinction of ſexes; and therefore theſe names are only calculated to miſlead. The [l] firſt differs from the ſecond in having the outer petals more acute and longer; and the middle lobe of the lip bifid and longer than the ſide ones: it is alſo a much larger plant, with broader leaves, uſually ſpotted. The ſecond [m] has the lip of the nectary crenulate, or ſlightly notched on the ſides, trifid, with the middle lobe emarginate, and the petals obtuſe and linear. The height of this ſeldom exceeds ſeven or eight inches; the leaves are half an inch broad; and the ſpike is cylindric, and has few flowers; the bractes are coloured, and a little longer than the germs; the petals forming the helmet converge, and are marked with green parallel lines; the middle of the lip is ſpotted, and the ſides are rolled back; the horn is equal to the germ, with the end emarginate;

[l] Orchis Maſcula *Lin.* Curtis, Lond. II. 62. Vaill. t. 31. f. 11, 12. Ger. 208. 1. Park. 1346. 1.

[m] Orchis morio *Lin.* Curtis, Lond. III. 59. Vaill. t. 31. f. 13, 14. Ger. 208. 2. Park. 1347. 4.

the

the moſt common colour of the corolla is deep purple, but it varies to roſe-coloured, and even white. The firſt is a foot, and even eighteen inches high; the leaves an inch and half broad; the ſpike handſome, long, and thin ſet with flowers; the bractes about the ſame length with the germs, purple and lance-ſhaped; the petals that form the helmet looſe, not converging, they are purple, with lines of the ſame colour; the edges of the lip are bent downwards, the colour pale purple, with deeper ſpots at the chaps; the ſpur is ſtraight, thick, as long as the germ, or longer, dilated and compreſſed at the end. The colour of the corolla varies, even to white. This grows in meadows; and the roots make excellent *Salep*. The ſecond affects open dry paſtures. Thus you have abundant means of diſtinguiſhing theſe two ſpecies of Orchis from each other; and the roots are a ſufficient mark of diſtinction from two others, no leſs common, which we ſhall examine preſently. In the mean time, there is a ſmall but pretty ſpecies with double bulbs, which we muſt not paſs by. It grows chiefly on dry expoſed chalk hills, and is called *Dwarf Orchis* [n]: the lip of the nectary is quadrifid, and white dotted with purple; the horn is obtuſe, and

[n] Orchis uſtulata *Lin.* Fl. dan. 103. Hall. t. 28. 2. Vaill. t. 31. f. 35, 36. Mor. t. 12. f. 20. Ger. 207. Park. 1345.

the

the petals are diſtinct. The height is from four to ſeven inches: there are ſeveral leaves next the ground, but few on the ſtem: the ſpike is ſhort and cloſe ſet; the bractes are ſhorter than the germ; the helmet is pointed, and of a deep purple on the outſide: within, the petals are marked with lines and dots of purple; the horn is a little bent, and not half the length of the germ.

Two very common ſpecies with palmate, or handed bulbs, are the *broad-leaved*[o] and *ſpotted Orchis*[p], generally found in moiſt meadows. The firſt has the roots rather palmate and ſtraight; the horn of the nectary conic, the lip three-lobed, and turning back on the ſides; the bractes large, and longer than the flowers, ſo as to give the ſpike a leafy appearance. The horn is ſhorter than the germ, bent and obtuſe. The colour of the corolla is purple, varying to roſe and white. The ſecond has narrower leaves, and a ſolid ſtem, whereas that of the firſt is hollow; it is alſo higher, and flowers later; the leaves of both are ſpotted with black, but this more generally; the bractes are ſmaller and narrower; the corolla of a paler purple; the lip of the nectary is deeper cut, the ſide lobes are

[o] Orchis latifolia *Lin.* Curt. Lond. V. 65. Mill. illuſtr. Fl. dan. 266. Hall. 32. 2. Vaill. t. 31. f. 1.—5. Ger. 220. f. 1, & 222. f. 3.

[p] Orchis maculata *Lin.* Hall. t. 32. 1. Vaill. t. 31. f. 9, 10. Ger. 220. 2. Park. 1357. 3.

notched,

notched, the middle one very narrow, quite entire, and drawing more to a point.

I ſhall mention only one ſpecies more of *Orchis*, and that alſo has palmate roots: it is found in paſtures, but by no means ſo common as the two laſt: you may call it *long-ſpurred*, or *ſweet Orchis*[q], and you will know it by the great length and ſlimneſs of the ſpurs: the lip is trifid, equal, ſlightly notched, and obtuſe; and the ſide petals ſpread out very wide. The ſtem is leafy, and grows to the height of eighteen inches; the bractes are ſharp pointed, and of the length of the germ; the corolla is purple, and all of one uniform colour; the ſmell is ſtrong, but, in ſome circumſtances, ſweet.

The ſecond genus of this natural tribe is the *Satyrium*, which, inſtead of the horn, or ſpur, has a ſhort, bag-form, or double-inflated nectary, at the back of the flower. This is a much leſs numerous genus than the laſt, having only eight known ſpecies. Of theſe I ſhall ſelect two; *Lizard Satyrion*[r], and *Frog Satyrion*, commonly called *Frog Orchis*[s]. The firſt is found in chalky paſtures, but rarely; and has been rendered Satyrium.

[q] Orchis conopſea *Lin.* Fl. dan. 224. Hall. t. 29. 2. Vaill. t. 30. f. 8. Ger. 222. 2.

[r] Satyrium hircinum *Lin.* Hall. t. 25. Mor. t. 12. f. 9. Ger. 210. 1. Park. 1348. 1.

[s] Satyrium viride *Lin.* Fl. dan. 77. Hall. t. 26. 2. Ger. 224. 9. Park. 1358. 9.

more rare by the diligence with which it has been ſought after, to tranſplant it into gardens, where it ſeldom continues long, this tribe being generally abhorrent of culture. It has double undivided bulbs; lance-ſhaped leaves; the lip of the nectary trifid, the middle lobe linear, oblique, extremely long, flaunting like a ribband, and ſeeming, as it were, bitten off at the end. It is a very large lofty plant, from eighteen inches to three feet in height; the leaves alſo are half a foot long and more, and three inches broad; the ſpike has many flowers, and, by age, grows very long and becomes bent; the bractes are ſlender, acute, greeniſh, and twice as long as the germs; the colour of the corolla is greeniſh without, and ruſty within, with purple lines and ſpots: the flower has a ſtrong goatiſh ſmell.

Frog Orchis is much more common in meadows. The bulbs of this are palmate, the leaves oblong and obtuſe; the lip of the nectary trifid, with the middle lobe obſolete, or ſo ſmall as to be obſcure. This is a much lower and ſmaller plant than the former, not being above ſeven or eight inches high: the radical leaves are broad and ovate; thoſe on the ſtem, which are few, lance-ſhaped: the ſpike is rather thin ſet with flowers: the bractes are lance-ſhaped, and longer than the germ: the helmet is almoſt cloſed, pale green, with a purple line dividing the petals; the lip is yellow,

yellow, hangs down ſtraight, and grows broader towards the end; the whole corolla becomes duſky red with age.

The third genus of the *Orchis* tribe is entitled *Ophrys*: it has no horn or bag at the back of the corolla, but one petal longer than the reſt, hanging down, and marked underneath with a longitudinal riſing, called the keel. This it is which in ſome ſpecies takes the form of an inſect ſo exactly, as to appear real at a certain diſtance. Ophrys.

One ſpecies, called *Common Twayblade* [t], or *Twyblade*, from its having always two leaves, and no more, is frequent in woods and buſhy paſtures. It has fibrous roots, two ovate leaves, and the lip of the nectary bifid. The ſtem is eighteen inches high, rather rough or hairy, and naked, except the two large leaves in the middle, between the root and the ſpike, which is ſometimes ſix inches long, and has forty flowers, thin ſet on ſhort peduncles; the bractes are very ſmall, broad, and ſharp-pointed; the germ is round, and thicker than in any other of the ſpecies; the corolla is of a greeniſh yellow.

The latter end of ſummer and beginning of autumn flowers the *Spiral Ophrys*, commonly called *Triple Ladies Traces* [u]; you

[t] Ophrys ovata *Lin.* Curtis, Lond. III. 60. Ger. 403. 1.

[u] Ophrys ſpiralis *Lin.* Curtis, Lond. IV. 59. Fl. dan. 387. Park. 1354. 3.

will find it on heaths and dry pastures. The root consists of oblong aggregate bulbs; the stem is a little leafy, the flowers are spiral, and all on one side of the stem; and the lip of the nectary is undivided and slightly notched. This is a small plant, seldom above five or six inches high, though in a less dry soil it will rise to a foot; it has four or five leaves next the ground; the spike is long and slender, having twenty flowers, white within and yellowish without; the bractes are not flat, but hollow, and longer than the germ; the three outer petals of the corollas are glued together; the lip is roundish and ciliate. It has a pleasant odour.

But the most interesting and admired species of this genus are the *Fly* and *Bee Orchises*, which agree in having two roundish bulbs, and a leafy scape or stem. Linnæus thinks the Fly and the two Bees [v] not to be specifically different, but in this I cannot agree with him. *Fly Ophrys* or *Orchis* [w] has the lip of the nectary quadrifid; in the *common Bee Orchis* [x] it consists of five lobes, which are deflex or bent downwards; and in the *green-winged Bee*

[v] Ophrys insectifera *Lin.*

[w] Orchis muscifiora *Halleri.* 1265. t. 24. 2. Ophrys insectifera myodes *Lin.* Oph. muscifera *Hudſ.* Vaill. t. 31. f. 17, 18. Ger. 213. 6. Park. 1352. 10.

[x] Orchis fuciflora *Hall.* Ophrys apifera *Hudſ.* Curtis, Lond. l. 66. Ger. 212. 4. Park. 1351. 5.

Orchis,

Orchis, now called *Spider Ophrys*[y], it is roundiſh, entire, emarginate, and convex. But beſides this character from the lip of the nectary, the *Fly* is a ſtiffer, ſtraighter plant than the *Bee*, not ſo leafy, and having the flowers thinner ſet; in other reſpects they are much alike, except in the corollas, which are widely different: that of the fly has the three outer petals ovate, entire, ſmooth, herbaceous, and ſpreading; the two inner linear and dark purple; the lip of the nectary oblong, dark purple above, and herbaceous underneath, with a blue ſpot or band below the upper lobes. *Bee Orchis* has the three outer petals ſpreading, oblong, and purple, marked with three green nerves; the two inner lateral ones linear, villous, and green; the lip of the nectary large, roundiſh, purple, and like velvet, the lobes deflex, with a double variegated yellow, ſmooth, ſhining ſpot at the baſe. *Spider Orchis* is a lower plant; the lip of the nectary is of a leſs cheerful colour, without any of the yellow that decorates the *Bee*, and both helmet and wings are green: the three outer petals are oblong and ſpreading, the inner linear and ſhorter; the lip of the nectary is large, roundiſh, entire, emarginate, convex, and appearing like velvet, duſky purple above,

[y] Ophrys inſectifera arachnites *Lin.* Oph. aranifera *Hudſ.* Vaill. t. 31. f. 15, 16. Ger. 212. 3.

with a green edge, and a double ſpot at the baſe; beneath it is herbaceous. Theſe three beautiful plants are found among graſs in a chalky ſoil, and form a ſucceſſion from April to Auguſt: the *Spider* comes firſt in April and May, the *Fly* next in June, and laſt of all the *Bee* in July and Auguſt.

I have been the more particular on this ſingular tribe of plants, becauſe, ſpurning culture, they are not liable to eſſential changes, or indeed to any that I know of, except in colour: you muſt alſo ſearch for them abroad, and conſequently unite exerciſe with ſtudy, which is one of the principal advantages of Botany; for I cannot allow you to gather plants by proxy, ſince you would thus loſe half the pleaſure of the purſuit, as well as the benefit: and why ſhould you not have as much enjoyment in ſearching for a beautiful plant, or finding an elegant flower, as the men have in looking for a hare, or ſhooting a partridge. I will only add, that ſhould you be ſo happy as to meet with the *Lady's Slipper* [z], you would be highly delighted with its ſingular, large, hollow, inflated nectary, the form of which has given occaſion to the name. Haller however obſerves, that it has more reſemblance to a wooden ſhoe in form,

Cypripedium.

[z] Cypripedium Calceolus *Lin.* Mill. fig. 242. Ger. 443. *Sowerby*'s Engliſh Botany, t. 1.

and therefore is unworthy the title of *Venus's Slipper*, which Linnæus has beſtowed upon it. Without entering into this important diſpute, I will obſerve to you, that the root is fibrous; the ſtem about a foot high, and leafy; the two firſt leaves ſmall, and keeping almoſt cloſe to the ſtalk; the reſt (from four to ſeven) ovate-lanced: one, or at moſt two flowers come out on the ſame ſtem, of which there are ſometimes ſeveral from the ſame root: the bracte is very large, as is alſo the germ: there are but four petals to the flower, ſpreading out almoſt at right angles to each other, and often convolute; their colour is purple; of the two outer petals, one ſtands up above the nectary, the other hangs down behind it; the two inner petals ſtand out ſideways, and are narrower: the ſlipper or lip of the nectary is yellow, ſpotted within, and marked longitudinally with ridges and furrows.

THE ORDER PENTANDRIA.

In the order *Pentandria* you will find the numerous and beautiful genus of *Paſſion-flower*. The flowers have three piſtils, a five-leaved calyx, five petals to the corolla, a radiate crown for a nectary; and the fruit is a berry on a pedicle. None of the ſpecies are European, but moſtly natives either of New Spain, the Braſils, or Paſſiflora.

the Weſt Indian Iſles; ſo that they require the protection of the conſervatory at leaſt, if not of the ſtove, except one or two, which will ſtand abroad in a ſheltered ſituation, with a little attention, in ſevere weather. I ſhall ſelect the ſpecies which you are moſt likely to meet with, rather than the rareſt. *Blue Paſſion-flower* [a], though a native of the Braſils, is ſeldom injured with us, except in very ſevere winters. Againſt a houſe it may be trained up to the height of forty feet, and throws out annually ſlender ſhoots, fifteen or ſixteen feet long: the leaves are palmate or handed, compoſed of five ſmooth, entire, obtuſe lobes, the middle one longeſt, the outer ſhorteſt, and often divided: they are petiolate; the petioles have two glands, and at their baſe is a ſtipule in form of a creſcent, and a long claſper, by which the ſlender ſhoots ſupport themſelves: the flower comes out at the ſame joint with the leaf, on a peduncle near three inches long; round the centre of it are two radiating crowns, the inner inclining towards the central column, the outer, which is longer, ſpreading flat upon the petals, and compoſed of innumerable threads, purple at bottom, but blue on the outſide. On the top of the central column ſits an oval germ, from whoſe baſe five awl-

[a] Paſſiflora cærulea *Lin.* Mill. illuſtr. Curt. magaz. 28. and Plate 30. of this work.

ſhaped

ſhaped ſtamens ſpread out horizontally, and theſe are terminated by oblong, broad, pendant anthers, which are eaſily moveable; from the ſide of the germ ariſe three ſlender, purpliſh ſtyles, diverging, and ending in obtuſe ſtigmas: the flower continues but one day, but there is a conſtant ſucceſſion from July till autumnal froſts ſtop them. The germ ſwells to a large, oval fruit, of the ſize, ſhape, and colour of the Mogul Plum, incloſing a ſweetiſh, but diſagreeable pulp, in which the oblong ſeeds are lodged.

Incarnate or *trilobate Paſſion-flower* [b] is a native of North America, and though the firſt ſpecies known among us, is not ſo common as the *Blue*. It differs from the former in having only three lobes to the leaves, which are ſerrate or toothed like a ſaw; the ſide lobes are ſometimes divided into two narrow ſegments: the petals of the corolla are white, with a double, purple fringe, ſtar, or glory: the fruit is as large as a middling apple, and when ripe is of a pale orange colour.

There is a ſort, called *Granadilla* [c] in the Weſt Indies, where the fruit is eaten. It has undivided, oblong leaves, hollowed next the petiole, which has two glands; the involucres are quite entire, as are alſo the

[b] Paſſiflora incarnata *Lin.* Mor. hiſt. ſ. 1. t. 1. f. 9.
[c] Paſſiflora maliformis *Lin.* Plum. amer. t. 82.

leaves

leaves about the edge. The corolla is large, with white petals, and a blue glory. The fruit is roundiſh, the ſize of a large apple, and yellow when ripe.

Another ſort, called *Water Lemon*[d] in the Weſt-Indies, has an agreeable acid flavour in the pulp of the fruit, which quenches thirſt, and is given there in fevers. It has undivided ovate leaves, quite entire about the edge; two-glanded petioles; and toothed involucres: the corolla is white with browniſh red ſpots, and the glory or crown is violet: the fruit is of the ſize and ſhape of a pullet's egg, and when ripe is yellow. But ſince the rarer ſpecies may not readily fall under your cognizance, I reſtrain my deſire of enlarging on ſo remarkable and beautiful a genus; and paſs on to a vulgar plant, which you will find in the laſt order, *Polyandria*, and with that I will cloſe our examination of this claſs, and my prate for the preſent.

THE ORDER POLYANDRIA.

Arum. This is the common *Arum*, *Wake-Robin*,

[d] Paſſiflora laurifolia *Lin.* Jacq. hort. 2. t. 162. amer. pict. t. 219.—P. alata is figured in Curtis's Magaz. 66. and P. lunata, is moſt elegantly figured by Mr. Sowerby, in a ſuperb and ſplendid work, begun by J. E. Smith, M. D. under the title of *Icones pictæ Plantarum rariorum.*

or

or *Cuckow-pint*[e], called alſov ulgarly *Lords and Ladies*. Early in the ſpring it puſhes up a one-leafed cowl-ſhaped ſpathe, under hedges and among buſhes; if you open this ſpathe, you diſcover a ſpadix, naked on the upper part, covered with germs at the bottom, and with anthers in the middle. This is diſtinguiſhed from the other ſpecies, which are many, by having no ſtem but that which bears the fructification, haſtate leaves that are quite entire, and the ſpadix club-ſhaped. Though it has the trivial name from the black ſpots upon the leaves, yet that is not a conſtant character, for oftentimes they are quite plain. As the plant advances, the ſpathe opens, and diſcovers the club, varying from yellowiſh green to fine purple or red; theſe gradually decay, and leave a head of round red berries, which, as well as the reſt of the plant, are very hot and biting. To this, with ſome others nearly allied to it, you would perhaps find it difficult to aſſign the proper claſs, unleſs, from the ſtrange and unuſual appearance of the fructification, you were led to ſearch for it in that now under conſideration. Theſe have not properly the ſtamens growing upon the ſtyle, but both are borne upon a receptacle lengthened out in manner of a ſtyle, and performing the

[e] Arum maculatum *Lin.* Curtis, Lond. II. 63. Mill. illuſtr. Mill. ic. t. 52. f. 1. Blackw. 228. Fl. dan. 505. Ger. 834. 1.

ſame

ſame office as the piſtil in the other genera. Linnæus obſerves that he might, and perhaps ought to have ranged ſuch plants under other claſſes; but he was deterred by the difficulty of aſſigning the number of ſtamens to each piſtil. Since he found a difficulty in removing them, you and I, dear couſin, will leave them quietly in the place which he has aſſigned them.

LETTER XXVIII.

THE CLASS MONOECIA.

May the 15th, 1777.

WE have hitherto, dear coufin, been converfant with fuch plants as bear perfect or complete flowers only, except in the clafs *Syngenefia*, wherein we found imperfect, and even neuter, flofcules among the perfect ones. But in the twenty-firft and twenty-fecond claffes, which we are now to examine, you will never find any complete or perfect flowers; on the contrary, if they have ftamens, there are no piftils, and if they have piftils, they are deficient in ftamens. This is the common character of thefe two claffes, and the only difference between them is, that in the clafs *Monœcia*, the ftaminiferous and piftilliferous flowers are found on the fame individual plant; whereas in the clafs *Diœcia* they are always on diftinct plants of the fame fpecies. It is fcarcely neceffary to add, that in both, the flowers which produce ftamens fall off without being followed by fruit or feed: and that the others, which have the germ, are fruitful.

The clafs *Monœcia*, which is the twenty-firft in the fyftem, has eleven orders, taking

ing their titles and characters from the foregoing classes; eighty genera, and three hundred and seventy species.

The third order, *Triandria*, contains several genera nearly allied to the Grasses in habit, leaves, and placentation, or having only a single lobe to the seed: they differ however in the culm or straw not being hollow, but filled with a spungy substance; and in having no corolla.

Typha. Since Haller thinks there is a natural connexion between the *Arum*, with which I finished my last letter, and the *Typha* or *Cat's-tail*, let us begin our examination with this. Having three stamens, it belongs of course to the order *Triandria*, and having the air of the Grasses, it ranges in the natural tribe of the *Calamariæ*, just mentioned. The flowers on both sides are borne on a cylindric *Ament*; the stamineous flowers surrounding the end of the stem; and those which have the pistils growing in the same manner below them, and very close set: there is no corolla to either: the first have an obscure, three-leaved calyx; in the second it consists of pappous or villous hairs, and these have one seed, sitting on a capillary down or bristle: such are the generic characters. The *greater*, or *broad-leaved Cat's-tail*, otherwise called *Reed-mace*[f], is known

[f] Typha latifolia *Lin.* Curtis, Lond. III. 61. Mor. hist. s. 8. t. 13. f. 1. Ger. 46. Park. 1204. 1.

known by its ſword-ſhaped leaves, and by having the two aments approximating. It is a large plant, being about ſix feet in height, with leaves three feet long and more, but not an inch wide; it is common in the water, on the banks of rivers, but eſpecially in moats, ponds, and marſhes. There is a ſmaller ſpecies [g], not ſo common, which has ſemi-cylindric leaves, and the two aments remote from each other; the ſtem of this is not above three feet high, and the leaves are much narrower, ſtiffer, and embrace the ſtem more.

Sparganium, or *Bur-reed*, approaches very near to *Typha*: but the flowers of each ſort are collected into a head, or roundiſh ament, thoſe which have ſtamens above, and thoſe which have piſtils below, on the ſame ſtem: neither have any corolla; both have a three-leaved calyx; the piſtilliferous flowers have a bifid ſtigma, and are followed by a ſingle juiceleſs drupe, containing one ſeed. *Erect* or *greater Bur-reed* [h] is common in the ſame ſituations with *Typha*, and few plants exhibit more plainly the character of the claſs *Monœcia*. The ſtem is erect, and about three feet high; the leaves are erect and Sparganium.

[g] Typha anguſtifolia *Lin.* Curtis, Lond. III. 62. Mor. hiſt. ſ. 8. t. 13. f. 2. Park. 1204. 2.

[h] Sparganium erectum *Lin.*—ramoſum *Hudſ.* Mor. t. 13. f. 1. Ger. 45. f. 1. Curtis, Lond. V. 66.—in V. 67. he figures Sp. ſimplex, as diſtinct from the ramoſum. Ger. 45. 2. Mor. f. 2.

three-

three-ſided, but the upper one flat: the ſtalk is generally branching.

Zea. *Mays*, otherwiſe called *Indian* or *Turkey Corn*[1], is of the ſame tribe. The ſtamineous flowers are borne in looſe ſpikes: their calyx is a two-flowered awnleſs glume; neither has the corolla any awn. The other flowers, which have one piſtil only, are in very cloſe ſpikes, below the former, and are incloſed with leaves. The glume both of calyx and corolla is bivalved: the ſtyle is filiform, very long, and pendulous: one ſeed follows each flower: the receptacle is oblong and hollowed, ſo that the ſeeds are immerſed half way into it, forming a very denſe ſpike. The Weſt Indian Mays has a ſtalk ten or twelve feet high; long, broad leaves; and ſpikes from nine inches to a foot in length, formed of gold-coloured grains. That which is cultivated in Italy, Spain, and Portugal, has more ſlender ſtalks, not more than ſix or ſeven feet high; the leaves narrower; the ſpikes ſhorter and more ſlender, with white grains. The North American Mays, which is the ſame with what is cultivated in Germany, does not riſe more than four feet in height; the leaves are ſtill ſhorter and narrower; the ſpikes not more than four or five inches long, with yellow and white grains mixed: the colour of theſe however varies; and

[1] Zea Mays *Lin.* Blackw. 547.

indeed the three diſtinctions are but varieties ariſing from ſoil and climate.

Carex, or *Sedge*, is a moſt numerous genus of the ſame order, and the ſame natural tribe. The flowers of both ſorts are borne on an ament or catkin, and each flower has a one-leafed calyx, and no corolla: the piſtilliferous flowers, which are generally in diſtinct aments below the others, have an inflated, three-toothed nectary, three ſtigmas, and a three-ſided ſeed incloſed within the nectary. Some few ſpecies have only one ſpike; many have ſeveral ſpikes, with both ſorts of flowers in each; but more have the ſtaminiferous and piſtilliferous flowers in diſtinct ſpikes. Theſe plants grow chiefly in marſhes, bogs, ditches, wet woods, and the banks of brooks and rivers; they are the graſs and fodder of fenny countries, and low ſwampy grounds [k]. Carex.

In this claſs, *Monœcia*, as well as in the next, you will find many trees. In the order *Tetrandria*—Birch, Alder, Box, Mulberry; in that of *Polyandria*—Oak, Cork, Evergreen Oak, Walnut, Hickery, Cheſnut, Beech, Hornbeam, Hazel, Plane;—and laſtly in that of *Monadelphia*—all the

[k] Carex pendula Curtis III. 63, riparia IV. 60, acuta 61, gracilis 62.—dioica Fl. dan. 369, capitata 372, arenaria 425, muricata 284, remota 370, caneſcens 285, limoſa 646, capillaris 168, panicea 443, veſicaria 647, hirta 379.—pauciflora Lightf. 6. 2, incurva 24, 1.—Many of the ſpecies are figured in Leers's excellent Flora Herbornenſis.

ſpecies of Fir and Pine, Cedar, Larch, Arbor Vitæ, Cypreſs.

Betula. *Alder* is one of the ſame genus with *Birch:* their common character is, that the flowers of both ſorts grow in aments or catkins, each ſeparate from the other; that the calyx is one-leafed and trifid; that each calyx in the ſtaminiferous ament includes three flowers, that have four-parted corollas: in the piſtilliferous aments there are only two flowers in each calyx, without any corolla; but theſe are followed by ſeeds winged with a membrane on both ſides, whereas the others drop from the tree, without leaving any mark behind them. In examining theſe, and the flowers in general of this and the following claſs, I muſt once for all inform you, that ſince many of them are cloſe ſet together in the ſame ament, you muſt carefully ſeparate one flower from the reſt, to avoid confuſion. You muſt alſo look for them very early in the ſpring, ſince moſt of the foreſt and timber trees flower before the leaf-buds expand.

Common Birch[l] has ovate leaves, drawn to a very narrow point at the end, and ſerrated, or ſharply toothed round the edge. Linnæus diſtinguiſhes the *Alder*[m] by its

[l] Betula alba *Lin.* Blackw. t. 240. Duham. t. 39. Ger. 1478. Evelyn's ſilva by Hunter, p. 218.

[m] Betula Alnus *Lin.* Duham. t. 15. Ger. 1477. 2. Evelyn's ſilva by Hunter, p. 233.

branching

branching peduncles: the ſeeds alſo are borne on a roundiſh *ſtrobile*, rather than an ament; and the leaves are roundiſh, crenate or obtuſely notched round the edge; they are of a dark green, with very prominent nerves underneath, and little ſpungy ſubſtances where they divide: the bark of the Alder is black, whereas that of Birch is white.

In *Box* both ſorts of flowers come forth together in bunches, from the axils of the leaves or branches, and ſit cloſe to the ſtem: the ſtaminiferous flowers have a three-leaved calyx, with two petals to the corolla, and the rudiment of a germ; the piſtilliferous flowers have a four-leaved calyx, three petals to the corolla, three ſtyles, and a three-celled capſule, terminated by three beaks, and having two ſeeds in each cell. Properly ſpeaking, there is only one ſpecies of box[n], varying a little in the ſhape of the leaves, and much in the ſize. **Buxus.**

Mulberry bears the ſtaminiferous flowers in an ament; the others in a ſeparate roundiſh head, which afterwards becomes a compound berry, with one ſeed in each protuberance; the firſt have a four-parted calyx; in the piſtilliferous ones it is four-leaved, and theſe have two ſtyles; neither have any corolla. *White Mulberry*[o], which is **Morus.**

[n] Buxus ſempervirens *Lin.* Blackw. 196. Ger. 1410.
[o] Morus alba *Lin.*

 the

the ſort commonly cultivated in France and Italy for feeding ſilk-worms, has ſmooth leaves, obliquely heart-ſhaped, and white fruit. *Black Mulberry*[p] has rugged, heart-ſhaped leaves: though cultivated for the fruit, yet the leaves are preferred to thoſe of the other for feeding ſilk-worms, and are uſed for that purpoſe in Perſia, from whence this tree originally came into the ſouth of Europe. White Mulberry is a native of China. Of another ſpecies[q], paper is made in Japan, from the bark; this has palmate leaves, and hiſpid fruit. *Fuſtick wood*[r] is alſo from a ſpecies of Mulberry: this has axillary thorns, and the leaves are oblong and more extended on one ſide than the other. This grows in the iſlands of the Weſt Indies, but in greateſt plenty at Campeachy: the wood is imported into Europe from both places for the uſe of the dyers, but the tree is too tender to ſupport our climate.

Quercus. In the order *Polyandria* the *Oak* leads the way. The ſtaminiferous flowers hang on a looſe ament or catkin, whilſt the piſtilliferous ones are ſeſſile in a bud: the calyx of the former is moſtly quinquefid, and the ſtamens are from five to ten in number: in the latter the calyx is one-

[p] Morus nigra *Lin.*

[q] Morus papyrifera *Lin.* Seba muſ. 1. t. 28. f. 3. Kæmpf. amæn. t. 472.

[r] Morus tinctoria *Lin.* Sloan. jam. 2. t. 158. f. 1.

leafed

leafed and quite entire, and there is one ſtyle, ſplit into five parts; but ſometimes only into two, three or four. The fruit, or acorn, is well known: it is an oval nut, covered with a tough ſhell, and immerſed at bottom into the calyx or cup.

We have two principal ſorts, or perhaps rather varieties [s] in England: one with the leaves on longer petioles, and the acorns ſeſſile, or on very ſhort peduncles; the other, having the leaves not ſo deeply, but more regularly ſinuate, the ſinuſes being oppoſite; they have ſcarcely any petioles: on the contrary the acorns grow on very long peduncles, are larger, and come out fewer together. There are ſome other variations in this noble tree, which being leſs conſiderable, do not attract our notice as botaniſts. Several ſpecies different from ours are found in North America; and ſome in the ſouthern countries of Europe.

Ilex or *Evergreen Oak* [t] has oblong-ovate leaves, of a lucid green above, but hoary underneath, ſtanding on long petioles, and continuing all the year; they vary much, ſome being quite entire, long and narrow; others broad, with the edges toothed and

[s] Linnæus makes them one, under the title of *Quercus Robur*, and deſcribes the ſpecies as having deciduous leaves, of an oblong form, but broader towards the upper part; the ſinuſes acute, and the angles obtuſe. Duham. t. 46.—48. Evelyn's ſilva by Hunter, p. 67. Ger. 1339.

[t] Quercus Ilex *Lin.*

 ſet

ſet with prickles, almoſt like thoſe of the Holly: the acorns are of the ſame ſhape with thoſe of the Oak, but ſmaller. The *grain-bearing Ilex* [u], which yields the *kermes* or ſcarlet grain, has ovate leaves toothed on the edge, and the indentures armed with prickles as in the Holly; they are ſmooth on both ſides: this is of ſo ſmall a growth, that it may be looked upon rather as a ſhrub than a tree. *The Cork-tree* [v] is a ſort of Ilex, with a fungous bark full of clefts or chinks, which is the principal as well as moſt obvious difference: in the air, and form of the leaves, it much reſembles the Evergreen Oak: the leaves however fall off in May, before the young ones come out, ſo that the Cork trees are bare for a ſhort time; which is not the caſe with the common Ilex. Moſt of the trees in this genus are much reſorted to by inſects, many of which form different ſorts of galls: but here we are ſtepping out of our province:—we will return to it again, by taking the *Walnut* under conſideration.

Juglans. This genus has the ſtaminiferous flowers thick ſet in oblong, cylindric catkins, under the lower leaves of the branches; they conſiſt of ſcales with one flower to each; the corolla is ſix-parted and the ſtamens are uſually eighteen, but vary in number from twelve to twenty-four. The piſtilliferous

[u] Quercus coccifera *Lin.*

[v] Quercus Suber *Lin.* Blackw. 193.

flowers

ſlowers come out cloſe to the branches, above the others, at the baſe of a petiole, generally in pairs: theſe have a quadrifid calyx, crowning the germ; a four-parted corolla; and two ſtyles: the fruit is a *drupe* containing a nut, with a furrowed ſhell, within which is a four-lobed, irregularly furrowed nucleus. *Common Walnut* [w] is diſtinguiſhed by having the component leaves oval, ſmooth, ſometimes a little toothed, and almoſt equal: there are many varieties in the fruit, and ſeveral diſtinct ſpecies in North America, one of which is the *Hickery* [x]. All the ſpecies have pinnate leaves, with a different number of leaflets; ours has from five to nine, and the odd leaflet is rather the largeſt. *Hickery* has ſeven lance-ſhaped leaflets, toothed on the edge, and the odd one ſeſſile.

Linnæus joins the *Cheſnut* and *Beech* in one genus, with this character: that the ſtaminiferous flowers, which are in catkins, have a quinquefid, bell-ſhaped calyx, and about twelve ſtamens: that the piſtilliferous flowers, which are produced from buds on the ſame tree, have a four-toothed calyx, three ſtyles, and a muricate, four-valved capſule, which before was the calyx, and contains two nuts. He obſerves that the ſtaminiferous flowers in the cheſnut are Fagus.

[w] Juglans regia *Lin.* Mill. illuſtr. Hunt. Evel. ſilva, p. 164.

[x] Juglans alba *Lin.* Cateſb. car. 1. 38.

 diſpoſed

diſpoſed on a cylindric ament, whereas thoſe of the Beech are in a ball. The catkins indeed of the former are very long, and the knots of flowers have near ten in each, and are diſtant from each other: the ſtamens are from five to eighteen, and have ſhort filaments: the piſtilliferous flowers are at the baſe of theſe, and are ſucceeded by two or three fruits cloſe together; their calyx has more frequently ſix ſegments than four; the fruit varies in the number of kernels and piſtils, but the moſt common number is ſix; and the kernels are convex on one ſide and flat on the other. The catkins of the Beech are roundiſh and looſe, with few flowers; the ſtamens are eight in number, on long filaments: and there are only two piſtilliferous flowers together, and each of theſe is ſucceeded by a roundiſh nut, containing three or four hard three-ſided kernels, which are commonly called *Beech maſt*. The ſpecific difference which Linnæus aſſigns to the *Cheſnut* [y] and the *Beech* [z], is taken from the leaves; which in the firſt are lance-ſhaped, ſawed with the teeth ending in points, and naked or ſmooth on the under ſurface; in the ſecond, ovate and obſcurely toothed, or rather waving on the edge.

[y] Fagus Caſtanea *Lin.* Mill. fig. pl. 84. Evel. ſilva by Hunter, p. 153. Ger. 1442.

[z] Fagus ſylvatica *Lin.* Evel. ſilva by Hunter, p. 131.

In

In the *Hornbeam* both ſorts of flowers are diſpoſed in catkins: both have a calyx conſiſting of one ciliate or fringed ſcale, and no corolla: the one has from eight to fourteen or ſixteen ſtamens; the other has two germs, with two ſtyles to each, and at the baſe of each ſcale of the ament or *ſtrobile* lies a ſeed, which is an ovate nut. In the *common Hornbeam*[a] the ſcales of the *ſtrobiles* are flat; and in the *Hop-Hornbeam*[b] they are inflated: ſuch is the ſpecific difference of theſe, which are the only known ſpecies. The leaves are wrinkled, marked with ſtrong nerves, of an ovate form, and ſharply toothed about the edge. Carpinus.

Hazel has the ſtaminiferous flowers on a long cylindric catkin, with one flower to each ſcale, which is trifid; it has from ſix to ten ſtamens; generally eight: the piſtilliferous flowers are remote from the others, ſeſſile and incloſed in a bud; the calyx is two-leaved and torn: each flower has two very long, red ſtyles; but you muſt obſerve that there are ſeveral flowers in the ſame bud, which you muſt therefore ſeparate for examination: the fruit, as you know, is an ovate nut. As uſual, neither of the flowers have any corolla. The *common Hazel nut* and *Filbert*[c] are ſuppoſed Corylus.

[a] Carpinus Betulus *Lin.* Evel. by Hunter, p. 158. Duh. t. 49. Ger. 1479.

[b] Carpinus Oſtrya *Lin.* Mich. gen. t. 104. f. 1, 2.

[c] Corylus Avellana *Lin.* Blackw. 293. Evel. ſilva by Hunter, p. 213. Duham. t. 77. Ger. 1438.

not

not to be ſpecifically different, and the ſpecies is characterized by the ſtipules, which are ovate, and end obtuſely; whereas thoſe of the *Byzantine* or *Spaniſh nut*[d], which Linnæus gives as a diſtinct ſpecies, are linear, and end acutely. Theſe do not arrive at the dignity of trees, but are only ſhrubs.

Platanus. The laſt tree I ſhall point out to you of this order is the *Plane*; which has the flowers of both ſorts in globular aments: the ſtaminiferous flowers have a few very ſmall ſcales for the calyxes, a corolla ſcarcely apparent, and anthers ſurrounding the filament: the piſtilliferous flowers have many very ſmall ſcales to the calyx; many petals to the corolla; ſubulate ſtyles, with recurved ſtigmas; and roundiſh ſeeds, terminated by a pointed ſtyle, and having a ſimple down adhering to their baſe. The two ſpecies of this tree, for there are no more, are well diſtinguiſhed by their leaves, which in the *Eaſtern* or *Aſiatic Plane*[e] are palmate; and in the *Occidental* or *Virginian*[f], lobate. The firſt was introduced early to Rome, and was the favourite tree of the Romans at their villas. All theſe trees are included in a natural tribe, called *Amentaceæ* by Linnæus, and *Juliferæ* by Haller and others;

[d] Corylus Colurna *Lin.* Seba muſ. 1. t. 27. f. 2.
[e] Platanus orientalis *Lin.* Ger. 1489. Park. 1427.
[f] Platanus occidentalis *Lin.* Cateſby car. 1. t. 56. Duham. arb. t. 25. Park. theat. 1421.

their

their character is ſufficiently obvious from their name, and what has been already ſaid in delivering the characters of the genera.

There remains ſtill a ſet of kindred trees, of the order *Monadelphia*, and of a natural tribe, entitled *Coniferæ* or *Cone-bearing*. Of theſe the Pine genus is chief: its generic characters are, that the ſtaminiferous flowers are diſpoſed in racemes, having each of them a four-leaved calyx; no corolla, but abundance of ſtamens terminated by naked anthers: the piſtilliferous flowers are on a cone; each ſcale or calyx has two flowers, without any corolla; one piſtil; and a nut furniſhed with a membranous wing. Pinus.

The whole genus may be divided into the *Pines*, having two or more leaves from the ſame ſheathing baſe, and the *Firs*, having the leaves quite diſtinct at the baſe. Of the firſt diviſion, the moſt known among us is the *Scotch Pine*[g], or, as it is vulgarly called, *Scotch Fir*: this has two leaves in a ſheath; and the primordial ones ſolitary and ſmooth. It is by no means peculiar to Scotland, but is found all through Denmark, Norway, and Sweden, in Switzerland, and moſt other parts of Europe, and even in the Weſt Indies. The *Pineaſter* or *wild Pine* of Italy, the ſouth of France and Switzerland, reſembles this, but the branches are wider diſtant, and more hori-

[g] Pinus ſylveſtris *Lin.* Mill. illuſtr. Evel. ſylva by Hunter, p. 274. Ger. 1356. 1.

zontal;

zontal; the leaves are larger, thicker, and longer, grow ſtraight, are of a darker green, and end obtuſely; the cones are ſeven or eight inches long: the leaves of the Scotch Pine are broader, grayiſh and twiſted; the cones ſmall, and of a light colour: the timber alſo is far preferable, yielding the beſt red or yellow deal. Linnæus, however, does not ſeem to have diſtinguiſhed them. The *Stone Pine* [h] has alſo double leaves, and the primordial ones ſolitary, but fringed; they are of a glaucous hue: the cones are thick, roundiſh, and end obtuſely; the ſcales are flat, and the nuts ſo large, that in the ſouth of France and Italy they think it worth while to break them, and ſerve the kernels up in deſerts. *Frankincenſe Pine* [i] has three leaves coming out of the ſame ſheath, and cones as large as thoſe of the Stone Pine, but more pointed, and with looſer ſcales, that open horizontally, and drop the ſeeds. The *Cembra Pine* [k] has five leaves in a ſheath; they are ſmooth, of a light green, long, and narrow; the cones are about three inches long, with cloſe ſcales, and large ſeeds eaſily broken. *Weymouth Pine* [l] has alſo five leaves in every

[h] Pinus Pinea *Lin.* Blackw. 189. Duham. arb. 2. 27.

[i] Pinus Tæda *Lin.*

[k] Pinus Cembra *Lin.* Gmel. ſib. 1. t. 39. Duham. 2. t. 32.

[l] Pinus Strobus *Lin.* Hunt. Evel. ſilva, p. 276.

ſheath,

ſheath, long and ſlender, but rugged on the edge; this tree grows remarkably ſtraight and tall, and the bark is very ſmooth. In North America it is called *White Pine*, and is excellent for maſts. The leaves of all theſe are linear and permanent; Linnæus calls this ſort of leaf *aceroſe*.

Linnæus includes the *Cedar of Lebanon*[m] and *Larch*[n] in this genus; others ſeparate them, becauſe the leaves are faſciculate, or come out in cluſters, ſpreading at top like a painter's bruſh: this circumſtance Linnæus gives for the ſpecific diſtinction, adding, that in the former they are acute, and in the latter obtuſe at the end; this is the only difference he mentions; the leaves of the Larch however are deciduous, thoſe of the Cedar permanent or evergreen: the character alſo of theſe two trees is totally different—the latter ſpreading its vaſt arms horizontally till the ends hang down with their own weight, and having a faſtigiate or flat top—the former having the branches decreaſing from the bottom upwards, and being therefore nearly pyramidal.

Of the *Firs* properly ſo called, *the Pitch-tree*, or *Norway Fir*[o], and the *Spruce*[p], are

[m] Pinus Cedrus *Lin.* Trew. Ehr. t. 1. Edw. av. t. 188.

[n] Pinus Larix *Lin.* Hort. angl. 11. Hunt. Evel. ſilva, p. 280.

[o] Pinus Picea *Lin.* Ger. 1363. Hunt. Evel. ſilva, p. 278.

[p] Pinus Abies *Lin.* Ger. 1354. Hunt. Ev. ſilva, p. 278.

the

the moſt common. The firſt has the leaves emarginate, or notched at the end: this is the tree from whence pitch is commonly extracted, and the wood of it is what we call white deal. The ſpruce has awl-ſhaped, pointed, ſmooth leaves, turned two different ways; the timber of this reſembles the other, and, when cut into boards, is called by the ſame name. *Silver Fir* is ſo named from the whiteneſs of the leaves underneath; they are emarginate, and in ſhape much reſemble thoſe of the Yew: a great deal of turpentine is made from this. *Balm of Gilead Fir*[q] has the leaves ſubemarginate, or but little notched at the end; they are dotted in a double line underneath. There are many varieties, eſpecially of the Spruce; but it would lead us too far to notice them.

Cupreſſus. I ſhall finiſh this knot of trees with the upright, the funereal *Cypreſs*, which has its ſtaminiferous flowers collected into an ovate ament, with one-flowered ſcales, and four ſeſſile anthers without filaments to each flower: the piſtilliferous flowers are in a roundiſh cone, eight or ten in number, one to each ſcale; theſe have many truncated points, hollow at the top, which are perhaps the ſtyles; under the ſcales of the cone lies an angled nut. *Common upright Cypreſs*[r] has imbricate leaves, with the leafing

[q] Pinus Balſamea *Lin.* Pluk. alm. 2. t. 121. f. 1.
[r] Cupreſſus ſempervirens *Lin.* Blackw. 127.

branches quadrangular: this takes naturally a cloſe pyramidal form, and when large has the fineſt effect imaginable near buildings. *Spreading Cypreſs* is only a variety of this, but grows to a very large ſize, and furniſhes the wood ſo famous for its durability, and reſiſtance to inſects. *Deciduous Cypreſs*[s] has the leaves in two ranks, and ſpreading: it is a native of America, and grows to a vaſt ſize. But it is time to deſcend from trees to herbs, and thus put an end to this long letter.

The *ſtinging Nettles*[t] are to be found in the order *Tetrandria* of this claſs; but ſuch vulgar ill-humoured plants may forgive your paſſing them by, where you have ſo many intereſting and even great perſonages to attract your notice. Urtica.

The immortal *Amaranth* however, having ſuperior elegance and beauty to boaſt, will not thus be paſſed unnoticed. It is of the order *Pentandria*, and having no corolla, is ranged by ſome in the natural tribe of apetalous flowers. The ſame raceme or bunch bears incomplete flowers of both kinds, each of them having a three or five-leaved calyx; the one bearing three or five ſtamens, the other three ſtyles, and a one- Amaranthus.

[s] Cupreſſus diſticha *Lin.* Cat. car. 1. t. 11.

[t] Urtica *Lin.*—pilulifera Mill. illuſt. Ger. 707. 1. Park. 440. 1.—urens Fl. dan. 739. Ger. 707. Park. 440. 2.—dioica Fl. dan. 746. Ger. 706. 2. Park. 441. 3.

celled

celled capſule opening horizontally, with one ſeed only lodged in it. The ſpecies are numerous: one of the moſt known is the *Amaranthus tricolor*, cultivated for the beauty of its leaves, which are variegated with green, yellow, and red: this is one of thoſe that have three ſtamens to the flowers, which grow in roundiſh heads, are axillary, and ſurround the ſtem; the leaves are broad lance-ſhaped. *Amaranthus bicolor* [u] has only two colours in the leaves, an obſcure purple and bright crimſon: this reſembles the other, but has lance-ſhaped pointed leaves. *Prince's Feather* [v] has five ſtamens to the flowers, which are produced in decompounded, cylindric, long, pendulous racemes, of a bright purple, and two feet or more in length. *Tree Amaranth* reſembles this, but is ſeven or eight feet high: the racemes are thicker, but not ſo long. *Bloody Amaranth* [w] has alſo five ſtamens: the racemes are compound and erect, the ſide ones very ſpreading; the leaves are ovate-oblong: this has purple ſtalks and leaves; the racemes are ſhort, and at the end of the ſtem there is a large cluſter of them placed croſſwiſe, with one upright in the middle: the flowers are bright purple at firſt, but grow darker. Thus I have

[u] Amaranthus melancholicus *Lin.*

[v] Amaranthus caudatus *Lin.*

[w] Amaranthus ſanguineus *Lin.* Mill. fig. 22.—cruentus Mart. cent. t. 6.

ſelected

ſelected the moſt ſpecious of this fine genus for your examination: your gardener will furniſh you with them from the hot-beds, when he raiſes his annual flowers.

From the order *Polyandria* I ſhall preſent you with two wild herbs—*Arrow-head* and *Burnet*. The firſt has many ſtaminiferous flowers, and a few with piſtils immediately below them: both have a three-leaved calyx, and a corolla of three petals: the one has about twenty-four ſtamens; the other many germs in a head, ending in very ſhort ſtyles, terminated by acute permanent ſtigmas. Our *common Arrow-head*[x] is eaſily diſtinguiſhed by its leaves ſhaped like the head of an arrow, and pointed: it grows in the water, has rounded white petals with purple claws, and bears an evident affinity to *Water-plantain*. Sagittaria.

Burnet has incomplete flowers of both ſorts in the ſame ſpike; thoſe with ſtamens below the others: they have a four-leaved calyx, and a four-parted corolla: the lower ones have from thirty to forty ſtamens; the upper, two piſtils, and a kind of berry formed from the tube of the corolla hardened. *Common* or *ſmaller Burnet*[y] is diſtinguiſhed from the other ſpecies by being unarmed or having no thorns; and the ſtems Poterium.

[x] Sagittaria ſagittifolia *Lin.* Fl. dan. 172. Ger. 416. 2. Park. 1247. 2.

[y] Poterium ſanguiſorba *Lin.* Curtis, Lond. II. 64. Ger. 1045. 1. Park. 582. 1.

being rather angular. This and the *Great Burnet*[z], though ſeparated ſo widely in the artificial ſyſtem, are evidently of the ſame natural genus: the calyx of the latter is two-leaved, and the number of ſtamens only four, and one piſtil; both in the ſame flower: it is alſo a much larger plant, with not ſo many pairs of leaflets: this grows in moiſt meadows: the other in dry, eſpecially chalky paſtures.

Ricinus. *Ricinus*, or *Palma Chriſti*, ranges in the order *Monadelphia*. The flowers have no corolla: ſome are furniſhed with many ſtamens, and theſe have a five-parted calyx; others have three bifid ſtyles, with a three-celled capſule, containing one ſeed in each cell; in theſe the calyx is three-parted. *Common Palma Chriſti*[a] has peltate, palmate leaves, toothed about the edge, of a glaucous hue underneath, and glands on the petioles. In the Weſt Indies there are ſeveral others, varying from this, and from each other; which are not, however, generally ſuppoſed to be diſtinct ſpecies. They call them *Agnus caſtus*, or *Oil-tree*, and extract from them an oil for their lamps; this is the *Caſtor Oil*, uſed in medicine. The common ſort grows in Sicily, and the other warm parts of Europe.

The order *Syngeneſia* of this claſs contains

[z] Sanguiſorba officinalis *Lin.* Fl. dan. 97. Mor. hiſt. ſ. 8. t. 18. f. 7. Ger. 1045.

[a] Ricinus communis *Lin.* Mill. fig. 219.

a ſet

a ſet of plants that belong evidently to the ſame natural tribe, entitled *Cucurbitaceæ*, or *Gourd plants*. They all agree in a one-leafed calyx, divided into five ſegments; a ſuperior, monopetalous corolla, divided alſo uſually into five; three filaments; one ſtyle, generally trifid: and a *pomum* for a fruit.

Momordica is diſtinguiſhed principally by the elaſtic burſting of the fruit, which in the common ſort is hiſpid; the ſtalks of this have no tendrils. From the property of throwing out the ſeeds with the juice, this plant has acquired the name of *Spirting Cucumber* [b]. Momordica.

Gourd has the ſeeds of the fruit with a tumid margin. *Long Gourd* [c] has the leaves ſlightly angular, downy, two-glanded underneath at the baſe; the flowers white, on long peduncles, and reflex at the brim; the fruit crooked, yellow when ripe, and the rind hard and woody, ſo that it will contain liquids; whence it is called *Bottle Gourd*. Cucurbita.

Pompion, corruptly called *Pumpkin* [d], is of this genus, and has lobate leaves, with ſmooth fruit, which will grow to the ſize of a peck.

The *Squaſh* [e], which is another ſpecies,

[b] Momordica Elaterium *Lin.* Pl. 31. of this work.
[c] Cucurbita lagenaria *Lin.* Mor. hiſt. ſ. 1. t. 5. f. 3.
[d] Cucurbita Pepo *Lin.*
[e] Cucurbita Melopepo *Lin.*

has alſo lobate leaves, erect ſtems, and the fruit flatted and knotty.

Warted Gourd[f] has likewiſe lobate leaves, and knobby fruit, covered with warts. Theſe differ much in their form and ſize.

Cucumis. But the moſt known and cultivated of theſe fruits are the *Melon* and *Cucumber*, which belong to another genus, called *Cucumis*, having the ſeeds of the fruit ſharp. *Melon*[g] has the angles of the leaves rounded, and the fruit covered with little ſwellings: it varies much, as you know, in the form of the fruit. *Cucumber*[h] has the angles of the leaves ſharp, and the fruit oblong and rugged[i]. All theſe having large flowers, with the parts very diſtinct, are proper to give you a juſt idea of this claſs; with theſe then I will finiſh, and releaſe you for the preſent.

[f] Cucurbita verrucoſa *Lin.*
[g] Cucumis Melo *Lin.* Blackw. 329.
[h] Cucumis ſativus *Lin.* Blackw. 4.
[i] This ruggedneſs is frequently loſt by culture

LETTER XXIX.

THE CLASS DIOECIA.

June the 1ſt, 1777.

THE twenty-ſecond claſs differs no otherwiſe from the preceding than in the diſpoſition of the incomplete flowers, namely on different individuals of the ſame ſpecies; this is its eſſential character, and this gave occaſion to its name—*Diœcia.* There being no difficulty then in underſtanding this, which indeed has been repeated ſeveral times before, let us go on without farther preface to the examination of ſuch plants as are moſt likely to fall in our way[k].

Such is the *Willow*, which is of the ſecond order—*Diandria.* Both ſtaminiferous and piſtilliferous flowers are produced in aments or catkins, on different trees; ſo that you will have double trouble in examining the flowers of this claſs; for when you have found one ſort, you will have to look about, and perhaps have ſome difficulty in finding the other. In ſo delightful a ſtudy however, you will not grudge a Salix.

[k] The genera in this claſs are fifty-five, and the ſpecies two hundred and nineteen.

little pains, after having already taken ſo much. The flowers of Willow have no corolla, and their calyx is nothing but the ſcales of the ament; there is a little honied gland in the centre of each ſtaminiferous flower: you will eaſily know the other aments, by the ovate germ in each little flower, gradually leſſening to a pair of ſtyles, ſcarcely diſtinguiſhable from it, but by the two erect, bifid ſtigmas, with which they are terminated; this germ becomes a one-celled, two-valved capſule, containing many ſmall ſeeds, crowned with a rough ſimple down. There are anomalies in this genus; for one ſpecies has one, another has three, a third has five ſtamens, and a fourth has complete flowers. From more than thirty ſpecies I ſhall ſelect the *White Willow* [l], which is a tree ſo common in watery ſituations: you will know it by the lance-ſhaped, acuminate leaves, toothed about the edges, pubeſcent, or villous, on both ſurfaces, and having the lower ſerratures glandulous: the leaves are very white underneath; and the catkins are ſhort and thick: it will grow to be a large tree, when it is not headed. Several ſpecies are commonly cultivated in Oſier-holts [m], but being al-

[l] Salix alba *Lin.* Blackw. t. 327. Ger. 1389. 1.

[m] Salix vitellina, amygdalina, purpurea, viminalis, &c. *Lin.*—Of theſe, S. purpurea is figured in Curtis Lond. n. 61. under the name of S. Monandra. For S. Triandra, ſee n. 62.

ways

ways kept down, in order to have a conſtant ſucceſſion of long, ſlender twigs, you will have little opportunity of examining their fructification. But one ſpecies being cultivated for its beauty, which fortunately depends upon the natural growth, you may ſtudy it at your leiſure: this is the *Weeping Willow* [n], known at firſt ſight by its long, ſlender, pendulous branches; the leaves are ſmooth, narrow, and linear, tending to lance-ſhaped. *Common Sallow* [o] has ovate leaves, wrinkled on the ſurface, which is villous above, and tomentoſe or nappy underneath, and ſlightly toothed or waved on the edges. There are ſeveral varieties of this vulgar ſpecies.

Miſſeltoe is of the order *Tetrandria*, its paraſitic quality you are well acquainted with, and that alone makes it generally obvious to every body: it is however no part of its character. The genus is determined by a four-parted calyx, and an anther growing to each part, without a filament, in the ſtaminiferous flowers; a four-leaved calyx ſitting on the germ; no ſtyle; and a berry incloſing one heart-ſhaped ſeed in the others; neither have any corolla. *Common* or *White Miſſeltoe* [p] is diſtinguiſhed from the reſt of the ſpecies by lance-ſhaped Viſcum.

[n] Salix babylonica *Lin.*

[o] Salix caprea *Lin.* Fl. dan. 245. Ger. 1390. 3.

[p] Viſcum album *Lin.* Mill. illuſtr. Duham. t. 104. Ger. 1350. 1. Park. 1393. 1.

leaves ending obtuſely, a dichotomous ſtalk, and axillary ſpikes of flowers.

In the next order *Pentandria*, we have Spinacia. Spinach, Hemp, and Hop. The firſt has a five-parted calyx in the ſtaminiferous flowers, and a quadrifid or four-cleft one in the others; theſe have four-cleft ſtyles, and one ſeed within the indurated calyx. Linnæus ſeparates the *garden*[q] ſort from the *Siberian*[r], by the ſeeds being ſeſſile, which in the latter are peduncled: of the former are ſeveral varieties: two remarkable ones, which perhaps may be diſtinct, the one having ſagittate leaves, and prickly ſeeds; the other rather ovate leaves, with ſmooth ſeeds.

Cannabis. *Hemp*[s] has a five-parted calyx in the flowers which bear ſtamens, but in the piſtilliferous ones it is one-leafed, entire, and gaping on the ſide: theſe have two ſtyles, and the ſeed is a bivalvular nut within the cloſed calyx, There is only one known ſpecies, and therefore until others are diſcovered, there is no occaſion for any ſpecific diſtinction.

Humulus. *Hop*[t] has a five-leaved calyx in the ſtaminiferous flowers; in the others it is one-leafed, obliquely expanding, and en-

[q] Spinacia oleracea *Lin.*

[r] Spinacia fera *Lin.* Gmel. ſib. 3. t. 16.

[s] Cannabis ſativa *Lin.* Mill. fig. pl. 77. Pl. 32.

[t] Humulus Lupulus *Lin.* Mill. illuſtr. Ger. 885. Park. 177.

tire;

tire; theſe have two ſtyles, and one ſeed within a leafy calyx: many of them are collected together to form what we call the Hop. In the three laſt genera the flowers have no corolla.

The order *Hexandria* has the *Tamus* or *black Bryony*, the flowers of which have a ſix-parted calyx and no corolla; the piſtilliferous flowers have a trifid ſtyle, and a three-celled berry below the flower, containing two ſeeds: our common ſpecies [u] has heart-ſhaped undivided leaves. Tamus.

The *Poplars* are in the order *Octandria*. Populus.
The flowers of both ſorts are here borne on ſimilar aments, conſiſting of ſcales torn on the edge, and each having one flower, without any petals, but a top-ſhaped nectary ending obliquely above in an ovate border; the piſtilliferous flowers have a quadrifid ſtigma, and are ſucceeded by a two-celled capſule, containing many downy ſeeds. *White Poplar* [v] has roundiſh leaves indented on the edges into angles, and downy underneath. *Great White Poplar*, or *Abele-tree*, is a variety of this, with larger leaves, more divided, and of a darker green. *Trembling Poplar*, or *Aſp* [w], has leaves like the former

[u] Tamus communis *Lin.* Mill. illuſtr. Mor. hiſt. ſ. 1. t. 1. f. 6. Ger. 871. Park. 178. 6.

[v] Populus alba *Lin.* Evel. ſilva by Hunter, p. 201. Duham. t. 36. Ger. 1486. 1. Park. 1410. 1.

[w] Populus tremula *Lin.* Blackw. 248. 2. Ger. 1487. 3. Park. 1411. 4.

in

in ſhape, but ſmooth on both ſides; theſe being ſet on long petioles that are flatted at the tip, tremble with the ſlighteſt breeze. *Black Poplar* [x] has rhomboid leaves, pointed and toothed; they are ſmooth on both ſides, of a light green; and the catkins are ſhorter than thoſe of the two former. *Carolina Poplar* [y] has very large heart-ſhaped leaves, obtuſely notched about the edges; and the ſhoots angled. *Tacamahaca* [z] is a ſpecies of Poplar, with oblong ovate leaves, toothed about the edges, white underneath, with a ſcarcely viſible down, and the veins forming a fine net-work: the ſtipules are remarkably reſinous.

Mercurialis. Of the order *Enneandria* there is an herb, frequent under hedges and in woods, called *Dog's Mercury* [a]: the flowers have a three-parted calyx, and no corolla; in ſome there are nine or twelve ſtamens, with globular, twin anthers; in others, on a diſtinct plant, two ſtyles, and a two-grained, two-celled capſule, containing one ſeed in each cell. The ſpecies here meant is diſtinguiſhed

[x] Populus nigra *Lin.* Mill. illuſtr. Blackw. 548. & 248. 1. Ger. 1486. 2.

[y] Populus balſamifera *Miller.* angulata. Duham. arb. 2. t. 39. f. 9.

[z] Populus balſamifera *Lin.* Cat. car. 1. 34. Duh. arb. 2. t. 38. f. 6. Mill. fig. t. 261.

[a] Mercurialis perennis *Lin.* Curtis, Lond. II. 65. Ger. 333. 1. M. annua. Curt. Lond. V. 68. Ger. 332. 1, 2.

from

from the rest by its very simple unbranched stem, and its rough leaves.

In the order *Monadelphia* you will find a genus of trees under the title of *Juniper*, including not only the Juniper properly so called, which is rather a shrub than a tree, but also the Savin, and American or Sweet Cedars, &c. The staminiferous flowers in this genus are borne on an ament, the scales of which form the calyx of each flower, having no corolla, but only three stamens: the pistilliferous flowers have a small, permanent, three-parted calyx, growing to the germ, which is below the flower; they have a corolla of three petals, three styles, and a three-seeded berry, with three tubercles of the unequal calyx on the lower part, and three little teeth at top from the remains of the petals. *Common Juniper* [b] has three spreading, pointed leaves, coming out together, that are longer than the berry. *Savin* has opposite, erect, decurrent leaves, with the oppositions boxed into each other along the branches; they are short and acute: this shrub spreads out much horizontally, rising little in height. There are several species of *Cedar* natives of America, *Bermudas Cedar* [d] is that which is imported for casing black lead in pencils, was for-

Juniperus.

[b] Juniperus communis *Lin.* Mill. illustr. Duham. t. 127. Ger. 1372. 1. Park. 1029. 1.

[c] Juniperus Sabina *Lin.* Blackw. 214.

[d] Juniperus bermudiana *Lin.* Herm. lugdb. t. 347.

merly

merly uſed for wainſcoting rooms, and now for ſhips in the Weſt Indies, the worms not attacking this kind of wood. The ſpecific diſtinction is from the leaves; the lower ones being threefold, the upper twofold[e], decurrent, ſubulate, ſpreading, and acute. Our plantations of ſhrubs have alſo the *Red Virginia*[f], *Carolina*, and *Barbadoes*[g] *Cedars*; and there are others which are natives of the ſouthern parts of Europe[h].

Taxus. The baleful *Yew*[i] is of the ſame order: the flowers have no corolla, nor, properly ſpeaking, any calyx, unleſs we allow the three or four-leaved bud to be ſuch: on ſome trees they will be found to have many ſtamens, terminated by peltate, eight-cleft anthers; on others, to have an ovate, pointed germ, ending in an obtuſe ſtigma without any ſtyle, the germ becoming a kind of berry, or rather ſucculent receptacle, with one ſeed in it, having the top naked: theſe flowers all come out from the axils of the leaves, which are linear, end in a ſharp point, and are ranged in a double row cloſe together along the mid-rib; the

[e] Miller ſays fourfold and imbricate.

[f] Juniperus virginiana *Lin.* Sloan. jam. 2. t. 157. f. 3.

[g] Juniperus barbadenſis *Lin.* Pluk. alm. 197. 4. Hort. angl. t. 1. f. 1.

[h] Juniperus thurifera, phœnicia, lycina, Oxycedrus *Lin.*

[i] Taxus baccata *Lin.* Evel. ſilva by Hunter, p. 257. Duham. t. 86. Ger. 1370. Park. 1412.

berry

berry is red, and mawkiſhly ſweet—not poiſonous, though the leaves certainly are ſo.

I will now finiſh our examination of this claſs, and cloſe this letter, with the ſingular genus of *Ruſcus*, the flowers of which have a ſix-leaved calyx, no corolla, but an ovate inflated nectary, perforated at top, in the centre of the flower: the ſtaminiferous flowers have no filaments, but only three anthers, ſitting on the top of the nectary, and united at the baſe, whence this genus is of the order *Syngeneſia*: the piſtilliferous flowers have one ſtyle, and a germ hid within the nectary, which becomes a globoſe, three-celled berry, containing two globoſe ſeeds. The common ſpecies, which we call *Butcher's Broom*, or *Knee Holly* [k], bears its flowers in the middle of the leaves, on their upper ſurface; theſe are of the ſhape and ſize of myrtle leaves, but ſtiffer, and end in prickly points; the berries are red, and almoſt as large as cherries: in another ſpecies [l] the flowers are produced on the under ſurface of the leaves: in a third [m] they are produced alſo underneath, but are protected by a leaflet, whereas in the other ſpecies they are naked: a fourth [n] flowers Ruſcus.

[k] Ruſcus aculeatus *Lin.* Mill. illuſtr. Blackw. 155. Duham. t. 59. Ger. 907. Park. 253.

[l] Ruſcus Hypophyllum *Lin.* Col. ecphr. 1. t. 165. f. 1.

[m] Ruſcus Hypogloſſum *Lin.* Col. t. 165. f. 2.

[n] Ruſcus androgynus *Lin.* Dill. elth. t. 250. f. 332.

from

from the margin of the leaves: and the *Alexandrian Laurel* °, which is a ſpecies of *Ruſcus*, from long racemes at the ends of the branches; the flowers of this are complete, and therefore the plant ought not to be found in this claſs, but ſince it is evidently of this genus naturally, Linnæus has left it with its own family, chooſing rather to violate the laws of his own arbitrary ſyſtem than thoſe of nature. The ſtalks of this are ſlender and pliable; the leaves are rounded at the baſe, but end in acute points; they are ſmooth, and of a very lucid green: the flowers are of an herbaceous yellow colour, and are ſucceeded by berries like thoſe of our Butcher's broom, but ſmaller. With this beautiful evergreen I leave you, dear couſin, till the next letter.

° Ruſcus racemoſus *Lin.* Mor. hiſt. ſ. 13. t. 5. f. 14.

LETTER

LETTER XXX.

THE CLASS POLYGAMIA.

June the 14th, 1777.

THERE are ſome perſons, dear couſin, who think the twenty-third claſs—*Polygamia*, might have been ſpared, and the plants compriſed in it[p] ranged in the other claſſes, according to the number, ſituation, proportion, &c. of the ſtamens. But let us take things as we find them, without enquiring too deeply into the merits, of what, after all, is of no great importance. The eſſence of this claſs conſiſts in having complete flowers, accompanied by one or both ſorts of incomplete ones, either on the ſame or different individuals. The latter circumſtance furniſhes the character of the three orders.

The firſt order of this claſs having the complete and incomplete flowers always on the ſame plant, is hence entitled *Monœcia*. You may perhaps remember, that ſome of the graſſes were ſaid to be of this order[q]; here alſo are the *Plantain-tree* and *Banana*[r]: *Valantia* or *Croſſwort*, which you Valantin.

[p] Genera 34, ſpecies 224.

[q] See letter XIII.

[r] Muſa paradiſiaca & ſapientum *Lin.* Trew. Ehr. t. 18—23.

may

may find in hedges and buſhy places, and will evidently perceive to be of a natural tribe[s] you have met with before: there is uſually one complete flower in this genus, accompanied on each ſide with an incomplete ſtaminiferous one; the former has the corolla four-parted, four ſtamens, a bifid ſtyle, and one ſeed; the latter have the corolla trifid in ſome ſpecies, quadrifid in others; three ſtamens in ſome, four in others, and an obſcure piſtil; none of the flowers have any calyx: frequently theſe plants produce incomplete flowers only, and therefore no ſeed; owing, I preſume, to their running ſo much at the root. Our wild ſpecies[t] is one of thoſe which have the incomplete flowers quadrifid, and it has two leaves to each peduncle, which ſupports about eight flowers, with yellow corollas; there are four leaves to each whorl, and they, with the whole plant, are covered with ſoft hairs.

Parietaria. *Pellitory of the Wall* has two complete flowers, with one piſtilliferous flower between them, within a ſix-leaved involucre; they have a four-cleft calyx, no corolla, one ſtyle, and one ſeed: the complete flowers are diſtinguiſhed by having four ſtamens; the other has none. Our common ſpe-

[s] *Stellatæ:* ſee letter XV.

[t] Valantia Cruciata *Lin.* Blackw. t. 76. Mor. hiſt. ſ. 9. t. 21. f. 1. Ger. 1123. 1.

cies

cies [u] has broad lance-ſhaped leaves, dichotomous or forked peduncles, and two-leaved calyxes: the piſtilliferous flowers are quadrangular and pyramidal.

Atriplex, or *Orach*, has ſuch affinity with Chenopodium or *Gooſefoot*, that, as Linnæus obſerves, if Orache had only complete flowers it would be a Gooſefoot; and if this had piſtilliferous flowers, it would be an Orache. Moſt of theſe are common weeds on dunghills, or on the ſea-coaſt. Atriplex.

Acer, or *Maple*, is a tree in which you may examine the character of the claſs and order at your eaſe. The flowers are produced in bunches; the lower ones complete, and thoſe which are towards the end ſtaminiferous: they have a quinquefid calyx, a corolla of five petals; the complete flowers have beſides all this one piſtil, and two or three capſules, joined at the baſe, flat, each terminating in a large, membranaceous wing, and containing one ſeed. The *Great Maple*, commonly called *Sycomore* [v], has five-lobed leaves unequally ſerrate, and the flowers in large racemes. *Common Maple* [w] has lobed leaves, obtuſe, and emarginate; generally they are divided half way into three lobes, Acer.

[u] Parietaria officinalis *Lin.* Curtis, Lond. IV. 63. Fl. dan. 521. Ger. 331. Park. 437.

[v] Acer Pſeudoplatanus *Lin.* Evel. ſilva by Hunter, p. 193. Duham. t. 9. Ger. 1484. 1. Park. 142. 1.

[w] Acer campeſtre *Lin.* Ger. 1484. 2. Hunt. Evel. ſilva, p. 183. and Pl. 33. of this work.

the ſide ones obtuſely ſemi-bifid, the middle one ſemi-trifid; the upper leaves rather cut into five lobes: the bunches of flowers are ſmaller. This tree grows much in hedges.

Mimoſa. The famous *Mimoſa* or *Senſitive* belongs to this firſt order of the claſs Polygamia. The flowers have a five-toothed calyx, a five-cleft corolla, and five or more ſtamens: the complete flowers have alſo one piſtil, and a legume for a ſeed-veſſel. This genus is very numerous, but all the ſpecies are not endued with the ſenſitive quality. That which is moſt common in the iſlands of the Weſt Indies, and in our ſtoves [x], has the ſtems armed with ſhort recurved ſpines; pinnate leaves compoſed of four or five pairs of leaflets, whoſe baſe joins at a point where they are inſerted into the petiole, ſpreading upwards like the fingers of the hand; the flowers come out from the axils on ſhort peduncles, in ſmall globular heads, the corollas are yellow; they are ſucceeded by ſhort, flat, jointed pods, with two or three orbicular, bordered, compreſſed ſeeds in each. Some ſpecies move much more readily than others; ſome drop the leaflets only, and others drop the petioles of the whole leaf alſo. The *true Egyptian Acacia* [y], and many other Acacias, having the ſame characters, are included in this genus: they

[x] Mimoſa pudica *Lin.* Comm. hort. 1. t. 29.
[y] Mimoſa nilotica *Lin.*

are

are too tender to flower much in our climate.

Three-thorned Acacia[z] is of another genus, and indeed of another order—*Diœcia*: for it has the ſtaminiferous flowers in a long, compact, cylindric ament, with ſome complete ones generally at the end of it; and, on a diſtinct plant, piſtilliferous flowers on looſe aments. The complete flowers have a quadrifid calyx, a four-petalled corolla, ſix ſtamens, one piſtil, and a legume: the ſtaminiferous flowers have a three-leaved calyx, a corolla of three petals, and ſix ſtamens: and the piſtilliferous flowers have a five-leaved calyx, a five-petalled corolla, one piſtil, and a legume. The common ſpecies is diſtinguiſhed from the other[a] by its large thorns, which have generally two ſmaller ones, coming out from the ſide: they are axillary, and are often produced in cluſters at the knots of the ſtem: the leaves are pinnate, and have ten pairs of ſmall leaflets. In America, its native country, this tree is called *Honey Locuſt*. Gleditſia

The *Aſh-tree* is alſo of this ſecond order: having on ſome trees complete flowers, on other piſtilliferous cnes, each frequently accompanied by the others; they have either a four-parted calyx or none, a corolla Fraxinus.

[z] Gleditſia triacanthos *Lin.* Duham. 1. t. 105. Hort. angl. t. 21.

[a] Gleditſia inermis *Lin.* Mill. fig. pl. 5.

 of

of four petals or none, and one piſtil: the complete flowers have alſo two ſtamens, and one lance ſhaped ſeed. *Common Aſh* [b] has pinnate leaves, with five pairs of leaflets, ſlightly ſerrate on the edge; the flowers have neither calyx nor corolla, and are produced in looſe bunches from the ſides of the branches. *Flowering Aſh* [c] has the leaflets ſerrate; the flowers are furniſhed both with calyx and corolla; and are in large looſe bunches at the ends of the branches. The *American* or *Carolina Aſh* [d], has the leaflets quite entire, and the petioles round.

Ficus. Of the third order—*Triœcia*, we have the *Fig*, which though it bears flowers that are viſible, yet conceals them within the fruit, and therefore may lead us well enough to the claſs *Cryptogamia*. What we call the fruit of the Fig Linnæus names the receptacle, or common calyx of the flowers; he deſcribes it as being top-ſhaped, fleſhy, converging, cloſed at the broad end with ſeveral ſcales, and having the inſide covered with little flowers, complete and incomplete; ſometimes in the ſame fruit, and ſometimes on different trees: the ſtaminiferous flowers have a three-parted calyx, and three ſtamens; the piſtilliferous flowers have a five-parted calyx, one piſtil,

[b] Fraxinus excelſior *Lin.* Evelyn's ſilva by Hunter, p. 145. Blackw. 328. Duham. t. 101. Ger. 1472.
[c] Fraxinus Ornus *Lin.* Mill. illuſtr. Hort. angl. t. 9.
[d] Fraxinus americana *Lin.* Cateſb. car. 1. 80.

and one roundiſh, flatted ſeed; neither of them have any corolla. Our *common* or *eatable Fig*[e] is diſtinguiſhed by its palmate leaves: the different fruits are but varieties ariſing from the ſame ſeed. The hiſtory and œconomy of this ſingular tree, as related by naturaliſts and travellers, will be an agreeable relaxation to you amidſt our dry botanical diſquiſitions.

[e] Ficus Carica *Lin.* Mill. illuſtr.

LETTER XXXI.

OF THE NECTARIUM OR NECTARY.

June the 21ſt, 1777.

HAVING now gone through all the claſſes of conſpicuous flowers, we ſhould regularly proceed to the laſt claſs of the ſyſtem, in which they are inconſpicuous; but having kept on a ſtraight courſe for a long time, we will now turn out of it, and take a view of the different appearances which the nectary puts on, in the ſeveral genera of plants wherein it is found.

Several of theſe have been curſorily mentioned as characters of the genus; and we have even hinted at the general uſe of the nectary [f]: but we ſhall now go farther, and ſay, that though this part of the flower has not hitherto been obſerved in two hundred genera [g], yet that in all probability it exiſts in all, if not as a diſtinct viſible part, as a gland or pore however, or a ſet of glands or pores, exuding that viſcid, ſweet juice, ſo uſeful ſecondarily for the nouriſhment of a great variety of inſects, and, at the ſame time doubtleſs primarily neceſſary to the fructification of the plant itſelf. For you will

[f] See letters IV. and XVII.

[g] Beſides the Graſſes.

will obſerve in monopetalous tubular corollas, that though they have no viſible nectary, yet there is a nectareous juice ſecreted into their tube[h], which is therefore probably provided with glands for this purpoſe, too minute to be ſeen with the naked eye, but which an accurate inſpection with glaſſes might perhaps detect. Polypetalous flowers with open calyxes, having no tube, or baſin for the reception of the nectareous juice, have in general a body deſtined to prepare and contain it, in order that it may be diſtributed to the ſurrounding parts of fructification, as it is wanted. In the compound and umbellate tribes of plants indeed no nectaries have been remarked, but then you remember, that the whole flower in both of them is ſo ſmall, that it is no wonder if a part ſo minute as the nectary frequently is in larger flowers ſhould eſcape our obſervation in theſe: we may preſume however that they abound in nectareous juice, ſince we obſerve that inſects are particularly fond of theſe tribes. No genus of the claſs Icoſandria has any diſtinct nectary; but then the calyx is one-leafed, and forms a commodious baſin for the reception of the nectareous juice, which is frequently very diſcernible in it. The verticillate tribe alſo is not mentioned by Linnæus as being fur-

[h] As particularly in the Honeyſuckle and Aloe.

[i] Didynamia Gymnoſpermia *Lin.*

 niſhed

niſhed with viſible nectaries; nor are they perhaps immediately neceſſary here, becauſe the corolla is monopetalous, and the monophyllous calyx forms a permanent tube: many genera however of this order have a gland in the bottom of the calyx, ſurrounding the baſe of the germ; this is large in the *Bugle*, and ſufficiently viſible in the *Dead Nettle*.

No appearance of the nectary is more common than this of glands. You have already ſeen[k] that they are conſiderable in ſeveral genera of the cruciform tribe; that they have furniſhed us with generic characters: and that they are even the cauſe of the claſſical character itſelf[l]. It has been juſt mentioned that they are found in the verticillate or labiate tribe: and many genera, diſperſed in various parts of the ſyſtem, have this glandular nectary. Thus *Plukenetia* (1080)[m] has four glands at the baſe of the filaments, as in the claſs *Tetradynamia*. *Cercis* (510) has a ſtyle-form gland under the germ. *Lathræa* (743) and *Orobanche* (779) have a gland at the baſe of the germ. *Caſſyta* (505) has three glands; *Echites* (299), and *Tabernæmontana* (301), have five; *Hernandia* (1049) has

[k] Letter XXIII.

[l] See letter II. IV. and V. compared with letter XXIII.

[m] The figures refer to the number of the genus in Linnæus's genera and ſyſtema.

ſix

ſix or four, ſurrounding the germ; and *Grielum* (1235) has a ſet of oblong glands, round the germ, uniting into a little crown. *Malpighia* (572) has two glands at the bottom and on the outſide of each leaf of the calyx: in *Baniſteria* (573) the caſe is the ſame, except that one foliole of the calyx has no glands, and therefore the whole number is eight; whereas in the other it is ten. *Reſeda* (608) has a gland ariſing from the receptacle between the ſtamens and the upper petal: and (*Croton* 1083) has five of them, fixed to the receptacle. *Aſtronium* (1111) has five glands in the diſk of the flower. *Cucurbita* (1091), or the gourd genus, has a ſingle, triangular, concave gland in the centre of the flower: and in the *Salix* (1098), or *Willow*, the ſituation is the ſame, but the form of it is cylindric.

Another very uſual form of the nectary is ſcales, which are in truth but flatted glands. *Monnieria* (850), and *Vicia* (873), or the Vetch genus, have one ſcale only, at the baſe of the germ. *Cuſcuta* (170), or *Dodder*, has four ſcales, at the baſe of the ſtamens. But many have five ſcales: as *Parnaſſia*[n] (384); at the baſe of the filaments in *Schrebera* (319), *Quaſſia* (529), and *Melaſtoma* (544); between the ſtamens in *Ireſine* (1113); at the baſe of the germ, in *Craſſula* (392), *Cotyledon* (578), and *Se-*

[n] See Plate 34. f. 3.

dum

dum (579); ſurrounding the receptacle, in *Samyda* (543); or at the baſe of the petals, in *Erythoxylon* (575), *Ranunculus*[o] (699), *Grewia* (1026), and *Kiggelaria* (1128). *Amaryllis* (406), and *Leontice* (423), have ſix ſcales; without the baſe of the filaments in the firſt, and inſerted into the baſe of the petals in the ſecond.

Not unfrequently does the nectary appear in the ſhape of valves, which are generally five in number; in *Plumbago* (213) placed at the bottom of the corolla, and incloſing the germ; ſurrounding the germ in *Achyranthes* (288); and covering the receptacle in *Campanula* (218) and *Roella* (219). *Aſphodel* (421) has ſix of theſe valves, inſerted into the baſe of the corolla, and forming a complete arch over the germ; a filament ſpringing from each of them[p].

In *Erythronium* (414) there are two callous tubercles at the baſe of each inner petal; in the *Laurus* (503) genus[q], three tubercles round the germ; and two round glands, on a ſhort ſtalk, near the baſe of each filament of the inner rank. In ſome ſpecies of *Iris* there are three dots[r] at the baſe and on the outſide of the corolla; in *Tamus* (1119) an oblong dot grows to the inſide of each diviſion of the calyx; and in another genus, *Swertia* (321), are ten of

[o] Plate 34. f. 4.
[p] Plate 34. f. 7.
[q] See letter XIX.
[r] Puncta.

theſe

these dots; two at the base of each division of the corolla, surrounded with bristles. In the *Hyacinth*[s] (427) there are three pores at the top of the germ: and in both the genera of *Fritillaria* (411), and *Uvularia* (412), there is an excavation at the base of each petal: in the *Crown Imperial* this is considerable, and generally exhibits a large drop of nectareous juice[t]. *Mercurialis* (1125)[u] has two subulate *acumens* or sharp points, one on each side of the germ; and *Vallisneria* (1097) has a *cuspis* on each petal.

You remember the beautiful appearance that the nectary made in some species of *Iris*[v] as a longitudinal villous line upon the petals: in the *Lily* (410) it is a pipe or tubulous line along the middle of each petal: and in *Frankenia* (445) it is a channel running along the claw.

In some genera the nectary takes the exact form of petals, and was always confounded with them until Linnæus pointed out the difference: this is the case with several plants of the first class[w], and with *Lecythis* (664) in the thirteenth; in all these it is of one petal only: in *Galanthus* (401), or *Snowdrop*, it consists of three parallel, notch-

[s] Our wild Hyacinth (H. non scriptus) has not these pores, or at least they are not visible to the naked eye.
[t] See Plate 34. f. 6.
[u] Letter XXIX.
[v] Letter XIV. See Pl. 34. f. 5.
[w] Letter XI.

ed,

ed, obtuſe, petal-like leaflets, forming a cylinder about half the length of the corolla. *Illicium* (611) has ſeveral awl-ſhaped folioles of the ſame length with the petals themſelves. *Cardioſpermum* (498) has a four-petalled nectary incloſing the germ; and in *Hartogia* (273), *Sauvageſia* (286), and *Helicteres* (1025), it is made up of five petals. *Andrachne* (1095) has five ſemi-bifid herbaceous folioles, leſs than the petals, and placed between them. All the *Graſſes*, *Rice* (448), and *Mays* (1042), agree in having a nectary of two minute, oblong leaflets. *Swietenia* (521), *Melia* (527), and *Melianthus* (795), have a one-leafed nectary, with a many-toothed mouth in the two firſt, and in the laſt within the loweſt diviſion of the calyx, to which it grows. In *Muſa* (1141) alſo, the nectary is one boat-ſhaped leaf, compreſſed, pointed, and inſerted within the boſom of the petal. Ten converging leaflets, incloſing the germ, form the nectary of *Zygophyllum* (530); each leaflet being fixed to the baſe of each filament. *Dalechampia* (1081) has a broad nectary, compoſed of many ovate, flat plates, in ſeveral rows.

I have mentioned before, that in tubulous corollas the nectareous juice is ſecreted into the tube: in many genera there is a horn or ſpur at the back of the flower, which anſwers this purpoſe of a recipient. Several plants have occurred in the courſe

of

of our examinations with a nectary of this form; as *Tropæolum* (466), *Larkspur* [x] (681), *Aconite* [y] (682), *Columbine* (684), *Antirrhinum* (750), *Fumitory* (849), *Violet* (1007), *Impatiens* (1008), and *Orchis* (1009): to these we may add *Pinguicula* (30), or *Butterwort*, *Utricularia* (31), and *Valerian* (44). In some species of Antirrhinum the horn is blunted, and becomes rather a bag; which is also its shape in the *Satyrium* genus (1010). The genera of this tribe are remarkable for their nectaries; in *Ophrys* (1011) it hangs down from the corolla, longer than the petals, and is keeled at the back part; in *Serapias* (1012) it is of the same length with the petals, ovate, gibbous below, and with an ovate lip; in *Limodorum* (1013) it is of the same length with the petals, of one leaf, concave, standing on a pedicle, and within the lowest petal; in *Arethusa* (1014) it is of one leaf, tubulous at the bottom of the ringent corolla, and connate with it; in *Cypripedium* (1015), or *Ladies-Slipper*, as you have seen before [z], it is very large and inflated; and in *Epidendrum* (1016) it is tubulous at the base, turbinate or top-shaped, with an oblique bifid mouth. Thus you observe that all the genera of this tribe have singular nectaries; whereas in the three classes with

[x] Plate 34. f. 2. [y] Plate 34. f. 1.
[z] Letter XXVII.

conjoined

conjoined filaments ſcarcely any are to be found[a]. The numerous genus of *Carex* (1046), or *Sedge*, has an inflated, permanent nectary, contracting above, and toothed at top, where it gapes, but continues to inveſt the ſeed; in *Ruſcus* (1139) alſo it is inflated and open at top, it is ovate, erect, and of the ſame ſize with the calyx.

In many genera the nectary takes the form of ſome well-known utenſil or other thing. Thus in *Staphylæa* (374), *Tinus* (504), *Winterana* (598), and *Urtica* (1054), or Nettle, it is *Urceolate* or *Pitcher-ſhaped.* In *Narciſſus*[b] (403), and *Pancratium* (404), it is *Funnel-ſhaped.* In *Epimedium* (148) it is *Cyathiform* or *Goblet-ſhaped.* In *Byttneria* (268), *Theobroma* (900), or *Chocolate*, *Ayenia* 1020), and *Kleinhovia* (1024), it is *Bell-ſhaped.* In *Ciſſampelos* (1138) it is *Wheel-ſhaped:* and in *Epidendrum* (1016), *Poplar* (1123), and *Gleditſia* (1159), it is *turbinate*, or ſhaped like a boy's top, narrow at bottom, and ſpreading out above. The moſt beautiful of theſe nectaries is the *Crown-ſhaped:* in *Dioſma* this is placed on the germ; in *Olax* (45), *Hamamelis* (169), *Nerium* (297) or *Oleander*, *Periploca* (303), *Silene* (567), and *Cherleria* (570), it terminates the tube of the corolla: but in the *Paſſion-flower* (1021) it is a triple crown or

[a] In *Monadelphia* and *Polyadelphia* only one in each; and in *Diadelphia* three. [b] See Plate 14. f. 2.

glory, the outer one longeſt, ſurrounding the ſtyle [c].

In *Garidella* (571), *Nigella* (685), and *Hellebore* [d] (702), the nectaries are bilabiate; the firſt has five, the ſecond has eight, and the third has an uncertain number. *Trollius* (700) has nine linear, flat, bent bodies, perforated at the baſe, on the inſide; and *Iſopyrum* (701) has five equal, tubulous, ſhort nectaries, with a trilobate mouth, inſerted into the receptacle, within the petals.

In *Arum* (1028) the nectaries reſemble the filaments of ſtamens, only that they thicken at bottom; they come out in two rows from the middle of the ſpadix. In *Peganum* (601) the filaments themſelves are dilated into nectaries at the baſe. In *Fevillea* (1118) they conſiſt of five compreſſed bent threads, placed alternately with the ſtamens. In *Trichilia* (528) the nectary is cylindric, and tubulous, formed out of the ten filaments, ſhorter than the petals, and with a five-toothed mouth.

You have obſerved that many nectaries already mentioned have an intimate connexion with the germ; it is a ſituation ſo common with this part of the flower, that ſome perſons have ſuſpected the ſole or principal uſe of it to be to ſupply and foſter the germ. Accordingly there are ſeveral other

[c] See Plate 30. [d] Plate 34. f. 8.

genera,

genera, in which it is thus placed. In *Mirabilis* (242), or *Marvel of Peru*, it is globoſe, permanent, and incloſes the germ; in *Ciſſus* (147), *Celoſia* (289), *Limeum* (463), and *Phyllanthus* (1050), it is a ring ſurrounding the germ: in *Cynanchum* (304) it is cylindric, with a five-toothed mouth; in *Apocynum* (305), *Aſclepias* (306), and *Stapelia* (307), it is made up of five bodies, which in the ſecond and third entirely conceal the ſtamens and piſtils, and in the third forms a double ſtar: all of them about the germ. In *Gualtheria* (551) it is made up of ten ſhort, awl-ſhaped, erect bodies, ſurrounding the germ, between the ſtamens.

It muſt not be diſſembled however, that whatever uſe theſe bodies may be of to the germ, when they adhere to it, or are near it; they are frequently found on other parts of the fructification. Many inſtances of this have already occurred, and to theſe we may add, that they are found on the petals in *Bromelia* (395), growing to each of the three, above the baſe; in *Berberis* (442), or the *Barberry*, in two roundiſh orange-coloured bodies at the baſe of each; in *Hermannia* (828), each petal having a little membrane, forming all together a cowled tube; in *Hydrophyllum* (204), and *Reaumuria* (686), in *laminæ* or plates growing to them; in *Myoſurus* (394), being five awl-ſhaped bodies. The nectary is found on the calyx in *Tropæolum* mentioned before, in *Monotropa* (536), in ſome ſpecies of *Biſcutella* (808), and in *Malpighia*, mentioned alſo before among thoſe which have glandular nectaries. This part is a globoſe gland on the exterior tip of the anthers in the *Adenanthera* (526), at the baſe of them in *Ambroſinia* (1238): and on the filaments in form of glands in *Dictamnus* (522), in form of ſcales in *Zygophyllum* (530), placed horizontally on the real filaments in *Commelina* (62); and in *Plumbago*, *Campanula*, and *Roella*, mentioned before. And, laſtly, the nectaries are not unfrequently placed on the receptacle; as in *Lathræa* (743), *Clutia* (1140), *Melianthus* (795), and ſome others: but theſe are ſo cloſe to the germ, which takes its riſe from the ſame baſe, that they may very well be ſuppoſed to be placed there for its uſe.

But what ſhall we ſay when we find the nectary, in the incomplete ſtaminiferous flowers, which have no germ; as in *Willow* (1098), *Aſtronium* (1111), *Ireſine* (1113), *Fevillea* (1118), *Poplar* (1123), *Rhodiola* (1124), *Kiggelaria* (1128), *Ciſſampelos* (1138), *Ruſcus* (1139), *Clutia* (1140), and *Ophioxylon* (1142). In all theſe caſes it certainly cannot be of any immediate uſe to the germ, which is not only on a diſtinct flower but on a different plant: this however being the moſt important part of the vegetable, ſince it is deſtined by nature to produce a new one of the ſame kind; and all the other parts of the flower being in ſome meaſure ſubſervient to this, whatſoever is immediately uſeful to theſe may fairly be ſaid to be mediately ſerviceable to the germ.

But let us return to our hiſtory of facts, and finiſh this dry diſcuſſion, which I ſhould not have troubled you with, if I could have directed you to any author where you might find the different forms and ſituations of the nectary regiſtered in one view [e].

Hitherto you have obſerved that this beautiful part of the flower is generally ſingle, though in many caſes formed of ſeveral portions: in ſome genera however it is double. Thus in *Krameria* (161), there are two nectaries, one above another; in *Stapelia*, as you have already ſeen, a double ſtar, both flat and quinquefid, the lower with linear diviſions torn at the end, ſurrounding the ſtamens and germs, the upper with acute, entire diviſions covering them: ſomething of the ſame kind is obſervable alſo in *Aſclepias*, the very ſingular ſtructure of whoſe flowers is particularly deſerving of your attention. *Paullinia* (497) alſo, and *Sapindus* (499) have two nectaries, very different from each other; the

• When I writ this letter, I entirely forgot that there was a diſſertation on the ſame ſubject printed in the 6th volume of the *Amænitates Academicæ*. The learned reader may compare that treatiſe with this.

one

one confifting of four petals inferted into the claws of the real petals, the other of four glands at their bafes. I may here obferve, that though the general ufe of the nectary, as the name implies, be to pour out the nectareous juice; yet it does not feem that all the bodies to which Linnæus has given the name ferve that purpofe: fuch may probably be the cafe in one of thefe nectaries of the genera before us, and perhaps of others, where this part is double. Laftly, *Clutia* (1140) has two fets of nectaries, one within the other; the outer of five three-parted, oblong bodies, placed in a ring within the petals, and of the fame length with their claws; the inner of five little glands, which are certainly melliferous at top: it is obfervable that in the piftilliferous flowers of this genus there are no glands or inner nectaries, and the outer ones are of the fame fize, and in the fame fituation, but differ in form, being roundifh and didymous, or twinned.

Concerning the form and variations in the other parts of the fructification, which furnifh the generic character of vegetables, enough is to be found in the elementary books[f]: of the leaves alfo, together with thofe other parts and circumftances, furnifhing characters for the differences of about ten thoufand one hundred fpecies, which

[f] Lee's Introduction; Rofe's Elements of Botany, &c.

is the whole number of plants at preſent arranged [g], there is no want of inſtruction in the ſame authors, tranſlated from Linnæus's original work. I ſhall only remark to you therefore, that a more minute attention and accurate obſervation of vegetables, diſcovered to Linnæus parts that former botaniſts had paſſed by unnoticed; and that his ſuperior ſagacity and genius enabled him to make a much more extenſive uſe of ſuch as were already known. The parts I now allude to, are what he calls *Fulcra*, props or ſupports of the plant. Among theſe the *arms* or *weapons*, that is, thorns and prickles; claſpers or tendrils; ſome ſorts of *pubeſcence*; and perhaps *glands*, in ſome few ſpecies had been noticed; but in a manner very looſe and imperfect: but the *ſtipule*, which is a ſcale at the baſe of the petioles; and the *bracte*, which is a ſcale or ſmall leaf next the flower, had not been ſo much as named; nor had any one thought of uſing theſe ſeven important though minute parts for diſtinguiſhing the ſpecies, a buſineſs to which they are ſo well adapted, both by their conſtancy and abundant variety.

He has alſo taken in other circumſtances very happily, beſides the mere *form*, to furniſh ſpecific differences, and for other pur-

[g] In the 14th edition of Syſtema Vegetabilium.—To theſe however a conſiderable number has been ſince added, from the South-ſea iſlands, and other places.

poſes;

poſes; ſuch as the mode and degree of *ramification* in leaves and branches, the *intorſion*, or manner of turning or bending in the ſtems; the *gemmation*, or various conſtruction of the buds; the *foliation*, or different folding of the leaves before they are expanded; the *infloreſcence*, or manner in which flowers are connected to the plant by their peduncles: all theſe, together with ſome others, which I have paſſed over, will occaſionally furniſh you with marks to diſtinguiſh plants from each other, even more certain in ſome caſes than the form itſelf, and therefore highly worthy of your attention; but I have already treſpaſſed on that too long, and will leave you to your leiſure and more important concerns.

LETTER XXXII.

THE CLASS CRYPTOGAMIA.

October the 4th, 1777.

I HAVE at length found time, dear cousin, to send you my last letter on the subject of Botany. I have not hastened it, because you have found full employment during the summer, either in examining such plants as had escaped you before, or in searching for their nectaries and other more minute parts. You have also by this time discovered, that the study or amusement which you have taken up, is not the affair of a single season.

As to the last and lowest class of vegetables—*Cryptogamia*, I shall at present touch it very slightly, because, though full of beauties, when examined with that attention which such small bodies require, it is much too difficult for our young cousin, and will probably be uninteresting even to you, unless you have already imbibed a greater passion for Botany than I wish you to have. The objects also of this class must be searched for in places, and at a season, by no means agreeable to your delicacy; and I will not have you risk your health, the most precious gift of heaven, even in pursuit

pursuit of the most delightful knowledge. Gentle exercise, such as a proper attention to the study of nature will induce you to take, accompanied with that cheerfulness, regularity, and temperance, for which you are so conspicuous, is your best security for a continuance of this blessing; and that you may enjoy it uninterrupted to a period yet distant, my good wishes shall not be wanting.

You are already acquainted with the meaning of the name *Cryptogamia*, and the character of the class[h]: you are also mistress of the four orders into which it is divided, together with their characters, such as they are[i]. I have only therefore to present you with a few of the most obvious species in each order, wherein the generic and specific characters are the least inconspicuous.

The number of genera in this class are fifty-one, of species eight hundred and fifty-eight.

FERNS.

The plants of the first order—the *Ferns*, are as large, and oftentimes as specious, as those of the foregoing classes: it is apparent also to the naked eye, that there is a fructification, though the parts of it are not

[h] See page 105. [i] See page 114, &c.

diſtinguiſhable. The general face of this, as it appears to the microſcope, has been already deſcribed [k].

In general the fructification in this order of Ferns is on the back of the leaves; that however is not univerſal. For inſtance, in the genus *Equiſetum*, or *Hórſetail*, it is in a ſpike, each ſeparate fructification being peltate, and gaping at its many-valved baſe: Hedwig has determined the flowers of the Horſetails and Adder's-tongue to be hermaphrodite. *Corn Horſetail* [l] has theſe ſpikes on a naked ſtem, and other leafy barren ſtems come up later in the ſeaſon. *Wood Horſetail* [m] has the leaves compound, or divided, and the ſpikes at the end of the ſame ſtems. A ſpecies common in ditches [n] has ſcarcely any leaves, and is perfectly ſmooth; in which circumſtance alone it differs from the *Shave-graſs* [o] uſed in poliſhing, which is rough.

Equiſe-tum.

Ophio-gloſſum.

Ophiogloſſum alſo, or *Adder's-tongue*, has the fructifications on a ſpike, in a jointed row along each ſide of it; when they are ripe, theſe joints gape tranſverſely. Our

[k] Letter X.

[l] Equiſetum arvenſe *Lin.* Curtis, Lond. IV. 64. Ger. 1114.

[m] Equiſetum ſylvaticum *Lin.* Ger. 1114. Hedw. theor. f. 1. 7.

[n] Equiſetum limoſum *Lin.* Ray. ſyn. t. 5. f. 2.

[o] Equiſetum hyemale *Lin.* Ger. 1113.

common

common ſpecies[p], which is found in moiſt meadows, may be known by the *frond* or leaf being ovate.

Oſmunda likewiſe has a ſpike diſtinct from the *frond*; it is branching, and each component fructification is globular. *Moonwort*[q], which grows on dry paſtures, has one naked ſtem, and one pinnate frond, forming the whole of this little Fern. *Flowering Fern*, or *Oſmund Royal*[r], a large ſpecies found on bogs, has bipinnate fronds, bearing the fructifications in a raceme at top. *Rough Spleenwort*[s] has lanceolate, pinnatifid fronds, with the diviſions confluent, quite entire and parallel: theſe are of two ſorts; the narrower being covered with fructifications on their backs, and the broader being barren. This therefore recedes from the character of the genus, in having a fertile frond inſtead of a ſpike, diſtinct from the barren one. Oſmunda.

The remaining genera have the fructifications invariably on the back of the fronds. In *Acroſtichum* they cover the whole diſk. In *Pteris* they are to be found only round Acroſtichum. Pteris.

[p] Ophioglоſſum vulgatum *Lin.* Fl. dan. 147. Mor. hiſt. ſ. 14. t. 5. f. 1. Ger. 404. Hedw. theor. f. 20—23.

[q] Oſmunda Lunaria *Lin.* Fl. dan. t. 18. Mor. hiſt. ſ. 14. t. 5. f. 1. Ger. 405.

[r] Oſmunda regalis *Lin.* Fl. dan. t. 217. Ger. 1131.

[s] Oſmunda Spicant *Lin.* Curtis, Lond. II. 67. Ger. 1140. Hedwig theor. f. 24—29. & Pl. 35. of this work.

the

the edge: the *common Fern* or *Brake*[t], which is ſo abundant in uncultivated grounds and woods, has ſupradecompounded, or triply-pinnate fronds, the leaflets pinnate, the lobes lance-ſhaped; the loweſt pinnatifid, and the upper ones leſs.

Aſplenium. *Aſplenium* has the fructifications in lines, that are frequently parallel. *Hart's-tongue*[u] has ſimple fronds, heart-tongued, that is drawn out into length, and hollowed next the petiole; quite entire, and the petioles ſhaggy: this grows on rocks and in ſhady places. There are ſeveral ſmaller ſpecies with pinnate or decompounded leaves, not uncommon on walls and rocks.

Polypodium. In *Polypody* the fructifications are in diſtinct roundiſh dots, placed in rows, and increaſing ſo much in ſize, as they advance to maturity, that they occupy the whole of the diſk in ſome ſpecies, and great part of it in others. *Common Polypody*[v] has pinnatifid fronds, the *pinnas* or lobes oblong, a little toothed and obtuſe; the root is ſcaly: this is common on trees, walls, and rocks. Many ſpecies that are generally called Ferns, from the diſpoſition of the fructifications, are of this genus: of theſe, that which is moſt common has vulgarly

[t] Pteris aquilina *Lin.* Blackw. t. 325. Ger. 1128

[u] Aſplenium Scolopendrium *Lin.* Curtis, Lond. I. 67. Ger. 1138.

[v] Polypodium vulgare *Lin.* Curtis, Lond. 1. 68. Ger. 1132.

 the

the name of *Male Fern* [w], and is found in woods, heaths, and on rocks, not covering the ground like the Brake, but in detached parcels: the fronds of this are doubly pinnate, the *pinnas* or lobes obtuſe, and crenulate, or ſlightly notched, and the ſtem chaffy.

Laſtly, *Adianthum* has the fructifications in terminal ſpots, under the margin of the frond, which is folded back. *True Maidenhair* [x], which is uſed, or ſuppoſed to be ſo, in the ſyrup of capillaire, is of this genus, and has decompounded fronds, the component leaves alternate, and the lobes wedge-ſhaped, lobate, and pedicelled. It grows, but rarely, on rocks and walls. Adianthum.

MOSSES.

The plants of the ſecond order—the *Moſſes*, have leaves like the more perfect vegetables, diſtinct from the ſtalk; and in this they differ from the Ferns, in which the ſtalk and leaf always, and the fructification often, are blended, to form the *frond*. They are perennial, and when ever ſo much dried up, will revive again with moiſture; as Haller experienced in ſome ſpecimens of Caſpar Bauhin's *Hortus Siccus*, which muſt

[w] Polypodium Filix mas *Lin.* Blackw. t. 323. Vaill. t. 9. f. 2. Mor. hiſt. ſ. 14. t. 3. f. 6. Ger. 1128.

[x] Adianthum Capillus Veneris *Lin.* Jacq. miſc. 2. t. 7. Ger. 1143.

have

have lain in a dry ſtate above a century. You know them by their air, or habit, as botaniſts uſually call it. A general idea of their fructification has been already given[y], as far as it is viſible to the naked eye; and we can only hope for a perfect account of it from a laborious examination with glaſſes of conſiderable magnifying powers[z].

The generic characters of the Moſſes are taken from the heads, which are either ſeſſile, or elſe the plant puſhes them up on a ſlender naked ſtem; this Linnæus calls the *Anther*, but I ſhall beg leave rather to name it the *Capſule*[a]: in four genera[b] it is naked, or not covered with a *calyptre* or *veil*; in the other ſeven it is.

Lycopodium. Sphagnum.

Lycopodium, or *Wolf's-claw Moſs*, has a two-valved, ſeſſile capſule, without any calyptre. *Sphagnum*, or *Bog-moſs*, has the capſule covered with a lid, and a ſmooth mouth. The *gray*[c] ſpecies is common on bogs, covering vaſt tracts of them; and is

[y] See letter X.

[z] This has now been done by *Hedwig* in his *Fundamentum Hiſtoriæ Naturalis Muſcorum Frondoſorum.* Lipſiæ 1782, quarto; and, *Theoria generationis et fructificationis Plantarum Cryptogamicarum*, Petrop. 1784, quarto; both with coloured plates of the parts of fructification much magnified.

[a] As Linnæus thinks it really is: (See *Genera*, p. 556,) and Hedwig has ſhown it to be.

[b] Lycopodium, Porella, Sphagnum & Phaſcum.

[c] Sphagnum paluſtre *Lin.* Fl. dan. 474. Dillen. t. 32. f. 1.

known

known not only by its hoary appearance, but by its deflected branches.

Polytrichum has a capſule covered with a lid, ſitting on a ſmall protuberant eminence, which is a kind of receptacle, and is called by Linnæus *Apophyſis*, by Haller the *Diſk*; the capſule is covered by a villous calyptre. There is a ſtar or roſe on a diſtinct individual, which has been generally taken for the piſtilliferous flower; Haller rather thinks it is only a kind of bud, from which new branches ſpring. The common ſpecies, called *Greater Golden Maidenhair* [d], is known by its ſimple ſtem, and the parallelopiped form of the capſule. This is a large ſort of Moſs, and abundant in woods, heaths, and bogs. Polytrichum.

The three remaining genera of Moſſes, which are alſo the principal and moſt numerous, are thus diſtinguiſhed. *Mnium* agrees with Polytrichum in having two ſorts of fructification; the one a lidded capſule, covered with a ſmooth calyptre: the other a ſtar or roſe, in the diſk of which are ſome globoſe little duſty bodies. *Bryum* and *Hypnum* have none of theſe ſtars or roſes: theſe have both a lidded capſule, covered with a ſmooth calyptre, and are diſtinguiſhed from each other by the ſtalk which ſupports the capſule being naked, and ariſing from a terminal tubercle in the

[d] Polytrichum commune *Lin.* Dillen. t. 54. f. 1. Ger. 1559.

firſt;

firſt; whereas in the ſecond it ſprings from the ſide of the branch, and is ſurrounded at bottom by a *perichætium*, ſcaly ſheath, or receptacle.

Mnium. One ſpecies of *Mnium*, whoſe filaments or capſular ſtalks are ſo ſenſible of moiſture, that it has obtained the name of *hygrometric*[e], has no ſtems; it has nodding turbinate or pear-ſhaped capſules, reflex four-cornered calyptres, and ovate leaves forming a head; they are of a yellowiſh green, and the filaments are an inch and half high, and red or orange-coloured.

Bryum. One of the moſt common ſpecies of *Brium* is the *hairy*[f], which covers the old thatch of cottages; this has the capſules rather erect, and the leaves ending in a hair, and recurved. *Apple-form Bryum*[g] has large ſpherical heads; and in the *Pear-form* ſpecies[h] they are obovate, covered with an awl-ſhaped calyptre; the ſhoots are ſtemleſs, and the leaves are ovate and awnleſs. *Brown Bryum*[i] has erect roundiſh capſules, with a pointed lid. This is a

[e] Mnium hygrometricum *Lin.* Fl. dan. 648. f. 2. Dillen. t. 52. f. 75. Mor. hiſt. ſ. 15. t. 7. f. 17.

[f] Bryum rurale *Lin.* Dill. t. 45. f. 12. Mor. t. 6. f. 1.

[g] Bryum pomiforme *Lin.* Dill. t. 44. f. 1. Mor. t. 6. f. 6.

[h] Bryum pyriforme *Lin.* Dill. t. 44. f. 6. Mor. t. 7. f. 16. & plate 36. of this work.

[i] Bryum truncatulum *Lin.* Curtis, Lond. II. 70. f. 2.

very

very ſmall Moſs, growing cloſe to the ground in thick tufts; the filaments are three or four lines high, and when the capſules have loſt their lid, they have a truncated appearance, whence their name.

Silky Hypnum[k], one of the moſt beautiful, and not the leaſt common of the genus, is known by its creeping ſhoots, its crowded erect branches, its awl-ſhaped leaves, and erect capſules. This grows both in dry places, ſuch as on walls, or trees; and in wet ones, as meadows: in the firſt, the leaves are narrow, and preſſed cloſe to the ſtalk; in the ſecond, they are broader, ſpreading, and ſhining, like ſilk: the capſules are long, round, enlarging a little at bottom, with a ſlender ciliated mouth, a ſcarlet beaked lid, and a pale calyptre; they are ſupported by a purple ſtalk, or filament, from half an inch to an inch in height, ſurrounded at the baſe by a ſhort thick ſcaly *perichætium*. This may ſerve as a ſpecimen of the numerous ſpecies of *Hypnum*; and we will now paſs on to the third order of the *Cryptogamia* claſs, containing the Hypnum.

ALGÆ.

Algæ or *Flags*, which are chiefly the *Lichens* or *Liverworts*, *Sea-weeds*, and

[k] Hypnum ſericeum *Lin.* Curtis, Lond. II. 69. Dillen, t. 42. f. 59. Mor. t. 5. f. 25

ſome

Marchantia. ſome few commonly called Moſſes, but having in reality the character of this order[l]. Of theſe laſt, *Common Marchantia*[m] may ſerve as an inſtance: it grows by ſtreams and fountains, in wet ſhady places, and on walls ſubject to a drip. There are two diſtinct fructifications in this genus, one ſtanding out from the plant on a peduncle, and conſiſting of a peltated calyx or receptacle, covered with ſmall one-petalled corollas underneath, each of which has one multifid anther or capſule; the other ſeſſile, ſhaped like a cup or bell, and containing many little roundiſh bodies, which ſome take for ſeeds. The ſpecies here pointed out is diſtinguiſhed by the common calyx being ten-cleft: it varies much in its appearance, and hence has its trivial name of many-form. This genus is evidently the connecting link between the Moſſes, and the *Lichens*, which we ſhall now examine.

Lichen. The genus of *Lichen* has a roundiſh, flattiſh, ſhining receptacle, or common calyx, ſeldom elevated; and a meal ſprinkled over the leaves. The receptacle affording a variety of forms, has ſuggeſted a ſubdiviſion of this otherwiſe unweildy genus, the leaf and manner of growth taking their parts in it. *Lichens* abundantly clothe the earth, rocks, and vegetables, eſpecially trees;

[l] See letter X.

[m] Marchantia polymorpha *Lin.* Dillen. t. 76. f. 6. Hedw. theor. f. 123—133.

in

in the form of meal, cruſt, leaf, or thread: age, ſoil, and ſituation, make ſo great a difference in their appearance, that numberleſs varieties have been advanced into ſpecies. The ſections of the genus are, 1. The *Tuberculate*, conſiſting of a cruſt adhering cloſely to the bark of trees, or ſtones, above which roundiſh tubercles riſe a little; theſe are rather irregular, a littled flatted at top, and without any rim round them. Sometimes they run into regular figures, and reſemble writing [n], or a map [o]. 2. *Scutellate*, or ſuch as have little ſhields, or roundiſh receptacles with a rim, and the diſk ſomewhat depreſſed, ariſing from a granulous cruſt more approaching to a leafy ſtructure than in the former ſection, and not adhering ſo ſtrongly. 3. *Imbricate*, compoſed of many ſmall leaves, generally in an orbicular form, lying over each other, the leaſt in the middle, and the largeſt on the outſide; from ſome of theſe ariſe little ſhields, and others have little mealy tubercles at the ends of the leaves. Nothing is more common than a yellow ſpecies [p] of this ſection, on trees, walls, and rocks; the leaflets of it are curled, deep yellow above, and aſh-coloured underneath; the ſhields are of a

[n] Lichen ſcriptus *Lin.* Dillen. t. 18. f. 1.

[o] Lichen geographicus *Lin.* Dillen. t. 18. f. 5.

[p] Lichen parietinus *Lin.* Dillen. t. 24. f. 76. Wall Liverwort.

 lighter

lighter yellow, grow brown with age, and are thick ſet towards the middle of the plant; other ſpecimens, inſtead of ſhields, have a yellow meal ſpread over them: the leaves by age become greeniſh, and then of a browniſh aſh-colour, warted and leprous. 4. *Leafy*, properly ſo called, conſiſting of one continued leafy ſubſtance, variouſly laciniate, cut or torn; theſe have generally large, wide ſhields, often on peduncles, either in the diviſions of the leaves, or on their edges. *Lungwort* or *Tree Lichen*[q], which hangs from old oaks, and beeches in woods, has very large jagged leaves, ſmooth, and ending obtuſely; the upper ſurface is wrinkled and pitted, the lower downy: the ſhields are of the ſize of a lentil, and placed on the edges of the leaves. 5. *Coriaceous* or *Leathery*: theſe are alſo leafy, but differ from thoſe of the fourth ſection in conſiſting of ſeveral leaves, of a tougher texture, broader, leſs ſharply laciniate, not branching, and generally adhering cloſer to the bodies on which they grow: the receptacles are very large, and from their reſemblance to the round ſhields of the ancients, called *peltæ*; they are generally on the edges of the leaves, and little or not at all notched on the edges. *Aſh-coloured Ground Liver-*

[q] Lichen pulmonarius *Lin.* Dillen. t. 29. f. 113. Ger. 1566.

wort

wort[r] is of this ſection: it is creeping, lobate, obtuſe, and flat; veined underneath, and villous, with a riſing *pelta* or target on the edge: this ſpecies is very common on the ground in woods, and on heaths, particularly on old ant-hills: the leaves are aſh-coloured, and white underneath. 6. *Umbilicate* or hollowed like the navel, and ſooty, or appearing black, or as if burnt. 7. *Cup-bearing*, conſiſting of a granulous cruſt, in proceſs of time unfolding into little leaves irregularly laciniate: from theſe ariſe a ſtipe or ſtem ſupporting hollow conical receptacles reſembling little tea-cups or drinking glaſſes, whoſe edge is often ſet with brown or ſcarlet tubercles. The different appearances of *Cup-moſs* are probably but varieties ariſing from the different age of the plant. 8. *Shrubby*, or reſembling ſhrubs or coral: theſe conſiſt of a leafy cruſt like the laſt, but they have no cups, only tubercles, and they are branched. The famous *Rhen-deer Moſs*[s] is of this ſection: it is perforate[t], very much branched, and the ſmall branches are nodding:

[r] Lichen caninus *Lin.* Fl. dan. 767. f. 2. Dillen. t. 27. f. 102. Mor. ſ. 15. t. 7. f. 1. This is the ſpecies formerly recommended againſt the bite of mad dogs, mixed with white pepper: but it is a remedy now exploded.

[s] Lichen rangiferinus *Lin.* Fl. dan. 180. Dillen. t. 16. f. 29.

[t] That is, there are little holes in the axils of the branches, as if made with a pin.

it grows on heaths and mountainous paſtures with us. 9. *Thready*, or conſiſting of mere round, ſolid ſtiff ſtalks or threads, frequently covered or incruſted with a meal, which is very inflammable, and terminating in dry globules, a little hollowed, and without any rim. Theſe moſt of them hang from the boughs of trees, and hence have the name of *Tree-moſs*. But this very numerous and widely diffuſed genus has already detained us too long.

The *Sea-weeds* are comprehended in three genera—*Ulva* or *Laver*, *Fucus* and *Conferva*. In the firſt, *Ulva*, the fructifications are in a diaphanous membrane, and the ſubſtance of the plant is membranaceous,
Fucus. at firſt bladdery, but afterwards leafy. *Fucus*, *Wrack*, or *Sea-weed* properly ſo called, has two kinds of bladders, the one ſmooth, hollow, and interwoven with hairs, the other ſmooth, filled with a jelly, in which are immerſed ſmall perforated grains, in each of which is ſuppoſed to be a ſeed: the texture of theſe plants is *coriaceous* or
Conferva. leathery. *Conferva* are compoſed of unequal tubercles, in very long capillary fibres, which are either continued or jointed. The two laſt genera will furniſh you with abundant amuſement whenever you are led to ſpend a little time on the ſea-coaſt; but the ſpecies are ſo numerous, that the examination of the ſpecific differences would carry

carry me into too wide a field: we will paſs on therefore to the laſt order of this laſt claſs of vegetable nature—the *Fungi* or *Muſhrooms*, which are univerſally known by their ſingular ſtructure and appearance; without branches, leaves, flowers, or any thing we can certainly call fructification, and ſcarcely any root. The *Agaric*, one of the principal genera in this order, is known by its horizontal manner of growing, and by having *lamellæ* or gills underneath [u]. The *Champignon* [v], or common eatable Muſhroom, is one of theſe, and has the following characters—the head is convex, ſcaly, white; and ſupported on a ſtipe or ſtalk; the gills are red; that which has white gills is only a variety of this, and though far inferior in quality, is not poiſonous. The *Chanterelle* [w], or little yellow Muſhroom, ſo common in the fairy rings on dry paſtures, is alſo ſtipitate, with the gills branched and decurrent. What is commonly called *Agaric* in medicine, and is uſed in ſtopping of blood, is of another genus. Agaricus.

Boletus, which grows horizontally like the laſt, but inſtead of gills, has pores on the under ſurface. Boletus.

[u] See plate 38. of this work.

[v] Agaricus campeſtris *Lin.* Mill. illuſtr. Fl. dan. t. 714.

[w] Agaricus Chantarellus *Lin.* Fl. dan. 264. Ger. 1580.

Phallus. *Morel*[x] is a fungus that is reticulate or netted all over the outſide or upper ſurface, and ſmooth beneath. The eſculent ſpecies has the head egg-ſhaped and cellular, the ſtipe or ſtem naked and wrinkled.

Lycoperdon. *Truffle* or *eſculent Puff-ball*[y], is a roundiſh fungus, filled with a mealy ſubſtance, taken for ſeed: this ſpecies is globular, ſolid, muricated, or rough on the outſide, without any root, and growing wholly under ground: the other ſorts are full of duſt, which they throw out when ripe, and are wholly above ground except their roots. *Common Puff-ball*[z] is roundiſh, and diſcharges its duſt by a torn aperture in the top; this varies much in form, and alſo in ſize, from a little ball to that of a man's head.

After all, the objects of this order are not univerſally allowed to be plants, but are ſuſpected, though ſeemingly without much reaſon, to be formed by animals, for their habitation, after the manner of Zoophytes or Corals. But this is a ſubject too difficult and nice for our diſcuſſion: and perhaps, after all, the *fungi* may prove to be one of thoſe links in the chain of nature, which unite the vegetable to the animal

[x] Phallus eſculentus *Lin.* Fl. dan. 53. Ger. 1583.
[y] Lycoperdon Tuber *Lin.* Michel. t. 102. Ger. 1583.
[z] Lycoperdon Boviſta *Lin.* Schœf. t. 190. Ger. 1582.

kingdom;

kingdom; and though they ſhould turn out to be the habitation of minute inſects, and to be formed for and by them, yet they may at the ſame time have the growth and texture of plants. Nature is full of theſe wonders, dear couſin; we are admitted to the view of a very ſmall portion of it only; there is little hope then that we ſhould be able to underſtand its relations fully, or to unravel all its myſteries.

AN

INDEX

OF THE

ENGLISH NAMES OF PLANTS.

Barley

ENGLISH NAMES.

Cyclamen

ENGLISH NAMES.

Orchis,

Rattan

INDEX

OF

LATIN NAMES.

Antirrhinum

Glechoma

Malva

Origanum

Prunella

LATIN NAMES.

Sauvageſia

NATURAL TRIBES, OR ORDERS OF

PLANTS,

Mentioned or explained in the foregoing Letters.

INDEX.

I N D E X

OF

T E R M S,

Casually explained in the Course of this Work.

INDEX

Mono-

Spadix

INDEX, &c.

FINIS.

Printed in the United States
By Bookmasters